Palgrave Studies in Green Criminology

Series editors
Angus Nurse
Department of Criminology and Sociology
Middlesex University
London, UK

Rob White
Department of Criminology
University of Tasmania
Hobart, Tasmania, Australia

Melissa Jarrell
Texas A&M University, Corpus Christi
Corpus Christi, Texas, USA

Criminologists have increasingly become involved and interested in environmental issues to the extent that the term Green Criminology is now recognised as a distinct subgenre of criminology. Within this unique area of scholarly activity, researchers consider not just harms to the environment, but also the links between green crimes and other forms of crime, including organised crime's movement into the illegal trade in wildlife or the links between domestic animal abuse and spousal abuse and more serious forms of offending such as serial killing. This series will provide a forum for new works and new ideas in green criminology for both academics and practitioners working in the field, with two primary aims: to provide contemporary theoretical and practice-based analysis of green criminology and environmental issues relating to the development of and enforcement of environmental laws, environmental criminality, policy relating to environmental harms and harms committed against non-human animals and situating environmental harms within the context of wider social harms; and to explore and debate new contemporary issues in green criminology including ecological, environmental and species justice concerns and the better integration of a green criminological approach within mainstream criminal justice. The series will reflect the range and depth of high-quality research and scholarship in this burgeoning area, combining contributions from established scholars wishing to explore new topics and recent entrants who are breaking new ground.

More information about this series at
http://www.palgrave.com/gp/series/14622

Ngoc Anh Cao

Timber Trafficking in Vietnam

Crime, Security and the Environment

palgrave
macmillan

Ngoc Anh Cao
People's Security Academy
Hanoi, Vietnam

Palgrave Studies in Green Criminology
ISBN 978-3-319-64279-6   ISBN 978-3-319-64280-2 (eBook)
DOI 10.1007/978-3-319-64280-2

Library of Congress Control Number: 2017957609

Cover illustration: Prisma by Dukas Presseagentur GmbH / Alamy Stock Photo

Printed on acid-free paper

This Palgrave Macmillan imprint is published by Springer Nature
The registered company is Springer International Publishing AG
The registered company address is: Gewerbestrasse 11, 6330 Cham, Switzerland

Contents

List of Abbreviations

ASEAN	Association of South East Asia Nations
CIFOR	Centre for International Forestry Research
CITES	Convention on International Trade in Endangered Species of Wild Fauna and Flora
CoC	Chain of Custody
EIA	Environmental Investigation Agency
EU	European Union
FAO	Food and Agriculture Organization of the United Nations
FC	Forest Company
FLEG	Forest Law Enforcement and Governance
FLEGT	Forest Law Enforcement, Governance and Trade
FPD	Forest Protection Department, Vietnam
FSC	Forest Stewardship Council
GDP	Gross Domestic Product
HSRP	Human Security Report Project
IMF	International Monetary Fund
INTERPOL	International Criminal Police Organisation
ITTA	International Tropical Timber Agreement
ITTO	International Tropical Timber Organisation
IUCN	International Union for Conservation of Nature
LOGGINGOFF	FLEGT Voluntary Partnership Agreement
LSITH	Large-scale Illegal Timber Harvesting
MARD	Ministry of Agriculture and Rural Development

MB	Management Board (Forest)
MNRE	Ministry of Natural Resources and Environment, Vietnam
MPI	Ministry of Planning and Investigation, Vietnam
MSITH	Medium-scale Illegal Timber Harvesting
NAV	National Assembly of Vietnam
NGO	Non-governmental Organisation
NTFPs	Non-Timber Forest Products
ODS	Ozone-depleting Substances
OECD	Organisation for Economic Co-operation and Development
SFEs	State-owned Forestry Enterprises
SSITH	Small-scale Illegal Timber Harvesting
TRAFFIC	Trade Records Analysis of Flora and Fauna in Commerce
UNDP	United Nation Development Programme
UNEP	United Nation Environmental Programme
UNODC	United Nations Office on Drugs and Crime
VND	Vietnam Dong (the Vietnamese currency)
WWF	World Wildlife Fund for Nature

List of Figures

List of Tables

1

Global Problem of Timber Trafficking

Environmental harm and crime have become an increasingly critical issue, of which every single person, organisation, state and global community as a whole should be aware. A great number of threats stemming from environmental problems are challenging humanity's security today. As a starting point, take just two examples that illustrate how environmental harm and crime threaten the living entities on Earth.

The first is the issue of climate change, indicated by the global average temperature increase, a rise in sea levels and altered weather patterns. As shown by Abbott (2008), climate change leads to a varied array of far-reaching devastating consequences, including socio-economic impacts: loss of infrastructure (e.g. over 20% of the gross domestic product (GDP) of the world economy may have to be paid annually due to inaction on climate change), resource scarcity (the demand for all three essential resources—food, water and energy—is already beyond the levels that can be sustained at the current time), mass displacement of people (by 2050, up to 200 million environmental refugees and one billion people may be displaced owing to natural disasters, conflict and large development projects); and the security consequences of inter-communal violence, civil unrest and international instability. Therefore, the Chief Scientific Advisor of the United Kingdom (UK), Sir David King, suggests that

N.A. Cao, *Timber Trafficking in Vietnam*, Palgrave Studies in Green Criminology,
DOI 10.1007/978-3-319-64280-2_1

climate change is a far greater threat to the world's stability than international terrorism (BBC News 2004), and The Lancet Commissions (2009: 1693) warns that climate change is "the biggest global health threat of the 21st century", putting "the lives and well-being of billions of people at increased risk".

The second example is a stimulating comparison by Lynch (2013: 47, 48) on the scale of victimisation between street crime and environmental or green crime in the United States of America (USA). Every year it is estimated that there are 25 million incidents of criminal victimisation in the USA, in which about 9 million offences involve violence. Meanwhile, every single day, 30% of the American population, or about 90 million people, are exposed to air pollution levels that infringe federal air quality standards. Thus, in the USA, the total number of air pollution violations in only one day is 10 times higher than the number of street crimes in one year.

Additionally, it is also estimated that there are approximately 284 trillion annual air pollution-related victimisation incidents. This figure is 31,536,000 times more than the number of incidents of victimisation from violent crimes as estimated by the US National Crime Victim Survey. The numbers of incidents of water pollution and hazardous waste sites are also 29,200 and 69,880 times respectively greater than street crime cases. Lynch (2013: 48) comes to the conclusion that "the fact that environmental violence incidents from these three offences alone are nearly 32 million times more frequent than the National Crime Victim Survey estimates should make us pause with concern".

In addition to its huge scale, green crime should alarm us today due to its illegal profits. Interpol (2012: 3) confirms that in fact, environmental crime is "currently one of the most profitable forms of criminal activity taking place throughout the world". By combining estimates from the Organisation for Economic Co-operation and Development (OECD), United Nations Office on Drugs and Crime (UNODC), United Nation Environmental Programme (UNEP) and International Criminal Police Organisation (INTERPOL), Nellemann et al. (2014) suggest that the monetary value of all forms of transnational organised environmental crime is worth between £70 and £213 billion annually. By 2005, it was estimated that while international trade in wildlife, fisheries and wild-sourced timber was worth over £332 billion annually (Engler 2008), illegal trade of wildlife and wildlife products may be valued up to £20 billion (Wyler and Sheikh 2008), making it one of the most lucrative illicit businesses, perhaps coming only after drug and arms trafficking (Schneider 2008).

Likewise, illegal trade in timber is worth between £30 and 100 billion annually or 10–30% of global wood trade[1] (Nellemann 2012). Profits from illegal trade in Ramin timber is reported to be even higher than drug smuggling (Khatchadourian 2008). The value of Merbau timber is only £120 per m^3 in an Indonesian port, but it is worth £2200 when retailed at an American store (EIA 2007b). Additionally, drugs and wildlife commodities are part of a joined global criminal economy, leading to an increasing probability that "the parallel trades will interact with each other" (South and Wyatt 2011: 556).

Regarding the detrimental impacts, green crime can be even more dangerous than other crimes. Wright (2011: 339) offers a marked comparison:

> The effects of TEC [transnational environmental crime] are not only damaging, but are often also definitive and long lasting, even permanent. Whereas there is a potentially unlimited supply of narcotics, there is a finite stock of an endangered species. Thus, once an animal is poached, it is gone forever. Likewise, once pristine rainforest is illegally logged it is irreplaceable and when ozone-depleting substances are released they cannot be stopped from damaging the atmosphere. Whereas trafficked persons can be offered support and counselling and drug users can enter rehabilitation, rectifying the effects of environmental crime can take a very long time, if they can be rectified at all.

As one of the typical green crimes, timber trafficking, as will be seen later in this chapter, is assumed to bring about a host of detrimental impacts on: (1) the economy by weakening economic efficiency, misallocating investments and stealing state revenue (Chan 2010; Contreras-Hermosilla 2001; EIA 2012; Guertin 2003; Rosander 2008; World Bank 2007); (2) on the natural environment by threatening biodiversity, intensifying deforestation and contributing to natural disasters (Goncalves et al. 2012; Interpol and World Bank 2009; Lawson and MacFaul 2010);

(3) on human well-being by threatening livelihood, employment, food and the physical safety of indigenous people living in forest areas (Boekhout van Solinge and Kuijpers 2013; Casson and Obidzinski 2007; Chan 2010; FAO 2007; Global Witness 2001); and (4) on political stability by creating conflicts between local communities with outsiders, challenging the proper operation of governments and provoking the expansion of organised and transnational criminal syndicates (Brack 2005; European Commission 2003; FAO 2007; Global Witness 2003; Human Rights Watch 2009). Because of its "alarming pace, level of sophistication, and globalised nature", trafficking in timber and other forest resources, as addressed by Nellemann et al. (2014: 4), has notoriously contributed to an "environmental crime crisis".

In brief, due to "the extraordinary extent and volume", various forms of green crime could be "the biggest crimes in the history of the world. No other crimes have threatened the existence of the entire planet" (Lynch and Stretesky 2014: 174). These crimes are "emerging as very serious global threats that cannot be underestimated any longer" (UNODC 2013: 1).

The enormity of green crimes and their associated issues have deeply concerned a variety of academic disciplines, including green chemistry, biology, medical science, epidemiology, environmental toxicology and economics, that have examined these practices and made "significant contributions" to our knowledge of the modern world (Lynch and Stretesky 2014: 176). Conversely, it seems that criminologists have protractedly neglected the criminological dimensions of environmental harm and crime, leading to a "woefully low" quantity of discussions on this topic (Zilney et al. 2006: 56). Lynch and Stretesky (2014: 176, 181) observe:

> Criminology has ignored the changing nature of the world around us, and has become less and less relevant to the problems found in the contemporary world… Criminology sleeps and dreams its long dream as if the world was not in crisis and the old routines practiced by criminology were sufficient.

The ignorance of criminologists may be one of the explanations for the fact that green crime often fails to prompt the required response by governments, the enforcement community and the public (Skinnider 2011). On international agendas, green crime has not been judged to be a priority in the discourse of international crime (Interpol and World Bank 2009). Indeed, among six key forms of criminality targeted by Interpol, green crime is considered as the one that has the lowest levels in terms of both illicit profit and the threat to public safety. Another example of the underestimation of environmental crime is that in Southeast Asia, "environmental crime is almost never mentioned in regional programmes and declarations on security and transnational crime" (Elliott 2007: 512).

Within the field of criminology, green crime in general already receives less attention than other categories of crime, but timber trafficking attracts even less concern from criminologists than other forms of green crime such as pollution, hazardous waste and wild animal trafficking. This is somewhat evidenced by the fact that among the eight most substantial collections on green crime to date, South and Brisman (2013), Spapens et al. (2014), White (2010) and White (2009), offer one essay relating to illegal timber trade, while Beirne and South (2006), Ellefsen et al. (2012), South and Beirne (2007) and Walters et al. (2013) present no chapter addressing this form of green crime.

Up to now, almost all intensive research projects on timber trafficking have been undertaken mainly by international organisations such as the Environmental Investigation Agency, Telapak, Global Witness, Interpol and the World Bank and by scholars in such fields as forestry, forest conservation, policy analysis and international development; whereas, as indicated by Boekhout van Solinge (2008, 2010b), Boekhout van Solinge and Kuijpers (2013) and Graycar and Felson (2010), the interest from criminologists on timber trafficking has been limited. Boekhout van Solinge and Kuijpers (2013) believe that on the basis of the ubiquity, high illicit profits and severe violence associated with timber trafficking, it would be a significant oversight if the crime was marginalised in the criminological agenda. Furthermore, it is predicted that placing timber trafficking into a green criminology framework may produce deeper and more fruitful insights into the crime (Boekhout van Solinge and Kuijpers 2013).

Another example of the neglect of international communities in regards to timber trafficking is that virtually each form of environmental crime is mainly targeted and intensively handled by a corresponding international treaty;[2] curiously, to date there is no single universal international instrument designed specifically to prevent and suppress timber trafficking (OECD 2012; Salo 2003). Although the Convention on International Trade in Endangered Species of Wild Fauna and Flora (CITES) lists 350 tree species for management, in practice many of the timber species listed by CITES are not being trafficked; whereas a large number of species excluded by CITES are being illicitly traded (Schloenhardt 2008). It is thus evaluated that the role of CITES in monitoring and controlling the illegal timber trade has been limited (Graycar and Felson 2010; Oldfield 2003a; Schloenhardt 2008). Added to CITES are several other international and regional initiatives,[3] but they too have had "little or no significant impact" (Goncalves et al. 2012: vii).

In a similar vein, at national level, timber trafficking has not received a high priority. Stewart (2014: 238) observes that "governments have, for the most part, failed to prioritise tackling this type of crime, resulting in a failure to enact the legislative reforms necessary or provide the financial and technical resources needed". The impacts of the responses from national and local authorities over timber trafficking are "far from satisfactory" (Reboredo 2013: 295), and still lag far behind the scope and enlargement of the crime (Nellemann et al. 2014).

Legal Timber Trade

Timber is a very important commodity. Dauvergne and Lister (2011: 1) note that "it is hard to get through a day without relying on timber. It is one of the world's most versatile natural resources in everything". The timber trade is, therefore, an important part of global commerce and has a long history. Tropical hardwood trade began centuries ago with a few valuable timber species such as teak, but since the sixteenth century, and particularly following the spread of the Industrial Revolution in Europe and colonial power in Asia and Africa, the trade has assumed a new level of diversity in the species of timber traded (Grainger 1993).

Currently, timber and its products travel commonly and globally, making it one of the most essential resources for people everywhere and one of the most popular commodities worldwide with no less than 5000 different wood products traded back and forth globally (Dauvergne and Lister 2011). In the early 1990s, it was estimated that global timber trade was valued at roughly £104 billion (TRAFFIC 2013). In the early 2000s, timber was the third largest commodity in international trade, resulting in one of the key resources of national revenue for many countries (Oldfield 2003b). By 2009, the Food and Agriculture Organisation of the United Nations (FAO) estimated that the trade's annual turnover was over £200 billion (TRAFFIC 2013).

Terminology and Definitions

The term *"illegal logging"* is probably most commonly used in both academic and policy discourses to denote a number of illegal activities related to forestry industries and forest ecosystems. There is not an internationally agreed definition of illegal logging and its use differs considerably in different contexts. Even what is legal or illegal also varies from country to country (Rosander 2008), or even from district to district within a country (Casson and Obidzinski 2007), leading the legality boundary to be "conceptually unsatisfactory" (Green et al. 2007). Some logging activities in Indonesia, for instance, may be judged as "illegal" by the central government, but "legal" by some district authorities (Casson and Obidzinski 2007). At the same time, logging without a government-approved management plan may be perfectly legal in the Southern USA, but might be illegal in the Brazilian Amazon (Seneca Creek Associates 2004).

Rosander (2008), Brack and Hayman (2001) and FERN (2002) provide relatively similar definitions of illegal logging in which illegal logging takes place when timber is harvested, transported, bought or sold in violation of national laws; whereas Interpol and the World Bank (2009) go further to encompass violations against both national laws in the origin country and ratified international treaties and conventions. The International Tropical Timber Organisation (ITTO 2013) differentiates *illegal logging* and *illegal trade in timber* in which the former mentions the removal of logs in a manner that is against the provision of relevant laws, while the latter (either domestic or international) is considered to be more multifaceted, concerning not only forest laws but also regulations on corporations, trading, banking, auditing, customs and taxes.

With the focus on cataloguing illegal logging, relying on the existing legal provisions, Goncalves et al. (2012) classify the crime into three broad categories: illegal products, illegal locations and illegal practices,[4] whereas based on the process of fraudulently legalising illegal timber, Richards et al. (2003) categorise timber harvesting into three different (although overlapping) groups: legal, legalised and clandestine harvesting.[5]

Additionally, some authors define and classify illegal logging by listing specific illicit behaviours. Seneca Creek Associates (2004), lists three groups of illegal conduct that appropriately rise to a "level of international concern", consisting of: (1) harvesting without authority or in excess of concession permit limits, (2) failing to report harvesting activity to avoid royalty payments or taxes, and (3) violations of international trading agreements such as CITES. Similarly, Nellemann (2012) lists 10 ways to illegally log, mainly including illegal cutting in protected areas; cutting without permits in remote areas, in conflict zones and in border areas; re-defining forest classification, using forged permits, exceeding legal concessions and laundering timber.

Since the term illegal logging was interpreted diversely in different countries, the term *"unauthorised harvesting"* was used during the Sixth Conference of the Parties of the Convention on Biological Diversity (Brack et al. 2002), while Graycar and Felson (2010) and Pendleton (2007) favour the term *"theft of timber"* and *"tree theft"*, respectively. Several publications, particularly those by non-governmental organisations (NGOs), occasionally employ the term *"unsustainable forest activities"* to imply the acts that originate from the customary rights of indigenous people and local communities (Schloenhardt 2008).

What is more, there are some other similar terms that somewhat confuse the readers in this topic, such as *"forest crime"* or *"illegal forest activities"* (Nellemann et al. 2014; Ravenel and Granoff 2004) or *illegal deforestation* (Boekhout van Solinge 2010a, b; Boekhout van Solinge and Kuijpers 2013). However, irrespective of different descriptions, generally, forest crime, illegal forest activities or illegal deforestation is often broader than illegal logging. These terms cover not only illegal harvesting of timber but also non-harvesting activities such as unauthorised occupation or conversion of forestlands, woodland arson, illegal transport, process and smuggle of forest products, transfer pricing, or other fraudulent accounting practices (Ravenel and Granoff 2004).

As can be seen from this section, there are indeed many terms and definitions relating to illegal logging. A summary by Guertin (2003) indicates that the current legal systems worldwide offer about 28 types of illegal activities in the forestry sector. The crux is that, taking all the terminological choices into account, a specific term needs to be chosen with a definition and the components that fit the present research's aim. As is evident, the term *illegal logging* is a tremendously open-ended term, often leading to substantially varied understanding of the practice. Some authors such as Smith (2002), Seneca Creek Associates (2004), Gutierrez-Velez and MacDicken (2008), Alemagi and Kozak (2010), ITTO (2013), Reboredo (2013) and Nellemann (2012) confine the meaning of illegal logging as the act of harvesting of timber only; whereas others such as Brack and Hayman (2001), Tacconi (2007a), Rosander (2008) and Schloenhardt (2008) argue that after harvesting, illegal loggers always connect with and conduct further illegal activities including buying, selling, processing or smuggling activities. It is argued in this research that a definition that excludes the subsequent activities may not be sufficient. Thus, the term illegal logging is not preferred in this research.

Other terms such as *forest crime, illegal forest activities, illegal deforestation, forest destruction* are either too broad or of little relevance to this research's interest, as they cover much broader targeted *objects* such as forestland, animals and even currency (in transfer pricing), but not timber, which is the chief object of concern in this research. Accordingly, this research uses the term *timber trafficking*[6] with a proposed definition:

> Timber trafficking is a form of environmental or green crime that involves the acts of harvesting, manufacturing, buying, selling, and smuggling timber and its products, which breach the relevant legal provisions enacted by local, national and international authorities.

This definition is supposed to be broad enough to cover illegal acts of harvesting as well as illicit after-harvesting activities including the smuggling, trade, processing and consumption of timber. It is also specific enough to avoid ambiguities of meaning because it focuses on timber as the targeted object of crime, and explicitly specifies the selected illicit activities involved. From this definition, it is worth clarifying some important points in this first chapter.

First, several illegal acts such as forest arson, illegal occupation of forestland and transfer pricing, that do not chiefly aim to obtain timber, are not considered as timber trafficking in this study. In Vietnam, there is a widespread form of violation in the field of forest protection namely "illegal deforestation" or "forest destruction" ("phá rừng trái phép"). Regardless of being destructive and illegal, such practice is not considered as a form of timber harvesting in this research because its foremost objective is not to harvest timber, but instead to appropriate forestland for cultivation by, for example, burning forests.

Second, timber trafficking encompasses both harvesting and after-harvesting acts. This means timber trafficking may occur inside or outside forests. Thus, if the term "illegal logging" is used in this research, it refers only to the unlawful harvest of timber. Illegal logging is, therefore, only a part of timber trafficking. Also, the definition implies that timber can be both internationally and domestically trafficked. However, to keep this research manageable, it focuses on illicit activities that are carried out inside the territory of Vietnam. That said, cross-border timber smuggling, though worthy of study, is beyond the research focus. Third, the legal provisions in this definition are not necessarily criminal laws, but they could be administrative laws and regulations. This socio-legal definitional approach will be justified and detailed in Chap. 2.

Scope and Trend of Timber Trafficking

Before 2000, comprehensive evaluations of the scale of timber trafficking were not conducted (FAO 2001), and consequently almost all data available was anecdotal or speculative (Kaimowitz 2003). Nonetheless, in the past decade or more, it is recognised that a number of studies have paid relatively thorough attention to estimating how pervasive and serious timber trafficking is at both global and regional levels. Some of the most intensive studies are Tacconi et al. (2003), (2007a), Tacconi (2007b) and Liddick (2011) who look at the global problem of illegal logging; Ravenel and Granoff (2004) who estimate the scale of timber trafficking in tropical countries; Rosander (2008) and Schloenhardt (2008) who both are interested in illegal timber trade in the East and Southeast Asia; Brack (2003, 2004, 2005, 2007) and Lawson and MacFaul (2010) who focus on measuring the effectiveness of the global response in controlling timber trafficking; and Solinge (2010a, b) who primarily investigates illegal deforestation in the Amazon.

In addition to these individual researchers, several research organisations have also undertaken studies in this field, with the most notable being the Seneca Creek Associates and Wood Resources International which investigates how timber trafficking affects the timber industry in the USA (Seneca Creek Associates 2004); the Environmental Investigation Agency, Telepak and the Centre for International Policy which conduct investigations in some logging hotspots such as Honduras (CIP and EIA 2005), Indonesia (Telapak and EIA 2007), China (EIA 2012), Mozambique 2013 (EIA 2013) and Laos and Vietnam (EIA 2011); and UNEP, Interpol and the World Bank which are concerned with law enforcement in the forestry sector (Interpol 2012; Interpol and World Bank 2009; Kendall 1998; Nellemann 2012; Nellemann et al. 2014; Stewart 2014; UNEP 2012). These institutions pay particular attention to the transnational scope of timber trafficking.

Despite varying degrees of scope and depth, collectively, these studies convey a deep concern about the expansive scope of timber trafficking worldwide. Indeed, the problem of timber trafficking is perceived as "common everywhere", ranging from the American, African, Asian to European continents; from developing to industrialised nations, and from boreal, temperate to tropical forests (Contreras-Hermosilla 2001: 3). Tacconi (2007a: 4) believes that "extensive illicit operations have been revealed whenever and wherever authorities have tried to find them". A study by Lawson and MacFaul (2010: xvii) concludes that timber trafficking "remains a major problem in all producer countries studied".

Timber trafficking is not confined within developing or producer countries, but also occurs in developed and consumer nations. A number of studies estimate that the scale of timber trafficking in Europe is also substantial (Brack 2005; FERN 2002; Smith 2004; Stahl 2010). FERN (2002), for instance, estimates that roughly 50% of the imports of timber from tropical forests into the European Union, and 20% of the imports from boreal forests, may be from illegal sources. In the UK, the Environmental Audit Committee (2009: 10) claims "it is possible that the UK is one of the world's largest importers of illegal timber and illegal timber products". It is estimated by Chatham House that in 2008, the UK imported a total of 1.5 million m^3 of illegal timber products, valued at over £1 billion (Bennett and Ares 2011).

Given the unfolding trend of timber trafficking, it is noteworthy that well-known international organisations concerned with timber trafficking provide differing evaluations of the trend of this crime. While Chatham House (The Royal Institute of International Affairs) believes that there is firm evidence to suggest that in recent years, the scales of timber trafficking "have declined significantly" in some important timber exporters (Lawson and MacFaul 2010: 91), in contrast, UNEP and Interpol assert that "it clearly spells out that illegal logging is not on the decline, rather it is becoming more advanced as cartels become better organised including shifting their illegal activities in order to avoid national or local police efforts" (Nellemann 2012: 5).

Even if there have been such drops in the proportion of illegal timber, it remains high. In addition to Indonesia and Brazil, other major timber-exporting countries witness a relatively high share of illegal timber: 20–50% in the Russian Far East, 22–35% in Cameroon, 59–65% in Ghana and 14–25% in Malaysia (Lawson and MacFaul 2010; Pye-Smith 2006). These figures are much smaller than those suggested by Nellemann et al. (2014), indicating that until recently in some main tropical countries, 50–90% of the timber is still suspected to be illegal.

When reviewing the existing literature on the scope of timber trafficking worldwide, it is readily apparent that the quantitative estimates of the scale, trends and value of the crime vary considerably. Four possible reasons account for these varying assessments. First, like other illicit markets, timber trafficking is essentially an unlawful business comprising clandestine activities occurring mainly in remote and isolated areas; it is, therefore, very hard to accurately gauge its scale. Second, as discussed earlier, the definitions and conceptualisations of timber trafficking are markedly uneven in different studies, organisations and countries. Third, the methodologies used to decide data collection and analysis techniques vary. Contreras-Hermosilla (2001), for example, observes that in many cases, research participants are reluctant to report illegal acts owing to either fear of revenge from offenders, or because they personally take part in the illegal activities.

Four, access to data on timber trafficking varies widely between different countries. Data collected from different countries is with "variable age and quantity", while in many countries, the data is neither available nor accessible (Smith 2004: 23). It is, hence, suggested that it is not possible to know the precise amount of timber trafficking (Seneca Creek Associates 2004), that much of timber trafficking remains either buried and undetected, or even laundered and recorded as legal (Contreras-Hermosilla 2001), and that the estimates provided should only be considered as very general assessments and, in some instances, as "best guesses" (Smith 2004: 23). Albeit the varying assessments, it is generally agreed that timber trafficking has become "more prominent, increasingly organised, sophisticated and transnational" (Stewart 2014: 241). These dimensions will be looked at in the next section.

The Sophistication of Timber Trafficking

Examining the nature and extent of timber trafficking, it is important to note that in many cases timber trafficking is a form of organised crime and/or involves organised crime networks (Boekhout van Solinge 2010b; FAO 2007; Pye-Smith 2006; UNDP 2013; Wyatt 2013a). Different studies address different aspects of timber trafficking as an organised crime including the involvement of many people (Graycar and Felson 2010), the significant degree of organisation and sophistication of unlawful enterprise and fraudulent practices (Stewart 2014), and the coordination of criminal networks in charge of harvesting, shipping, processing and trading timber (Interpol and World Bank 2009).

Additionally, there is strong evidence to underpin the argument that timber trafficking is a form of transnational crime. Nellemann et al. (2014) indicate that timber trafficking, together with animal trafficking; illegal fisheries and illegal mining and dumping of toxic waste are the main forms of transnational environmental crime. Timber trafficking includes cross-border illicit activities, that have been moderately well documented, between China and Russia (EIA 2012; Wyatt 2013a), Laos and Vietnam (EIA 2011), Liberia and its neighbouring countries (Global Witness 2003) and Southeast Asia and the Pacific (UNODC 2013). For example, some illegal timber originating from Southeast Asia, mainly from Indonesia and Malaysia, is sent directly to consumer countries, while some is smuggled to processing countries, mainly Vietnam and China, before reaching final markets (UNODC 2013). It is observed that the more countries that are involved in the trafficking chain, the harder it is to trace the timber origin, and the easier it is for the offenders to abuse the inconsistencies between different national legislation and international treaties (Interpol and World Bank 2009).

Another aspect demonstrating the sophistication of timber trafficking is that the offence has intimate links with other criminal behaviour. The most obvious link is its connection with corruption, which is a well-researched topic. There is widely agreed consensus that in the control of timber trafficking, corrupt behaviour can span the entire chain, involving various officials as well as their family members and close associates, making it tremendously challenging to pre-empt, detect and investigate the crime. Seneca Creek Associates (2004: 8) illustrate the status of the corruption involvement in timber trafficking worldwide.

In addition to corruption, in some parts of the world, timber trafficking is believed to have a connection with a number of other serious criminal activities such as arms and diamond trafficking (Boekhout van Solinge 2008), drug and arms trafficking, kidnappings and car robberies (CIP and EIA 2005; Wells et al. 2007), and money laundering (Interpol and World Bank 2009). However, unlike the connection with corruption, such links tend to be under-researched.

Consequences of Timber Trafficking

Relevant literature suggests that the detrimental impacts of timber trafficking may be severe and far-reaching. The impacts can be initially grouped into four different categories: economic, social, environmental and political.

Economic Impacts

For the majority of studies on timber trafficking that attempt to evaluate its damaging impacts, economic impacts are always the top concern. These studies emphasise that economic development of both the country in which it occurs and the country which is the final destination is negatively impacted by timber trafficking and other trades associated with the crime. However, it should be noted that similarly to the calculations of the monetary value of timber trafficking, calculating the actual economic consequences caused by timber trafficking is "bound to be imprecise" (Richards et al. 2003: 285). Richards et al. (2003: 285) note that much estimation relies on second-best or indirect methods as they lack "expensive, intensive and dangerous forest-based research involving close contact with loggers and other forest users".

Owing to the lack of empirical data to enable a comprehensive assessment of the impacts of illegal forest acts on the economy, Contreras-Hermosilla (2001) employs an inferential approach by analysing the economic impacts of some illegal practices in other sectors, and then applying these results to the forestry sector. Contreras-Hermosilla (2001: 15) concludes that "there is no reason to believe that consequences in the forestry sector would be at variance with these findings". He then elaborates a number of possible economic impacts caused by illegal forest activities including timber trafficking. These impacts are the erosion of economic efficiency, reduction and misallocation of investments in sustainable forest management, stealing of government revenue, and loss of global technical and financial assistance. Added to these is the creation of an unfair playing field for legitimate forest industries (Rosander 2008), the undermining of consumer confidence since their purchased products are made from timber "harvested in an environmentally responsible manner" (Guertin 2003: 7) and developing countries being denied "the opportunity to realise the economic value of preserving their forests as part of the global solution to stopping climate change" (Chan 2010: 9).

More specifically, it is argued from various studies that timber trafficking is the factor that causes a reduction in the global price of timber products by almost 7–17%. The most specific figure is around 16% as provided by (Lawson and MacFaul 2010). In the early 2000s, every year the crime deprived governments of revenue of £10–15 billion (World Bank 2002, 2007), increasing to £30 billion in 2012 (Nellemann 2012). Inside developing countries alone, this loss reaches £15 billion (EIA 2012; Stewart 2014).[7] It is evident that although there are varying estimations, generally the monetary cost of timber trafficking is considerable for governments and societies worldwide.

Social Impacts

Research on timber trafficking suggests that globally, the crime may threaten the livelihoods, employment and food resources of around one billion forest-dependent people worldwide (Lawson and MacFaul 2010).[8] This is because illicit logging lessens forest quality and leads to outright deforestation, thus depriving local inhabitants of vital resources such as agricultural implements, construction materials, medicines and fuelwood (Chan 2010; FAO 2007; Global Witness 2001).

Additionally, indigenous loggers may be heavily exploited. Indeed, participating in illicit timber harvesting, local loggers work in extremely onerous and dangerous conditions, but constantly receive the least profits, are dependent on logging coordinators, and are vulnerable to overexploitation (Casson and Obidzinski 2007; Chan 2010). A study by EIA (2007a: 5),citing a 2005 study by the International Labor Organisation, estimates that in the Peruvian Amazon, 33,000 people, mainly involved in illegal logging of mahogany and cedar, were working under forced labour conditions.

Furthermore, violence is never far away from timber trafficking operations. The world's largest tropical rainforests in Brazil, Peru, Congo and

Indonesia have seen "significant deforestation-related violence" (Boekhout van Solinge and Kuijpers 2013: 201). It is not rare to see cases of timber trafficking that involve murder, violence, threats and atrocities against indigenous forest-living people, journalists and local environmental activists (Boekhout van Solinge 2008; EIA 2012; Interpol and World Bank 2009; Nellemann 2012).[9] When local people, for example, criticise illegal forestry activities, the involved parties, at times including corrupt government officials, react with threats or even violence (Kaimowitz 2003).

Environmental Impacts

Compared to the economic and social dimensions, detrimental impacts of timber trafficking on the natural environment attract less, but currently increasing, attention from scholars. It is an overall evaluation by EIA (2012: 2) that "illegal logging and the trade in stolen timber are among the most destructive environmental crimes occurring today", directly threatening vital forests of the world. The notable point is that the environmental impacts can be extensive, not simply upon the surrounding regions, but have global implications (Brack 2004). There are probably three significant ways in which timber trafficking may negatively affect the natural environment, especially forest ecosystems.

At the outset, illegal logging particularly, illegal large-scale harvesting, is judged as a "significant contributor" to (Interpol and World Bank 2009: 4), and "among the most important drivers" of (Lawson and MacFaul 2010: 1), severe deforestation. Significant loss of forests as a direct result of illegal timber harvesting is documented across regions of the world including South America (Boekhout van Solinge 2010a), Central America (Richards et al. 2003), Southeast Asia and Asian Pacific (UNODC 2013), the African Congo Basin (Ruiz Pérez et al. 2005) and Central and Eastern Europe (Hirschberger 2008).

Globally, during the 1990s, 12.5 million hectares of forests disappeared (Liddick 2011), and recently more than 100 million m^3 of timber are still being illegally harvested annually, resulting in the degradation and possible eventual destruction of five million hectares of forest each year (Lawson and MacFaul 2010), which means that an area with the size of a football field is cleared by illegal loggers every two seconds (Goncalves et al. 2012). Once massive areas of forest are lost, inevitably the green functions of these forests, particularly carbon sequestration, are substantially hampered. It is calculated in a variety of studies that the clearance and degradation of the world's forests account for 12–25% of human-induced greenhouse gas emissions (Greenpeace 2007; Solomon 2007), and that deforestation is one of the heaviest contributors to climate change (Gorte and Sheikh 2010; Stern 2007).[10]

Second, timber trafficking poses a dramatic threat to biodiversity, particularly in already overlogged areas, watershed zones and biodiversity hotspots (Rosander 2008). The current high rate of timber trafficking means populations of inhabitants and wild animals in the forest are "rapidly shrinking" (Boekhout van Solinge 2010b: 34). In the Riau province in Indonesia, a notorious hotspot of illegal logging, 65% of its forests have disappeared over the past 25 years, while at the same time, there has been an 84% decline in elephant populations, and 70% decline in the number of tigers (WWF 2008).[11]

Third, timber trafficking could be one of the major facilitators of natural disasters such as flash flooding, landslides, forest fires and soil erosion. In addition to the function of providing habitats for more than two-third of the world's terrestrial species, forests serve as the means of reducing and filtrating water flows (Stewart 2014). Once forests are severely damaged, these functions are substantially weakened, which is an underlying facilitator of flash flooding and landslides (Reboredo 2013). Given the forest fire incidents, Pye-Smith (2006) claims that there is "clear evidence" to show that fires in Far East forests have become worse in recent years, and that the worst affected zones seem to be those that are most severely logged.

In addition, illegal operations during the harvesting, transportation and processing of timber are believed to generate considerable pollution (Greenpeace 2007). Carbon emissions, for instance, resulting from constructing roads and infrastructure for logging would be 2.5 times greater than emissions from the selective logging itself (Greenpeace 2007). There also exists evidence that wildlife trafficking may introduce invasive species and related diseases that threaten native wildlife and humans (Karesh et al. 2007; Wyatt 2013c). However, these aspects have not received much attention in research on timber trafficking.

Political Impacts

Current literature on timber trafficking reveals a number of possible political impacts. To begin with, so-called "conflict timber", where the illicit profits from timber trade are used as a major source of finance for conflicts, has been recognised in many countries such as Liberia, Democratic Republic of the Congo, Sierra Leone, Cambodia, Côte d'Ivoire and Myanmar (Brack 2005; European Commission 2003; FAO 2007; Global Witness 2001, 2003). For instance, the UN Security Council's effort to cut funding from timber trade, which sustains conflict in Liberia, is considered as "an effort to undercut funding for the war" (Brack 2005: 30). Likewise, in West Kalimantan, Indonesia, some green activists warn that timber trafficking is sometimes employed as a "political weapon" to fund election campaigns and to attack the opposition figures' reputation and finances. This technique has successfully eliminated one major rival from district office (Human Rights Watch 2009: 41). Additionally, transnational timber trafficking may trigger political rows, occasionally supplemented by military deployments or shows of force (Elliott 2007).[12]

In the long term, timber trafficking may undermine the legitimacy of the state. When, via corruption, timber trafficking touches the political or military domain, "the political infrastructure and stability of a country can be jeopardised" (Interpol and World Bank 2009: 15). Since timber trafficking and corruption are "twin brothers" (Reboredo 2013: 299), if timber trafficking cannot be minimised, it "can lead to a population's disaffection and loss of confidence in law enforcement authorities and their capabilities and, by extension, a loss of confidence in national institutions" (Interpol and World Bank 2009: 15).[13]

Positive Aspects

In conjunction with the harmful consequences generated by timber trafficking, as indicated in a number of studies, the crime is occasionally presumed to bring about several economic and social benefits. Pendleton (2007) and Tacconi et al. (2003) observe some positive outcomes for different parties, including local governments who receive large amounts of monetary fines and other revenue sources stemming from timber trafficking activities, for the poor and unemployed who are provided with jobs and income, and for consumers who benefit from lower prices of timber products.[14] Additionally, serving as a means for community members to "actively guard its normative boundaries", timber trafficking perhaps makes a contribution to community cohesion (Pendleton 2007: 22). Timber trafficking may also create some forms of small businesses and supplementary work, such as cooking and washing laundry for a small remuneration (Casson and Obidzinski 2007).

Given the benefits of timber trafficking, much literature is concerned that albeit visible in the short term, the benefits would be dwarfed by harmful impacts in the longer term. This can be particularly clearly seen in some countries in West Africa and Southeast Asia where "the collapse of the forest industry" has already occurred (Kaimowits 2007: 114).

Overview of Timber Trafficking in Southeast Asia and Vietnam

In the 1990s, 80% of all legal timber exports worldwide came from Southeast Asia[15] (Grainger 1993). In parallel to the legal timber trade, over the past decades, the region has played a manifold role in the global illegal timber trade (Schloenhardt 2008). South-East Asia now covers only 7% of the world's old-growth forests (UNODC 2010), but together with the Amazon and Central Africa, it is one of the world's three most critical hotspots wherein "the vast majority of deforestation and illegal logging takes place" (Nellemann 2012: 6). Research by Elliott (2007) indicates that this region, containing a number of heavily forested countries, is the main provider and processor of illegal timber.

The Southeast Asian countries also witness high proportions of illegal timber in the total amounts of harvested timber, possibly reaching 90% in Cambodia, 60–80% in Indonesia, 50% in Myanmar, 45% in Laos, 35% in Malaysia and 20–40% in Vietnam (DFID 2007; Interpol and World Bank 2009). Lawson and MacFaul (2010), however, suggest slightly lower figures of illegal timber namely 40–61% in Indonesia and 14–25% in Malaysia. In 2010, about 40% of wood-based products from Southeast Asia coming to the EU were believed to originate from illegal timber, while a half of timber volumes imported to China from this region may be illicit (UNODC 2010). A recent report on major transnational organised crimes in Southeast Asia by UNODC (2013) estimates that the value of trafficking in wood-based products in the region is worth £17 billion annually, making it even more lucrative than trafficking in heroin (£16.3 billion) and in methamphetamine (£15 billion). It is fair to say that timber trafficking has reached alarming levels in the region (UNODC 2013).

Some important conditions are believed to coexist, and their coexistence contributes to the flourishing of timber trafficking in this region. Indeed, while it is convincing that illegal logging tends to thrive in areas with weak forest governance and transparency (EIA 2012), in most countries in Southeast Asia, sustainable management in, and respective certification schemes for, the forestry sector are "almost absent" (Reboredo 2013: 295). In addition, the illicit trade in this region enjoys the abundant advantages created by the explosion of the legal timber trade and the problems of corruption, fraud, poor regulation and lax enforcement (UNODC 2013).

In Southeast Asia, Vietnam is an important location for illegal trade in both timber and non-timber forest products (NTFPs). Concurrently, among the Association of Southeast Asian Nations (ASEAN) countries, Vietnam should be seen as "a central importance in the ASEAN initiative" since it is an indispensable corridor of illegal wildlife trade between ASEAN countries, particularly Cambodia and Laos, and it has growing connections with other countries such as Malaysia and China (Lin 2005: 201). In terms of timber trafficking, Interpol and the World Bank (2009: 7) list Vietnam as one of the four countries in Southeast Asia (together with Cambodia, Indonesia and Malaysia) that are most affected by the crime.

Timber trafficking began to be widely noticed by the Vietnamese public in the 1990s, particularly after the notorious case of logging Tanh Linh Forest, in which 40 defendants were found guilty and subject to a total of 225 years' imprisonment (McElwee 2004). It currently still remains a major criminal problem in Vietnam (Forest Trends 2013; Department of Legal Affairs—MARD 2012). To the Vietnamese public, timber traffickers have become a deep concern; they are called "Lam Tac", which translates literally as "forest hijackers" (Sikor and To 2011: 692).[16]

Despite the significance of the problem, there are only a handful of studies on timber trafficking in Vietnam. These mainly take economic or legislative perspectives where the principal aim is to investigate the legal provisions to manage the forestry industry; none of them set as their principal objective to formulate a comprehensive typology of the crime and to assess material, physical and mental impacts of timber trafficking on the Vietnamese, especially poor forest-based inhabitants. The most notable research on illegal timber in the Asian-Pacific by Schloenhardt (2008: 84) indicates that unlike in some other countries in this region, such as Indonesia, Malaysia, China and Cambodia, "there were no reliable reports about the levels and patterns of timber trafficking in Vietnam". Thus, Vietnam would be an appropriate location to conduct an empirical green criminological examination on timber trafficking in the country, which would help better comprehend the regional and global problem of timber trafficking.

Methodology and Methods

To obtain a detailed understanding of timber trafficking in Vietnam, a research project has been carefully planned and conducted which considers fieldwork locations, research participants and methods of collecting data.

Locations Chosen for Doing Fieldwork

The locations for doing fieldwork were determined so as to maximise the diversity and deepth of the research findings and, where possible, to improve the representativeness of the findings as regards the problem of timber trafficking in Vietnam as a whole. With regard to the fieldwork locations chosen, a set of four main "criteria" was formed beforehand. Specifically the chosen provinces should have:

- A large forest area adequate to support the observation of the forest governance of provincial authorities and the reliance of the local communities on the forests.
- A large number of forest offences that would guarantee a wide variety of incidents of timber trafficking.
- Serious cases of timber trafficking in terms of number of offenders, consequences of the crime, and the sophistication of criminal activities, which ensure deeper insights into the crime,
- Located in various regions each with different natural, economic, social and cultural conditions, which would ensure an assortment of diverse insights into the problem of timber trafficking in Vietnam.

Based on these criteria, five Vietnamese provinces in three regions were selected as research sites for conducting the majority of the data collection: Bac Kan in the North, Thanh Hoa and Quang Binh in the Centre, and Kon Tum and Gia Lai in the Central Highlands. Vietnam contained over 13 million hectares of forest in 2011, meaning that on average, each of Vietnam's provinces contained around 200,000 hectares of forest. The five chosen provinces had a total of 2.7 million hectares of forest (540,000 hectares/province), accounting for 21% of the entire forest area of Vietnam (MARD 2012). Besides, all of the chosen provinces have national parks where illegal logging currently and frequently takes place.

Moreover, in 2011, an average of 450 cases of forest offences were detected in each province of Vietnam (FPD 2012). Meanwhile, the five selected provinces recorded more than 5600 cases (about 1100 cases/province), making up some 20% of the total cases discovered nationwide (FPD 2012). The selected provinces also consistently suffered from serious cases of timber trafficking as reflected through various channels including media coverage and official reports.[17]

In addition to these provinces, interviews were also conducted with NGO staff and environmental police officers at central departments in Vietnam's capital, Ha Noi. Another interview with a timber trader and an observation were also conducted in Bac Ninh province, which was one of the largest centres of timber processing in Vietnam. Taken together, the seven selected provinces were believed to be "conspicuously good examples" (Thomas 2011: 3) of how timber trafficking is taking place and how it is affecting Vietnamese society.

Selection of the Research Participants

To collect interview data with the highest degree of relevance and detail possible, interviewees chosen for this research were those who interact with, or are believed to hold a deep understanding of, timber trafficking in Vietnam. Accordingly, seven different cohorts of interviewees were selected with the specific reasoning for this selection detailed in Table 1.1.

Having provided the details of locations and participants for this research's fieldwork, the next section will justify and describe the three main techniques used to collect both primary and secondary data. These are semi-structured interviews, secondary data collection and direct observation.

Table 1.1 Seven cohorts of interviewee chosen and the justifications for the choice

No.	Cohorts of interviewee	Justifications for the selection
1	Environmental police officers	In charge of preventing timber trafficking, collecting systematic information and conducting initial investigations about timber trafficking
2	Investigative police officers	In charge of intensively conducting formal criminal investigations on timber trafficking
3	Forest protection officers	Exclusively vested in protecting, managing and developing forests
4	Local authorities	One of the forest owners and also representatives of the local residents
5	Local residents	Reside inside or near forest areas, directly impacted by timber trafficking
6	Timber traders	Buy, process and sell timber and timber products
7	NGO officers	From Forest Trends, WWF, TRAFFIC, IUCN and Forestry Science Institute who are in part interested in the control of timber trafficking in Vietnam

Data Collection

Semi-structured Interviews

After formulating a detailed fieldwork plan and making contact with the selected participants between June and July 2013, 41 semi-structured interviews were conducted between August and September 2013 with 11 environmental police officers, 6 investigative police officers, 7 forest protection (Kiem Lam) officers, 3 local authorities, 6 local residents, 3 timber traders and 5 NGO staff. The list of interviews, which includes the times each interviewee was conducted, is detailed in Appendix A. Appendix A also defines the codes for each of the interviews. These codes will be used in the following chapters to identify information provided by the corresponding interviewee. The code "1EP01", for example, means that it was the first ("1") interview that was carried out on 2nd August 2013 with the first ("01") environmental police officer ("EP"); while the code "2FO01" refers to the 2nd ("2") interview conducted on 3rd August 2013 with the first ("01") forest protection officer ("FO").

Although semi-structured interviewing has practical obstacles such as being time-consuming, costly and susceptible to interviewer prejudice and varying eloquence of interviewees (Bryman 2012; Henn et al. 2009; Punch 2005), it was considered that its intrinsic strengths made it appropriate for this research project. This is because semi-structured interviewing could this research require very detailed, multifaceted and fresh information on how timber trafficking is occurring, how it is affecting various aspects of Vietnamese society and how it can be better controlled. Additionally, due to the sensitivity and complexity of timber trafficking in Vietnam, it was envisaged that it would be extremely difficult to receive frank and detailed responses from interviewees without the creation of a cordial, trusted rapport via face-to-face conversations with them. The establishment of such a rapport would be guaranteed through the use of in-depth semi-structured interviewing. To construct the desired rapports, two strategies were adopted. From the outset, it was vital for me to make sure that the interviewees trusted their interviewer by reassuring them that the information they provided would be used solely for scientific purposes, not for, for instance, an undercover criminal investigation in which some of them might be implicated. Furthermore, the interviewees were informed that they would benefit from the research findings because this study would propose solutions that may be effective for reducing the crime and its detrimental impacts.

Subsequently, during the conversations, I tried to create friendly atmospheres by enthusiastically engaging with the interviewees. Indeed, I ensured that I was a keen listener, showing genuine interest in the participant's information and displaying sympathy for their situation. As I was engaged with the participant's circumstances, interests and concerns, which were pertinent to this research's themes, the participants were more willing to share information. As a governmental official, for example, I was talking with police and Kiem Lam officers about the limitations of current legislation and working mechanisms in public sectors including the forestry sector. As an academic researcher, I was sharing with NGO experts an interest in green criminological research. Having born and raised in a rural community, I was able to comprehend and talk about the hardships and challenges facing the poor indigenous villagers.

With the only exception of the interview in English with IUCN staff, all other participants were competent in, and comfortable with, speaking Vietnamese; as a result, all other interviews were conducted in Vietnamese. During the conversations with some highland residents, when I was occasionally unclear about their intended meaning due to their local accent, I asked them to repeat their comments to make sure that I fully understood their meaning.

During interviewing, the main contents of conversations revolved around three broad themes that were corresponding to three research questions:

1. The current typologies of timber trafficking in Vietnam: overall evaluation of the scale and trends of timber trafficking; main illegal activities (harvesting, smuggling, trading and processing); the demographics of actors involved in each step (age, gender, occupation and ethnicity); techniques used; time; locations; the planning for carrying out the crime, the connection with other criminal offenses such as fraud, assault, corruption, violence against on-duty officers and illegal use of weapons.

2. The impacts of timber trafficking in Vietnam: the ways timber trafficking threatens the livelihoods, food security, physical and psychological health, traditional culture of local people and their community; and positive aspects.
3. The facilitators of timber trafficking in Vietnam and suggestions to better control the crime: issues relating to demand for timber; legal provisions; effectiveness of law enforcement; insufficient investment in finance, personnel and equipment, and positive elements that contribute to curb timber trafficking.

The above-mentioned information was largely derived from the reviewing of existing literature on timber trafficking. However, during the interviews, the possibility that new, relevant information may be discovered from the interviewees was always kept in mind. Thus, the interviewees were encouraged to expand their responses when it was felt beneficial to do so. That is to say, depending on what and how interviewees responded, some types of question such as follow-up, specifying, structuring, and interpreting questions as suggested by Kvale (1996) were used to obtain the desired information.

Once the new ideas or arguments, for example, ones that lead to the classification of small, medium and large scale illegal logging, have initially emerged; in the next interviews, attention was paid to enrich and validate these ideas. Likewise, certain issues, which had been discussed at length and probably adequately addressed in earlier interviews, would not be talked in detail in the next interviews. The adjustment of the topics discussed once again demonstrated the merit of flexibility when doing semi-structured interviews (Bernard 2013; Bryman 2012; Harding 2013).

Furthermore, depending on the type of interviewee, the focus of each interview was adjusted. For example, police officers may be asked all question types, whereas questions relating to impacts on human security were predominant in the conversations with forest-based indigenous people. Similarly, green NGO officers are often specialists in policy reform within the forestry sector, while timber traders hold a deep understanding of the economic consequences of the crime. The questions asked were, therefore, directed to match the particular expertise of the interviewees. In addition, in order to validate information supplied by the interviewees, especially ambiguous responses, it was frequently necessary that this information was restated and clarified before further questions.

The length of the interviews was determined both by the extent of interviewee's understanding relevant to this research and by their responsiveness during the interviews. There were some interviews that lasted more than two hours as the respondents were very keen to discuss the topic, and they provided very detailed and penetrating information. In contrast, some participants were hesitant to comment in detail on the research subjects, either showing an unwillingness to engage fully with their interviewer or lacking understanding or experiences pertinent to this research. In such cases, the interviews finished after about 20–30 minutes. The locations for doing interviews were chosen based on the preference of the interviewees. Most interviews with NGO staff and law enforcement officers were carried out in their workplaces; the remainder took place in coffee shops. All interviews with local people were undertaken at their homes.

Secondary Data

This research collected a wide range of documents and statistics apposite to the project, mainly including:

- Nine criminal case records on timber trafficking including investigative conclusions issued by investigative police agencies and verdicts declared by criminal courts. These documents were official judgments from procedure-conducting bodies on criminal cases of timber trafficking in Vietnam. As stated in Article 10 (Determination of the Facts of Criminal Cases) and Article 63 (Matters to be Proved in Criminal Cases) of the 2003 Criminal Procedure Code of Vietnam (NAV 2003), these records must match the requirement to (1) "determine the facts of criminal cases in an objective, versatile and full manner including evidence of guiltiness, evidence of innocence, and evidence of the circumstances aggravating and extenuating the criminal responsibility of the accused or defendants" and to (2) clarify "the nature and extent of damage caused by the criminal acts". In this research, these criminal case files provided details on a number of recent cases of timber trafficking, including who were criminals, where, when and how timber trafficking took place and what consequences resulted from the crimes.

- Five annual reports from 2008 to 2012, one consolidated report synthesising 40 years of operations by the Forest Protection Department at national level, and 32 annual reports and statistics from forest protection agencies, police forces and authorities at provincial level. These reports provided this research with detailed statistics showing the scope and trends of timber trafficking. They also evaluated the advantages, disadvantages and effectiveness of forest protection work in particular the combatting of timber trafficking in Vietnam as a whole and in the provinces surveyed in this research in recent years. Additionally, several papers evaluating the socio-environmental impacts of forestland conversion projects, DVDs produced by Vietnam National Television on the consumption of timber products in Vietnam were collected.
- The conference proceedings titled "Preventing and Combating Violations of Forest Protection Laws" organised by the Department of Environmental Police in May 2012. The proceedings consisted of 32 papers presented by environmental police forces and forestry research institutions nationwide. These papers provided multidimensional information upon the forest offences across Vietnam.
- The quantitative database system available online at the "Operational Report" section of the Forest Protection Department website (http://www.kiemlam.org.vn/), which provided monthly and yearly statistics for the different categories of forest offences, including the numbers of violations and violators, the amounts of seized forest products and the areas of destroyed and converted forests, in all 63 provinces and six centrally managed national parks in Vietnam.

Overall, these official sources of secondary data played a major role in this research, providing notable benefits, uppermost of which was an opportunity to acquire all related understanding and information that have been released by the Vietnamese state agencies. Furthermore, since operations in the sector of criminal justice system are heavily monitored and shaped by policies and documentation, a secondary analysis, as suggested by Crow and Semmens (2008), was truly worthwhile. However, one potential shortcoming of using the aforementioned documents was that as these documents were not specifically created for this research's purpose, they might provide incomplete information.

In addition, this research looks for data in both English and Vietnamese languages from a wide array of domestic and international NGOs which were interested in green crime, timber trafficking and certain aspects of human security. Key organisations consulted included Illegal Logging Portal, FLEGT Voluntary Partnership Agreement (LOGGINGOFF), Centre for International Forestry Research (CIFOR), Convention on International Trade in Endangered Species of Wild Fauna and Flora (CITES), the Trade Records Analysis of Flora and Fauna in Commerce (TRAFFIC), Environmental Investigation Agency (EIA), Forest Trends, International Union for Conservation of Nature (IUCN), World Bank, Interpol, UNDP and FAO. The publications from these institutions provided overviews and figures relevant to some of the research issues particularly the policy framework of the Vietnamese forestry sector.

Newspaper Articles

In this research, because "Vietnam has an extremely lively press tradition" (Sumrall 2009: 24), over 200 newspaper articles relevant to the research areas have been collected and analysed. The articles were selected primarily from the most prestigious Vietnamese newspapers, with some from English-language media. Some of the newspapers consulted most in this research are listed in Table 1.2.

Key words and phrases, mainly in Vietnamese, related to different aspects of timber trafficking were typed into the Vietnamese version of the Google search engine https://www.google.com.vn/. These key words may be *"khai thác, buôn bán, vận chuyển, chế biến gỗ trái phép"* (illegal harvesting, dealing, smuggling, processing of timber), *"buôn lậu gỗ"* (illegal trade in timber), *"phá rừng trái phép"* (illegal forestation or illegal logging), *"lâm tặc"* (forest hijackers/illegal loggers), *"lâm tặc chống/móc ngoặc kiểm lâm"* (forest hijackers fight against/in partnership with forest

Table 1.2 Some newspapers whose articles were used in this research

Original Vietnamese name	Literal English name	Affiliated institutions	Circulation
Tuổi Trẻ	Youth/Juvenility	The Ho Chi Minh Communist Youth Union in Ho Chi Minh City	500,000 copies per day in 2008 (H. Nhut 2008)
Thanh Niên	Youth/Juvenility	The Vietnam National Youth Federation	2 million per week (Thanh Nien News 2015)
Nhân Dân	The People	The Communist Party of Vietnam	220,000 per day (Nhan Dan 2004)
Dân Trí	Knowledge of the People	The Studying Encouragement Society of Vietnam	Not found
VNExpress	Swift News in Vietnam	The Ministry of Science and Technology	Not found
Công An Nhân Dân	People's Public Security	The Ministry of Public Security	Not found

protection officers), *"lợi dụng thủy điện và cao su khai thác gỗ"* (abusing hydropower and rubber to log), and *"hậu qủa của phá rừng"* (consequences of illegal logging).

Alternatively, relevant articles could be found by using shorter key words such as *"lâm tặc"* (illegal loggers) *"gỗ lậu"* (illegal timber) and *"kiểm lâm"* (forest protection) entered in the search facility of the online newspapers consulted. For example clicking the word *"lâm tặc"* into the search facility of the Doi Song & Phap Luat (Life and Law) newspaper's website yielded about 30 journals and news dated March–September 2014 mainly addressing the illegal logging and serious incidents of Lam Tacs assaulting Kiem Lams. Meanwhile a similar number of journals and news articles were found using the same search facility by clicking the word *"gỗ lậu"* talking mostly about illegal cases of timber transport and trade. Some results were repetitive between the two searches. At times searching on English versions of major Vietnamese newspapers, particularly Tuổi Trẻ (http://tuoitrenews.vn/), Thanh Niên (http://www.thanhniennews.com/), Nhân Dân (http://en.nhandan.org.vn/) and Vietnamnet (http://english.vietnamnet.vn/) also yielded useful results

There was another method of obtaining newspaper articles pertinent to this research, which was available at the official website of the FPD. The website provided an invaluable section, "Daily Newspaper Review" (http://www.kiemlam.org.vn/diembao.asp), which selected and presented a large number of substantive articles from popular Vietnamese newspapers pertaining to forest protection issues. Every weekday, around 15–20 newspaper articles relevant to the field of forest protection and governance were selected and listed under three categories: "Forest Protection"; "Forest Management, Utility and Development"; and "Observation of the World". A total of over 1000 articles were collected through this method. The majority of these articles are relevant to this research, all were studied, and several are cited in this book. The "Daily Newspaper Review" is a highly valuable source of information for any research on forest offences in Vietnam.

As suggested by Mautner (2008: 32), media sources for social scientists, particularly newspapers and magazines, "represent a key data source". The up-to-date, diverse data from newspapers had two important roles to play in this research. First, it was used in particular for the purpose of examining sensitive aspects of timber trafficking such as the involvement of corruption and the illegal abuse of forestry policies, which some interviewees may hesitate to talk about in detail.[18] As shown in this chapter, timber trafficking is a crime that provokes a great deal of outcry from the Vietnamese media (Sikor and To 2011). In the field of wildlife trafficking, the media in Vietnam have been carrying out "a truly commendable job" with courageous journalists providing many stories, commentaries and investigative articles on the state of smuggling (Roberton 2013: 1). This is because in a "media-saturated society", while the media reflect the debates in criminal justice and make important contributions to the accountability of criminal justice bodies, the public consider media output as an "influential source" of crime information (Mawby 2011: 225–226).

In recent years, regarding the fight against corruption in Vietnam, for example, media in the country has played a "significantly increased role", being recognised by the urban Vietnamese as the second most trusted institution to lead this fight (Martini 2012: 8). Furthermore, as 97% and 92% of the Vietnamese population in urban and rural areas, respectively are now literate, the written journalism industry in Vietnam reaches a particularly broad audience (Earl 2013). This means that the Vietnamese press can reflect broad public attitudes on many major issues including timber trafficking and its connection with corruption in the country. Additionally, as observed by Mautner (2008: 32), newspaper data is easy to collect and already in written form, so it is not necessary to carry out time-consuming transcriptions.

The second function of the information extracted from the media in this research was that it helped to supplement and validate information from primary sources. Indeed, because the topic of timber trafficking in Vietnam was under-researched in academia, much of the information provided by the research interviewees was not addressed in existing literature. Data from the media, together with criminal case files and official documents, played an important role in augmenting and upholding the substantive patterns of timber trafficking, its impacts and the solutions offered to reduce the crime presented in this research.

Direct Observations

During the fieldwork, direct observations on how timber trafficking is occurring and affecting the forest-based communities were carried out. Guided and accompanied by police officers, some observations and "sidewalk activities" as suggested by Yin (2009: 109) on some forest areas and three national parks currently affected by timber trafficking in Vietnam were conducted. These parks are Ben En in Thanh Hoa, Ba Be in Bac Kan and Phong Nha Ke Bang in Quang Binh. From these observations, to some extent the severe impacts of timber trafficking on forests could be witnessed. The visits also enabled an appreciation of the great difficulties and risks involved in the process of logging and carrying timber out of the forests. When following several local people in Kon Tum province, the vehicles carrying logs coming to Vietnam from Laos were observed at the Bo Y International Border Gate between the two countries. This border gate was always one of the busiest points for importing timber from Laos to Vietnam. From the gate, people showed unofficial small routes that local porters use to carry timber from Laos to Vietnam.

During fieldwork in Quang Binh province, there was an informative and interesting observation on the arrest of a lorry driver and the seizure of the lorry's load of about 14 m^3 (approx. 14 tonnes) of Trac timber, an endangered timber species. The arrest was carried out by investigative police in collaboration with the Kiem Lam force in the province. The truck that departed from the Central Highlands was supposed to unload in Bac Ninh province in the North. It was covered by a large number of sacks of sweet corn, while the core contained wood furniture and several hundred planks of Trac timber. It took over half a day for several manual workers employed by the police force to finish unloading the timber. The arrest showed clearly a technique currently used to transport illegal timber in Vietnam.

In addition, one day was spent on visiting the Phu Khe Thuong timber market and the Dong Ky timber processing village in Bac Ninh province. Phu Khe Thuong is believed to be a unique timber market in Vietnam where massive quantities of various timber species, especially endangered species, are publicly sold and bought.

Finally, two days were spent living in a local household in Ngoc Hoi district, Kon Tum province. Ngoc Hoi is the only district in Vietnam sharing a border with both Laos and Cambodia. Over half of its population is ethnic minorities whose occupation is inexorably linked to forests. During these two days, I had a chance to observe the living conditions, working styles and culture of local people there. This gave me a greater understanding of how dependent on forests the indigenous people are and how the forest loss has affected their economic, community and personal security.

Conclusion

This chapter begins with the indication that green crime has been recognised as one of the most profitable and fastest growing sectors among all international criminal industries. Timber trafficking is a notable form of green crime, but like other types of green crime, it has been frequently overlooked by criminologists, which prompts the execution of this research with its aim and research questions clarified in the second section of this chapter.

As introduced in the next sections, there are many different illicit activities constituting timber trafficking; as a result, terms and definitions of the crime vary significantly and widely worldwide. However, the term *"timber trafficking"* is preferred in this study because it reflects a variety of illicit interrelated acts including the harvesting, transportation, trade, processing and consumption of timber. Evaluations of the extent of timber trafficking also differ. Some claim that the scope of the crime has been recently decreasing particularly in hotspot countries. However, others assert that because the crime is conducted in a more sophisticated manner, many illicit activities are often not explored and calculated by official criminal statistics. This means that the declining figures of timber trafficking in these statistics do not necessarily indicate that the actual scale of timber trafficking has been declining. Both camps, however, acknowledge that the magnitude of timber trafficking is still extensive and serious. This magnitude is demonstrated by very high figures in all aspects: illicit profits, illegal timber volumes, forest areas destroyed, and high proportions of illicit timber volumes compared to the total timber production. Furthermore, timber trafficking has become a sophisticated, transnational and/or organised crime with inseparable links to other examples of major criminality especially corruption.

Potential impacts of timber trafficking on the environment and society have been reviewed in this chapter. These economic, social, environmental and political consequences are believed to be abundant and predominantly detrimental, which suggests that timber trafficking should gain a firmer foothold on the green criminological agenda. However, it is observed that these consequences are mentioned in the literature mostly in an ad hoc manner, without a comprehensive framework to systematically identify and evaluate such impacts.

This chapter also shows that Southeast Asia is one of three tropical regions worldwide in which illegal logging, particularly in Indonesia, Cambodia and Malaysia, takes place on a large scale. This region has distinctive socio-legal contexts such as a rapid economic transition, rich tropical forests, relatively low capacity of forest governance, and diverse political and legal systems. All of these are conducive to the existence of timber trafficking. Within Southeast Asia, Vietnam is recognised as an important location of timber trafficking, which has created a large domestic public outcry. Irrespective of this, timber trafficking in Vietnam attracts little attention from international academic communities, especially criminologists, leading to limited understanding about the crime. It is, therefore, clear that an empirical research project that takes into account a green criminological perspective to investigate the problem of timber trafficking in Vietnam is essential.

Finally, this chapter has shown the rationale for the two major choices made in this study comprising fieldwork locations and research participants. Five provinces in three different regions containing large forest areas, a high number of violations of forest protection laws and serious cases of timber trafficking, in combination with seven different relevant cohorts of participants, were selected to conduct the fieldwork. This chapter then detailed the three main methods of collecting data: semi-structured interviews, documentary and secondary data and observations. Forty-one semi-structured interviews were conducted. Numerous official documents including criminal case files, reports from law enforcement authorities and conference papers, as well as hundreds of newspapers articles were collected and analysed.

Notes

1. There are other estimates. The Environmental Audit Committee (2006) suggests that worldwide the illicit earnings from this crime are around £10–15 billion annually, accounting for about 10% of global timber trade (Goncalves et al. 2012). Seneca Creek Associates (2004) estimate that the annual value associated with production of suspicious wood products is worth £23 billion.
2. The main international environmental agreements are the 1973 Washington Convention CITES designed for controlling wildlife trafficking, the 1987 Montreal Protocol on Substances that Deplete the Ozone Layer targeting illegal trade in ozone-depleting substances (ODS)

and the 1989 Basel Convention on the Control of Trans-boundary Movement of Hazardous Wastes and Other Wastes and their Disposal targeting illegal dumping and transport of various kinds of hazardous waste.

3. They are Forest Law Enforcement and Governance (FLEG); Forest Law Enforcement, Governance and Trade (FLEGT); and International Tropical Timber Agreement (ITTA). Like CITES, these frameworks are also devoid of enforceable mechanisms without penalties and sanctions for the countries that exploit timber resources in unsustainable manners (Salo 2003; European Commission 2003; Schloenhardt 2008).
4. Illegal products include the harvesting of protected tree species and the felling of trees below an allowable size; illegal locations refers to the harvesting of timber in locations where logging is prohibited or requires valid permits; and illegal practices denotes the failure to comply with laws and regulations on the harvesting operations and the behaviour of forestry concession-holders (Goncalves et al. 2012).
5. The latter two groups (legalised and clandestine) conform to "illegal logging". Legalised production is where the timber is fraudulently legalised at the logging locations (by, for example, adding illegal timber logged in other locations), in transport (transporting more than permitted volumes) or in processing. Legalised production is approved by official documents, forest charges are paid, and then it is calculated in official statistics. Clandestine timber, by contrast, avoids all certifications and charges, and remains unregistered (Richards et al. 2003).
6. The term *"timber trafficking"* is occasionally used in a handful of studies (EIA and Telapak 2001; Peters 2000), though these studies do not provide a definition of timber trafficking. The term *"timber trafficking"* tends to be used less often than *"wildlife trafficking"* (Prieksat 2009; Wyatt 2012, 2013b, d). In this research, timber exists in different forms, most commonly being logs, wood planks and wood products.
7. In Indonesia, timber trafficking costs the government no less than £2 billion every year due to corruption, uncollected taxes, unacknowledged subsidies and general poor management of resources (Chan 2010). Likewise the annual costs are approximately £5.3 million for Cameroon (SGS Trade Assurance Services 2002), tens of millions of US dollars in Mozambique (EIA 2013) and £560–640 million in the Bolivian, Brazilian and Peruvian Amazon (Gutierrez-Velez and MacDicken 2008).

8. It is estimated that over 90% of the 1.2 billion people living in extreme poverty rely on forests for some part of their livelihood (World Bank 2007).
9. In Rosita, Nicaragua, armed groups threaten loggers and traders to earn money for living (Richards et al. 2003). In Indonesia, on average, incidents in timber trafficking causing deaths, injuries, and property destruction happen twice a day (Reboredo 2013). In the Para state of Brazil alone, during the period from 1971 to 2004, 772 human rights and forest activists were murdered, while hundreds are also facing threats of murder as a result of their combat against coalitions of loggers, farmers and cattle ranchers (Boekhout van Solinge and Kuijpers 2013).
10. The project of Goncalves et al. (2012) reckons that over the last decade, a 22% drop in illegal logging would result in the prevention of over one billion tonnes of emissions of carbon dioxide. It is thus recommended that cutting deforestation and illegal logging is "the fastest, most effective and least controversial means" to trim global emissions of climate gases (Nellemann 2012a: 13).
11. In Honduras the unregulated harvesting of mahogany timber ranging between 30,000 and 50,000 m^3 annually is believed as the factor that leads the species to near extinction outside of protected areas (Richards et al. 2003). Timber trafficking in central Africa is threatening the survival of populations of the great apes, including gorillas and chimpanzees (Brack 2007).
12. Disputes over consignments of illegal logs and cross-border incursions of illegal logging do exist between China, Indonesia and Malaysia (Elliott 2007).
13. In Papua Province, Indonesia, for example, timber barons make use of local police and military personnel to threaten community leaders to accept completely unfair compensation for their timber (EIA 2007a).
14. For example, during the past decade, while much of the timber trade in Suifenhe, a Chinese town on the Russian border, is believed to be illegal or at least involves some illegal acts such as corruption, the trade is almost the entire driving force behind Suifenhe experiencing a spectacular economic transformation (Pye-Smith 2006).
15. There are some advantages that explain why at that time the worldwide timber importers preferred timber from the region, including open gov-

ernment policy, previous links between politics and trade, infrastructure and especially the quality and quantity of the region's timber (Grainger 1993).

16. The implications of the term are clear in the Vietnamese language: timber traffickers damage the common heritage and act against the public interest (Sikor and To 2011). Their practices were rendered further illegitimate when the Vietnamese media repeatedly reported conflicts between timber traffickers and forest protection officers and violence against the authorities, leading to many deaths and injuries (Sikor and To 2011).
17. The two main reports referenced for this task were "Forty years of the construction and development of Vietnam Kiem Lam" by the FPD (2013) and "Current status of forest protection and solutions to improve the work of forest protection" by the Department of Legal Affairs—MARD (2012).
18. In the fieldwork, a senior NGO interviewee also suggests that: "If you go to interview a chairman of people's committee, of course, he will never say 'I'm guilty'. He is not going to show you his bank accounts. You have to infer what is going on and I think there are a lot of data points that allow you to infer what is going on. So I think that it is realistic and that is best you could do, but you can draw pretty strong inferences" (19NG03).

References

Abbott, C. (2008). *An uncertain future: Law enforcement, national security and climate change*. London: Oxford Research Group.

Alemagi, D., & Kozak, R. A. (2010). Illegal logging in Cameroon: Causes and the path forward. *Forest Policy and Economics, 12*(8), 554–561.

BBC News. (2004). Global warming—Biggest threat. *BBC News*. Retrieved June 2, 2015, from http://news.bbc.co.uk/1/hi/sci/tech/3381425.stm

Beirne, P., & South, N. (2006). *Green criminology*. Aldershot: Ashgate.

Bennett, O., & Ares, E. (2011). *Illegally logged timber: EU and UK legislation, Standard Note: SN/SC/5858*. Retrieved from file:///C:/Users/w10036832/Downloads/sn05858.pdf

Bernard, H. R. (2013). *Social research methods: Qualitative and quantitative approaches*. Los Angeles, CA: Sage.

Boekhout van Solinge, T. (2008). Eco-crime: The tropical timber trade. In D. Siegel & H. Nelen (Eds.), *Organized crime: Culture, markets and policies* (pp. 97–111). New York: Springer.

Boekhout van Solinge, T. (2010a). Deforestation crimes and conflicts in the Amazon. *Critical Criminology, 18*(4), 263–277.

Boekhout van Solinge, T. (2010b). Equatorial deforestation as harmful practice and criminological issue. In R. White (Ed.), *Global environmental harm*. Cullompton, UK: Willan.

Boekhout van Solinge, T., & Kuijpers, K. (2013). The Amazon rainforest—A green criminological perspective. In N. South & A. Brisman (Eds.), *The Routledge international handbook of green criminology*. London: Routledge.

Brack, D. (2003). Illegal logging and the illegal trade in forest and timber products. *International Forestry Review, 5*(3), 195–198.

Brack, D. (2004). The growth and control of international environmental crime. *Environmental Health Perspectives, 112*(2), A80–A81.

Brack, D. (2005). Controlling illegal logging and the trade in illegally harvested timber: The EU's Forest Law Enforcement, Governance and Trade initiative. *Review of European Community & International Environmental Law, 14*(1), 28–38.

Brack, D. (2007). *Illegal logging*. London: Energy, Environment and Development Programme—Chatham House.

Brack, D., & Hayman, G. (2001). *Intergovernmental actions on illegal logging: Options for intergovernmental action to help combat illegal logging and illegal trade in timber and forest products*. London: Chatham House.

Brack, D., Marijnissen, C., & Ozinga, S. (2002). *Controlling imports of illegal timber: Options for Europe*. London: The Royal Institute of International Affairs and FERN.

Bryman, A. (2012). *Social research methods*. Oxford: Oxford University Press.

Casson, A., & Obidzinski, K. (2007). From new order to regional autonomy: Shifting dynamics of illegal logging in Kalimantan, Indonesia. In L. Tacconi (Ed.), *Illegal logging, law inforcement, livelihoods and the timber trade*. London: Earthcan.

Chan, A. (2010). *Illegal logging in Indonesia: The environmental, economic and social costs*. Minneapolis: BlueGreen Alliance.

CIP & EIA. (2005). *The illegal logging crisis in Honduras: How U.S. and E.U. imports of illegal Honduran wood increase poverty, fuel corruption and devastate forests and communities*. Washington, DC: Center for International Policy & Environmental Investigation Agency.

Contreras-Hermosilla, A. (2001). *Law compliance in the forestry sector: An overview*. Washington, DC: World Bank.
Crow, I., & Semmens, N. (2008). *Researching criminology*. Maidenhead: Open University Press.
Dauvergne, P., & Lister, J. (2011). *Timber*. Cambridge: Polity.
Department of Legal Affairs—Mard. (2012). Conference paper: The status of forest protection and solutions to enhance the work of forest protection. In *Preventing and combatting violations against legislations on forest protection* (pp. 18–29). Hanoi: The Department of Preventing and Combating Environmental Crimes (in Vietnamese).
DFID. (2007). *Crime and persuasion: Tackling illegal logging, improving governance*. London: Department for International Development.
Earl, C. (2013). Saigon style: Middle-class culture and transformations of urban lifestyling in post-reform Vietnamese media. *Media International Australia, 147*, 85–97.
EIA. (2007a). *AttentionWal-Mart shoppers: HowWal-Mart's sourcing practices encourage illegal logging and threaten endangered species*. McKinleyville, CA: Environmental Investigation Agency.
EIA. (2007b). *No question asked—The impacts of US market demand for illegal timber and the potential for change*. Washington, DC: Environmental Investigation Agency. Retrieved from http://www.illegal-logging.info/sites/default/files/uploads/eia_no_questions_asked.pdf
EIA. (2011). *Crossroads: The illicit timber trade between laos and Vietnam*. London: Environmental Invetsigation Agency.
EIA. (2012). *Appetite for destruction: China's trade in illegal timber*. London: Environmental Investigation Agency.
EIA. (2013). *First class connecions—Log smuggling, illegal logging, and corruption in Mozambique*. London: Environmental Investigation Agency.
EIA & TELAPAK. (2001). *Timber trafficking: Illegal logging in Indonesia, South East Asia and international consumption of illegally sourced timber*. London and Bogor, Indonesia: Environmental Investigation Agency and Telapak.
Ellefsen, R., Larsen, G., & Sollund, R. (2012). *Eco-global crimes: Contemporary problems and future challenges*. Farnham: Ashgate Publishing Ltd.
Elliott, L. (2007). Transnational environmental crime in the Asia Pacific: An un(der)securitized' security problem? *Pacific Review, 20*(4), 499–522.
Engler, M. (2008). The value of international wildlife trade. *Traffic Bulletin, 22*(1), 4–5.
Environmental Audit Committee. (2006). *Government response to the Committee's Second Report of Session 2005–06 on Sustainable Timber—Third Special Report*

of Session 2005–06. Environmental Audit Committee—House of Commons. London: The Stationery Office.

Environmental Audit Committee. (2009). *Reducing greenhouse gas emissions from deforestation: No hope without forests, fifth report of session 2008–09, report, together formal minutes, oral and written evidence*. London: The Stationery Office.

European Commission. (2003). *Communication from the Commission to the Council and the European Parliament: Forest Law Enforcement, Governance and Trade (FLEGT): Proposal for an EU action plan*. Brussels: Commission of the European Communities.

FAO. (2001). *Illegal activities and corruption in the forest sector*. The Food and Agriculture Organization of the United Nations. Retrieved from ftp://ftp.fao.org/docrep/fao/003/y0900e/

FAO. (2007). *State of the world's forests 2007*. Rome: Food and Agriculture Organization of the United Nations.

FERN. (2002). *Illegal logging, and the global trade in illegally sourced timber; a crime against forests and peoples*. Forests and the European Union Resource Network. Retrieved February 15, 2013, from http://www.sinkswatch.org/sites/fern.org/files/pubs/ngostats/logging.pdf

Forest Trends. (2013). *Small-scale illegal logging in Vietnam: Implications for FLEGT and REDD+*. The Forest Trends.

FPD. (2012). *Report on the operational performance in 2011 and the working programme in 2012*. Hanoi: The Forest Protection Department (in Vietnamese).

FPD. (2013). *Forest protection force of Vietnam: Forty years of construction and development (1973–2013)*. Hanoi: The Forest Protection Department (in Vietnamese).

Global Witness. (2001). *The logs of war, the timber trade and armed conflict*. London: Global Witness.

Global Witness. (2003). *The usual suspects: Liberia's weapons and mercenaries in Côte d'Ivoire and Sierra Leone—Why it's still possible, How it works and How to break the trend*. London: Global Witness.

Goncalves, M. P., Panjer, M., Greenberg, T. S., & Magrath, W. B. (2012). *Justice for forest: Improving criminal justice efforts to combat illegal logging*. Washington, DC: World Bank.

Gorte, R. W., & Sheikh, P. A. (2010). *Deforestation and climate change*. Congressional Research Service.

Grainger, A. (1993). *Controlling tropical deforestation*. London: Earthscan Publications.

Graycar, A., & Felson, M. (2010). Situational prevention of organized timber theft and related corruption. In R. V. Clarke & N. Tilley (Eds.), *Situational prevention of organised crimes* (1st ed., pp. 81–92). Hoboken: Taylor and Francis.

Green, P., Ward, T., & McConnachie, K. (2007). Logging and legality: Environmental crime, civil society, and the state. *Social Justice, 34*(2), 94–110.

Greenpeace. (2007). *Carving up the Congo—Destruction by numbers logging roads to climate ruin*. Amsterdam.

Guertin, C. É. (2003). *Illegal logging and illegal activities in the forestry sector: Overview and possible issues for the UNECE Timber Committee and FAO European Forestry Commission*. Quebec, Canada: Quebec Wood Export Bureau.

Gutierrez-Velez, V. H., & MacDicken, K. (2008). Quantifying the direct social and governmental costs of illegal logging in the Bolivian, Brazilian, and Peruvian Amazon. *Forest Policy and Economics, 10*(4), 248–256.

H.Nhut. (2008). Tuổi Trẻ và những ngủời bạn đồng hành [Tuoi Tre and the fellow]. *Tuoitre*, 19/6/2008. Retrieved June 2, 2015, from http://tuoitre.vn/Chinh-tri-Xa-hoi/264325/Tuoi-Tre-va-nhung-nguoi-ban-dong-hanh.html#ad-image-0

Harding, J. (2013). *Qualitative data analysis from start to finish*. London: Sage.

Henn, M., Weinstein, M., & Foard, N. (2009). *A critical introduction to social research*. London: Sage.

Hirschberger, P. (2008). *Illegal wood for the European market: An analysis of the EU import and export of illegal wood and related products*. WWF-Germany.

Human Rights Watch. (2009). *"Wild Money": The human rights consequences of illegal logging and corruption in Indonesia's forestry sector*. New York: Human Rights Watch. Retrieved from http://www.hrw.org/sites/default/files/reports/indonesia1209web_0.pdf

Interpol. (2012). *Environmental crime, It's global theft*. Interpol—Environmental Crime Programme. Retrieved from http://www.interpol.int/Crime-areas/Environmental-crime/Environmental-crime

Interpol & World Bank. (2009). *Chainsaw project: An Interpol perspective on law enforcement in illegal logging*. Rural Development and Natural Resources—World Bank & Interpol.

ITTO. (2013). *What are illegal logging and illegal trade?*. Bangkok, Thailand: International Tropical Timber Organisation. Retrieved April 24, 2013, from http://www.itto.int/feature06_01/

Kaimowits, D. (2007). Forest law enforcement and rural livelihoods. In L. Tacconi (Ed.), *Illegal logging: Law enforcement, livelihoods and the timber trade* (pp. 110–138). London: Earthscan.

Kaimowitz, D. (2003). Forest law enforcement and rural livelihoods. *International Forestry Review, 5*(3), 199–210.

Karesh, W. B., Cook, R. A., Gilbert, M., & Newcomb, J. (2007). Implications of wildlife trade on the movement of avian influenza and other infectious diseases. *Journal of Wildlife Diseases, 43*(3), S55–S59.

Kendall, R. E. (1998). Interpol alert. *Our Planet.* Retrieved June 2, 2015, from http://www.ourplanet.com/imgversn/105/kendall.html

Khatchadourian, R. (2008). The stolen forests—Inside the covert war on illegal logging. *The New Yorker.* http://www.newyorker.com/magazine/2008/10/06/the-stolen-forests

Kvale, S. (1996). *InterViews: An introduction to qualitative research interviewing.* Thousand Oaks, CA: Sage.

Lawson, S., & MacFaul, L. (2010). *Illegal logging and related trade indicators of the global response.* The Royal Institute of International Affairs.

Liddick, D. (2011). *Crimes against nature: Illegal industries and the global environment.* Santa Barbara, CA: Praeger.

Lin, J. (2005). Tackling Southeast Asia's illegal wildlife trade. *Sybil, 9*, 191.

Lynch, M. J. (2013). Reflections of green criminology and its boundaries: Comparing environmental victimization and considering crime from an eco-city perspective. In N. South & A. Brisman (Eds.), *The Routledge international handbook of green criminology* (pp. 43–57). London: Routledge.

Lynch, M. J., & Stretesky, P. (2014). *Exploring green criminology: Toward a green criminological revolution.* Farnham: Ashgate Publishing Ltd.

MARD. (2012). *National forest status declaration.* Development, M. O. A. A. R. Hanoi: Ministry of Agriculture and Rural Development.

Martini, M. (2012). *Overview of corruption and anti-corruption in Vietnam.* Anti-Corruption Resource Centre (U4), Transparency International & Chr. Michelsen Institute.

Mautner, G. (2008). Analysing newspapers, magazines and other print media. In R. Wodak & M. Krzyżanowski (Eds.), *Qualitative discourse analysis in the social sciences* (pp. 30–53). Basingstoke, UK: Palgrave Macmillan.

Mawby, R. C. (2011). Using media to understand crime and criminal justice. In P. Davies, V. Jupp, & P. Francis (Eds.), *Doing criminological research* (2nd ed., pp. 223–244). Los Angeles, CA; London: Sage.

McElwee, P. (2004). You say illegal, I say legal: The relationship between 'illegal' logging and land tenure, poverty, and forest use rights in Vietnam. *Journal of Sustainable Forestry, 19*(1-3), 97–135.

NAV. (2003). *The criminal procedure code of Vietnam*. Hanoi: The National Assembly of Vietnam.

Nellemann, C. (2012). *Green carbon, black trade: Illegal logging, tax fraud and laundering in the worlds tropical forests. A rapid response assessment*. Birkeland Trykkeri AS, Norway: Interpol Environmental Crime Programme—United Nations.

Nellemann, C., Henriksen, R., Raxter, P., Ash, N., & Mrema, E. (2014). *The environmental crisis—Threats to sustainable development from illegal exploitation and trade in wildlife and forest resources*. United Nations Environment Programme and GRID-Arendal, Nairobi and Arendal. Retrieved from http://www.grida.no/

Nhan Dan. (2004). Báo Nhân Dân: Ngủời bạn tin cậy đồng hành cùng bạn đọc [Nhan Dan Newspaper: The trusted friend accompanying the readers]. *Nhan Dan*, 14/03/2004. Retrieved June 2, 2015, from http://www.nhandan.com.vn/bandoc/item/19795802-nguoi-ban-tin-cay-dong-hanh-cung-ban-doc.html

OECD. (2012). *Illegal trade in environmentally sensitive goods*. Paris: Organisation for Economic Co-operation and Development.

Oldfield, S. (2003a). Chapter 12: regulation of the timber trade. In S. Oldfield (Ed.), *The trade in wildlife: Regulation for conservation*. London: Earthscan.

Oldfield, S. (2003b). *The trade in wildlife: Regulation for conservation*. London: Earthscan.

Pendleton, M. R. (2007). *The social basis of illegal logging and forestry law enforcement in North America*. In L. Tacconi (Ed.), *Illegal logging, law inforcement, livelihoods and the timber trade*. London: Earthcan.

Peters, J. L. (2000). Illegal trafficking of timber in Cambodia. *Colorado Journal of International Environmental Law and Policy, 11*, 102.

Prieksat, J. (2009). *International trade and trafficking in wildlife: The value of biodiversity cites, and going further*. International Environmental Law.

Punch, K. (2005). *Introduction to social research: Quantitative and qualitative approaches*. Thousand Oaks, CA: Sage.

Pye-Smith, C. (2006). *Logging in the Wild East: China and the forest crisis in the Russian Far East*. Forest Trends & Rights and Resources.

Ravenel, R. M., & Granoff, I. M. E. (2004). Illegal logging in the tropics: A synthesis of the issues. *Journal of Sustainable Forestry, 19*(1–3), 351–366.

Reboredo, F. (2013). Socio-economic, environmental, and governance impacts of illegal logging. *Environment Systems and Decisions, 33*(2), 295–304.
Richards, M., Wells, A., Gatto, F. D., Contreras-Hermosilla, A., & Pommier, D. (2003). Impacts of illegal ity and barriers to legality: A diagnostic analysis of illegal logging in Honduras and Nicaragua. *International Forestry Review, 5*(3), 282–292.
Roberton, S. (2013). Corruption along the Vietnam-China border permits massive smuggling of endangered animals. *Policy Innovation*, 7/8/2013. Retrieved June 2, 2015, from http://www.policyinnovations.org/ideas/briefings/data/000260
Rosander, M. N. (2008). *Illegal logging: Current issues and opportunities for Sida/SENSA engagement in Southeast Asia*. Bangkok, Thailand: RECOFTC & Sida.
Ruiz Pérez, M., Nguinguiri, J.-C., Nzala, D., Toirambe, B., Yalibanda, Y., Ezzine De Blas, D., et al. (2005). Logging in the Congo Basin: A multi-country characterization of timber companies. *Forest Ecology and Management, 214*(1), 221–236.
Salo, R. S. (2003). When the logs roll over: The need for an international convention criminalizing involvement in the global illegal timber trade. *Georgetown International Environmental Law Review, 16*(1), 127.
Schloenhardt, A. (2008). *The illegal trade in timber and timber products in the Asia-Pacific region—Research and Public Policy Series No. 89*. Canberra: Australian Institute of Criminology.
Schneider, J. L. (2008). Reducing the illicit trade in endangered wildlife. The market reduction approach. *Journal of Contemporary Criminal Justice, 24*(3), 274–295.
Seneca Creek Associates, LLC. (2004). *"Illegal" logging and global wood markets: The competitive impacts on the US wood products industry*. American Forest & Paper Association.
Sikor, T., & To, P. X. (2011). Illegal logging in Vietnam: Lam Tac (forest hijackers) in practice and talk. *Society & Natural Resources, 24*(7), 688–701.
Skinnider, E. (2011). *Victims of environmental crime: Mapping the issues*. International Centre for Criminal Law Reform and Criminal Justice Policy.
Smith, W. (2002). The global problem of illegal logging. *ITTO Tropical Forest Update*, 12/1.
Smith, W. (2004). Undercutting sustainability: The global problem of illegal logging and trade. *Journal of Sustainable Forestry, 19*(1–3), 7–30.
Solomon, S. (2007). *Climate Change 2007: The physical science basis: Contribution of working group I to the Fourth Assessment Report of the Intergovernmental Panel Climate Change*. Cambridge: Cambridge University Press.

South, N., & Beirne, P. (2007). *Issues in green criminology: Confronting harms against environments, humanity and other animals*. Cullompton, UK; Portland, OR: Willan.

South, N., & Brisman, A. (2013). *The Routledge international handbook of green criminology*. London: Routledge.

South, N., & Wyatt, T. (2011). Comparing illicit trades in wildlife and drugs: An exploratory study. *Deviant Behavior, 32*(6), 538–561.

Spapens, T., White, R., & Kluin, M. (2014). *Environmental crime and its victims: Perspectives within green criminology*. Ashgate Publishing.

Stahl, J. (2010). The rents of illegal logging: The mechanisms behind the rush on forest resources in Southeast Albania. *Conservation and Society, 8*(2), 140.

Stern, N. H. (2007). *The economics of climate change: The Stern review*. Cambridge: Cambridge University Press.

Stewart, D. (2014). Project leaf, and interpol's work on illegal logging and forest crime. In T. Spapens, R. White, & M. Kluin (Eds.), *Environmental crime and its victims: Perspectives within green criminology* (pp. 236–247). Surrey: Ashgate.

Sumrall, K. A. (2009). *Confronting illegal wildlife trade in Vietnam: The experience of education for nature—Vietnam*. Degree of Master of Science, University of Michigan.

Tacconi, L. (2007a). The problem of illegal logging. In L. Tacconi (Ed.), *Illegal logging: Law enforcement, livelihoods and the timber trade* (pp. 1–16). London: Earthscan.

Tacconi, L. (2007b). Illegal logging and the future of the forest. In L. Tacconi (Ed.), *Illegal logging: Law enforcement, livelihoods and the timber trade* (pp. 275–289). London: Earthscan.

Tacconi, L., Boscolo, M., & Brack, D. (2003). *National and international policies to control illegal forest activities*. Report for the Ministry of Foreign Affairs, Government of Japan. Bogor, Indonesia: CIFOR.

Telapak & EIA. (2007). *The thousand-headed snake: Forest crimes, corruption and injustice in Indonesia*. Bogor, Indonesia.

Thanh Nien News. (2015). About us. *Thanh Nien News*. Retrieved June 3, 2012, from http://www.thanhniennews.com/about/about-us-27720.html

The Lancet Commissions. (2009). *Managing the health effects of climate change*. Lancet and University College London Institute for Global Health Commission.

Thomas, G. (2011). *How to do your case study: A guide for students and researchers*. Los Angeles, CA; London: Sage.

To, X. P. (2012). Natural forests and its suffering from timber need [Rừng tự nhiên chịu sức ép từ nhu cầu sử dụng gỗ]. *Con Nguoi & Thien Nhien*, 6/11/2012. Retrieved June 3, 2015, from http://www.thiennhien.net/2012/11/06/rung-tu-nhien-chiu-suc-ep-tu-nhu-cau-su-dung-go/

TRAFFIC. (2013). *Timber trade*. Retrieved March 6, 2015, from http://www.traffic.org/timber-trade/

UNDP. (2013). *Justice index—Assessment of distributive justice and equality from a citizen-based survey in 2012*. Hanoi: United Nation Development Programme.

UNEP. (2012). Organized crime trade worth over US£30 billion responsible for up to 90 % of tropical deforestation. *UNEP Press Release*, 29/9/2012. Retrieved June 3, 2015, from http://www.unep.org/newscentre/default.aspx?DocumentID=2694&ArticleID=9286

UNODC. (2010). *The globalization of crime—A transnational organized crime threat assessment*. United Nations Office on Drugs and Crime.

UNODC. (2013). *Transnational organized crime in East Asia and the Pacific*. Bangkok, Thailand: United Nations Office on Drugs and Crime—Regional Office for Southeast Asia and the Pacific.

Walters, R., Wyatt, T., & Westerhuis, D. (2013). *Emerging issues in green criminology: Exploring power, justice and harm*. Houndmills, Basingstoke, UK: Palgrave Macmillan.

Wells, A., Gatto, F. D., Richards, M., Pommier, D., & Contreras-Hermosilla, A. (2007). Rural livelihoods, forest law and the illegal timber trade in Honduras and Nicaragua. In L. Tacconi (Ed.), *Illegal logging: Law enforcement, livelihoods and the timber trade* (pp. 139–167). London: Earthscan.

White, R. D. (2009). *Environmental crime: A reader*. Cullompton, UK: Willan.

White, R. (2010). *Global environmental harm: Criminological perspectives*. Cullompton, UK: Willan.

World Bank. (2002). *A revised forest strategy for the World Bank Group*. World Bank Group.

World Bank. (2007). Weak forest governance costs US£15 billion a year. *News Release No. 2007/86/SDN*. Retrieved June 3, 2015, from http://web.worldbank.org/WBSITE/EXTERNAL/COUNTRIES/AFRICAEXT/EXTAFRSUMESSD/EXTFORINAFR/0,,contentMDK:21055716~menuPK:2493506~pagePK:64020865~piPK:149114~theSitePK:2493451,00.html

Wright, G. (2011). Conceptualising and combating transnational environmental crime. *Trends in Organized Crime, 14*(4), 332–346.

WWF. (2008). Destruction of Sumatra forests driving global climate change and species extinction. *WWF*. Retrieved June 3, 2015, from http://worldwildlife.org/press_releases/destruction-of-sumatra-forests-driving-global-climate-change-and-species-extinction-wwf

Wyatt, T. (2012). *Green criminology & wildlife trafficking: The illegal fur and falcon trades in Russia Far East*. Saarbrücken Lap Lambert Academic Publishing.

Wyatt, T. (2013a). The Russian Far East's illegal timber trade: An organized crime? *Crime Law and Social Change, 1*(61), 15–35.

Wyatt, T. (2013b). Uncovering the significance of and motivation for wildlife trafficking. In N. South & A. Brisman (Eds.), *The Routledge international handbook of green criminology*. London: Routledge.

Wyatt, T. (2013c). *Wildlife trafficking—A deconstruction of the crime, the victims and the offenders*. Basingstoke, UK: Palgrave Macmillan.

Wyler, L. S., & Sheikh, P. A. (2008). *International illegal trade in wildlife: Threats and US policy*. DTIC Document.

Yin, R. K. (2009). *Case study research: Design and methods*. Thousand Oaks, CA: Sage.

Zilney, L. A., McGurrin, D., & Zahran, S. (2006). Environmental justice and the role of criminology an analytical review of 33 years of environmental justice research. *Criminal Justice Review, 31*(1), 47–62.

2

Convergence between Green Criminology and Security Studies: the Framework to Examine Timber Trafficking

The primary aim of this chapter is to develop an overarching conceptual framework that helps orient the course of investigating timber trafficking. To achieve this aim, two main tasks need to be carried out. First, literature on green criminology will be examined, which will clarify the specific perspectives from green criminology that are relevant and then employed throughout this research to conceptualise timber trafficking, its causes, offenders and victims. This task also helps to identify some potential gaps in knowledge in which further green criminology research would be interested. As will be seen, one of these gaps is the lack of conceptual frameworks that can be adopted to assess the consequences of green crime in an inclusive and systematic manner. This is the motivation to conduct the second task of this chapter, which is to reconcile the chosen perspectives in green criminology with a "compatible" approach in the discipline of security studies with a view to formulating the desired overarching framework.

N.A. Cao, *Timber Trafficking in Vietnam*, Palgrave Studies in Green Criminology,
DOI 10.1007/978-3-319-64280-2_2

Green Criminology and Conceptualisation of Timber Trafficking

A scholarly branch has recently developed within criminology, established to specifically examine various forms of harm and crime against the environment, which have long been either overlooked by mainstream criminology or subsumed within the study of other criminal types such as corporate or organised crime (South and Beirne 2006). This branch is frequently termed "green criminology".

> Green Criminology can be defined as a framework of intellectual, empirical and political orientations toward primary and secondary harms, offences and crimes that impact in a damaging way on the natural environment, diverse species (human and non-human) and the planet. (Ruggiero and South 2013: 360)

The term was first proposed by Lynch (1990)—an American criminologist—as he envisions that radical criminology could benefit from an association with environmental movements. Over the past two decades, the criminological discourse has witnessed the rapid growth of green criminology, in which a number of in-depth projects, focusing on both obtaining empirical data and developing theoretical frameworks, have been conducted. A number of well-known criminologists in the West have made significant efforts to develop it as "an independent and unique branch of the criminological field" (Cao and Wyatt 2013: 2), and "a staple at most international criminology conferences" (White 2013c: 8).

There are different semantics relating to green criminology,[1] including "green perspectives" (South 1998), "conservation criminology" (Gibbs et al. 2010; Halsey 2013; Herbig and Joubert 2006), "environmental criminology" (Clifford and Edwards 2012), "eco-global criminology" (Ellefsen et al. 2012; White 2009, 2010b, 2011), "eco-critical criminology" (Lynch and Stretesky 2007), "eco-crime" (Walters 2010) and "eco-global crime" (Ellefsen et al. 2012). Despite the different terminological usages, these perspectives share a common spirit that challenges conventional perceptions in criminology about how to conceptualise the nature and extent of environmental harm and crime.

In addition, these terms—with a somewhat different focus—can be seen as "distinct sub-areas or perspectives" in the scholarly area of green criminology as a whole (White 2013a: 23). Henceforward, this research will consistently use the term "green criminology" as favoured by many green criminologists. Doing so, however, does not imply that other terms are undervalued.

Considering the aims of green criminology, despite the different suggestions of green criminologists from different countries,[2] South (1998: 212) initially points out that its aims are "both modest and ambitious". The reason is that rather than claiming to put forward a "theory" of green criminology, it humbly highlights the significance of a "green" consideration as a "sensitising perspective" (South 1998: 212). In other words, a green criminology is more of a guideline than a set of axioms, rules or rigid theoretical structures (Lynch and Stretesky 2003). However, green criminology is "ambitious" because it challenges the status quo of orthodox criminology that is limited within criminal law and criminal justice (Lynch and Stretesky 2014), and it highlights the area of green issues, known globally as prominent issues in the twenty-first century, that have been ignored by mainstream criminology (South 1998).

Some may raise the fundamental question as to why green criminology is needed. The key answer is that although there are a range of crimes and criminal justice activities directly reflecting environmental issues (Potter 2010, 2014), overall mainstream criminology tends to either overlook or show essential limitations and irrelevance in terms of identifying and evaluating the nature, causes and actual seriousness of these issues. As noted in the introductory section of Chap. 1, Lynch (2013) and Lynch and Stretesky (2014), for example, estimate that in the USA, green victimisations due to air pollution exposure, water pollution and toxic waste exposure are about 33.6 billion times, 2100 times and 20,000 times respectively more likely to happen than victimisation incidents induced by violent street crime. However, traditional criminology, despite a recent "greening" in the studies of some green crime types (White 2013a: 26), ignores or is simply unable to identify these problems.

This is possibly because of the traditional criminologists who take criminal justice stipulated in criminal laws as the only valid approach to justice and the vast majority of criminological works are delivered within this frame of reference (Lynch 2013: 49). Green criminology that employs a variety of approaches to justice would be a promising solution to deal with this limitation of traditional criminology. Indeed, green criminology offers three broad frameworks of analytical thinking and activism efforts, whereby research can follow to address environmental harm and crime. The first framework is *environmental justice* which focuses on evaluating the access to specific natural resources determined by geographical regions and the impacts of environmental threats on specific populations defined, for example, by class, occupation, gender, age and ethnicity classifications (Bryant 1995; Bullard 1996; Schlosberg 2007; Walker 2012; White 2008a, 2011). Concerning the element of natural resource access, environmental justice focuses on assessing the *equity* of the access to environmental resources across social and cultural divides. Who has access to the benefits of natural resources and why? What factors prevent all people from equally sharing in the environment? These are the main questions with which the concept of environmental justice is concerned (Walters 2010).

The second justice framework—*ecological justice*—focuses on examining the relationship between humans and the rest of the natural environment such as plants and creatures which live in the biosphere. It views humans as one component of a complex ecosystem and all living organisms are interconnected; and therefore, environmental issues are essentially transboundary (White 2007, 2008a, 2011). Using the argument that "humans routinely discriminate against nonhuman animals with our everyday usage of speciest language" (Beirne 2007: 62), the third framework of *species justice* pays attention to addressing animal rights with the concept of speciesism (Beirne 2007, 2009; White 2008a; Wyatt 2012).

In parallel to these justice frameworks, green criminology employs a range of ideological and philosophical positions concerning the relationship between humans and the natural world. It employs not only the conventional position of *anthropocentrism*, which has been overwhelmingly used in traditional criminology, but also *ecocentrism* and *biocentrism* which have been largely ignored in orthodox criminology.[3]

The employment of the wide arrays of justice frameworks and ecophilosophies allow green criminology to produce more exhaustive evaluations of the systemic and particularistic nature of such harm, which is crucial to formulate more environmentally friendly legislation and respond more effectively to the harm and crime (Eman et al. 2009; White 2008a). For these reasons, the creation of green criminology is imperative and may also be, as Lynch and Stretesky (2014: 275) argue, "a criminological revolution". The following sections will discuss revolutionary approaches in green criminology in order to conceptualise green crime.

Definition of Green Crime

Redefining, or at least offering a variety of ways for defining green or environmental crime, is a key mission of green criminology, being the first effort to distinguish it from orthodox criminology. Natali (2013: 77) indicates:

> The extremely delicate knots we are untying in relation to the fine-tuning of a definition of environmental or green crime have, for some time, been *at the centre* of the debate within green criminology.

At the outset, it is, however, noted that there is "no clear definition of environmental crime" largely agreed across countries and "this has led to an indiscriminate use of the term, contributing to confusion as to its meaning" (Skinnider 2011: 16). There are a variety of ways to categorise various conceptual approaches which aim to define environmental crime. Table 2.1 details a suggestion by Skinnider (2011) who categorises four different perspectives (legalist, socio-legal, ecocentric and biocentric) to define environmental crime. Other green criminologists use different terms for these definitional methods. For instance, instead of "legalist", Gibbs et al. (2010: 125), White (2011: 21) and Lynch and Stretesky (2003: 229) use the terms "strict legalist", "conventional criminological conceptions" and "corporate perspective", respectively to refer to the method that merely uses criminal law to define green crime.[4]

It can be first observed from Table 2.1 that overall the range of criminalised activities widens following the order of the four approaches. This means that while the first legalist approach only criminalises a very limited number of environmentally harmful activities, in the biocentric approach, the scale of acts defined as crime is largest because it counts any

Table 2.1 Different approaches to define green crime, adapted from Skinnider (2011: 16–18)

No.	Perspectives	Definitional foundations	Definitions of environmental crime
1	Legalist	Criminal laws	"Violation of criminal laws designed to protect the health and safety of people, the environment or both"
2	Socio-legal	All forms of legal provisions including criminal laws and administrative civil laws and regulations	"Any illegal activity or formal rule-breaking, whatever form the rule might be"
3	Ecocentric	Environmental and ecological impacts	"Those acts that have identifiable environmental damage outcomes and originated in human action but that may or may not violate existing rules and environmental regulations"
4	Biocentric	Species justice	"Intentional or negligent human activity or manipulation that impacts negatively on the earth's biotic and abiotic natural resources, resulting in immediately noticeable or indiscernible natural resource trauma"

human acts that disrupt a biotic system. This range of criminalised activities is demonstrated in Fig. 2.1. However, this pattern has occasional exceptions because in reality there are some criminalised acts, such as growing cannabis, that may not be captured by ecocentric and biocentric approaches as these planting activities can be environmentally friendly (White 2008b).

Concurrently, it is also a general evaluation in green criminology that, in defining environmental crime, the more freedom from criminal law an approach has, the friendlier to the environment the approach may be. This is to say that although relying on criminal law to determine what is an environmental crime may be precise and "value-free", as argued by mainstream criminologists (Lynch and Stretesky 2003: 229), the definitional legalist methodology shows salient problems.

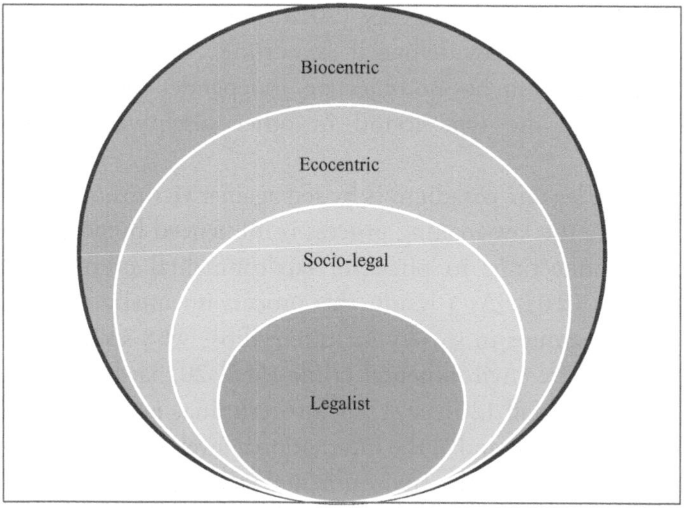

Fig. 2.1 Scope of criminalised acts in different definitional approaches in green criminology

First, since "crime" in criminal law has "no ontological reality" (Hall 2013a: 14), a definitional approach that focuses only on criminal law results in criminologists not creating an objective and independent definition of crime and substituting the legal definition of crime for a scientific definition (Lynch and Stretesky 2014). Additionally, there are some fundamental differences between green crime and other types of crime (Spapens 2014), but criminal law fails to recognise these differences (Mares 2010). UNODC (2012: 1), for instance, compares the difference in criminalising forest crime and poverty crime:

> Most property crimes, such as robbery, theft, arson and vandalism, are criminalised because they inflict harm on people or man-made property by creating uncertainty, diminishing confidence, and harming commerce and economic growth. All of these reasons apply for criminalising the same acts against natural resources. However, there is an additional dimension to the fight against wildlife and forest crime; legislation to protect wildlife and forests also aims to ensure the sustainability of natural resource systems. This sets a different dynamic for wildlife and forest law enforcement.

Therefore, Lynch and Stretesky (2014: 5) argue that if a crime is a crime, since criminal law defines it as a crime, the "clearly tautological identification of crime has no objective, independent point of reference or definition of the type found in other disciplines identified as sciences".

Second, the legalist paradigm is biased against victims of green crime. This is because the law-making process is influenced by politicians who have close connections to powerful environmental criminals (Simon 2000; White 2010a). As a result, this process unequally represents the interests of all groups in society including those who are powerless but heavily affected by environmental crime (Hall 2013a, b; Mares 2010). Investigation by Hall (2013a: 137) into criminal justice, both within domestic jurisdictions and at the international level, unveils that "in most criminal justice systems, the environmental victim still faces considerable barriers to recognition and even greater challenges in terms of their participation in the justice process". For the crime against non-human animals, for example, much criminal law largely considers animals as human property, failing to recognise the nature of the crime and its victims[5] (Nurse 2013a, b; Wyatt 2012).

Third, the legalist view would curb further worthy criminological inquiries, thereby disregarding some important responsibilities of criminologists. Indeed, while "a vast majority of ecological damage is allowed under the law" (Stretesky et al. 2014: 6), if remaining a corporate standpoint, how would criminology interpret environmental harm that falls outside the criminal law? There are many practices such as selecting residential land of ethnic minority communities as waste sites, overselling pesticides to farmers, clear-felling of old-growth forests, ecocide, and animal abuse and cruelty that would be perceived as standard and "above the law" businesses (Brisman 2012; White 2011). It is thus difficult for orthodox criminologists to obtain an understanding about these practices.

Additionally, in reality, in most countries, violations against the environment are typically regulatory or administrative offences that are included in various environmental bills rather than as provisions in criminal codes (Skinnider 2011). In New South Wales, Australia, for example, the majority of environmental violations are civil, regulatory and non-criminal

(Westerhuis 2013). Likewise, in Vietnam, although green crime shows an upward trend, the number of environmental violations handled by criminal laws is very minimal, accounting for about 1.1% of the total number of violations against environmental laws and regulations (Nguyen 2010). In the field of illegal logging, while much illegal and destructive logging is conducted by, or with the complicity of, states, which concurrently provide the formal definition of "the crime of illegal logging", it would be problematic if criminologists "use the criminals' own definition of 'crime' to study this crime" (Green et al. 2007: 119). A four-year study that assesses the role of criminal laws and criminal justice systems in controlling illegal logging in four heavily forested countries (Brazil, Indonesia, Mexico and the Philippines) concludes that the cumulative probability of an incident of illegal logging being charged with a crime is only under 0.082% (Akella and Cannon 2004).

The inherent limitations of the legal-procedural, corporate or legalist methodology bring about a motivation for green criminology to develop new trends, such as socio-legal, ecocentric and biocentric methodologies, in the course of defining green crime. These diverse perspectives, as claimed by Gibbs et al. (2010: 127), "encourage discussion about how to move beyond a legalistic definition of environmental crime". Stretesky et al. (2014: 2), for example, define green crime as "acts that cause or have potential to cause significant harm to ecological systems for the purpose of increasing or supporting production". Green crime in this definition encompasses: (1) acts that are already defined as crime by the state in the form of administrative and regulatory law, and (2) acts that are not criminalised but scientific evidence suggests that they may cause significant ecological destruction (Stretesky et al. 2014).[6]

This research does not use a legalist approach to define timber trafficking. Instead, by employing a socio-legal method, as partly indicated in Chap. 1, timber trafficking in this research is a form of green crime defined as any acts that violate both criminal and administrative legal provisions designed to maintain the sustainable operation in harvesting, transportation, processing of, and trade in, timber. These illicit acts are thus considered as the crime of timber trafficking.

It is hoped that the socio-legal approach will help this research minimise some of the inherent drawbacks of the legalist approach. For example, the range of logging activities defined as timber trafficking in this research would be far larger than those which concern only behaviours stipulated in the Vietnamese Penal Code.[7] A given act of illegally logging a certain volume of timber that does not meet the criminal level, defined by criminal laws, but meets the level of administrative laws or regulations, is still considered as timber trafficking in this research. Still, a logging operation that is authorised by competent bodies would be seen as timber trafficking in this research if the logging permit is obtained unlawfully, such as via corruption and forged documents; or this logging operation does not entirely comply with specific terms of the logging regulations such as exceeding the approved volume of timber and/or the approved areas of harvesting.

Due to the employment of a socio-legal approach that is concerned only with illegal behaviours, entirely legal activities are beyond the scope of this research, though these legal activities are indeed worthy of examination in green criminological research. It is believed that focusing on illegal practices helps keep this research manageable. Moreover, illegal practices do have significant harmful impacts on the environment, and are thus also worthy of investigation in a green criminological study. In fact, in some cases, illegal logging may be more environmentally detrimental than legal logging because legal logging operations, if properly conducted, only harvest the volumes of timber permitted by competent authorities. Moreover, the legal logging must strictly follow all logging procedures designed to minimise possible harm to the environment.[8] Conversely, illegal logging operations, if not effectively controlled, may remove vast, if not limitless, amounts of trees. Furthermore, illicit logging practices use harvesting methods that are often most ecologically destructive.[9] Meanwhile, in the context of Vietnam, where there is a clear dearth of understanding about timber trafficking, it is perhaps necessary to focus on illegal activities first, and then concentrate on legal practices in subsequent examinations.

The Aetiology of Green Crime

As mentioned previously, the embryonic idea of green criminology was not to build a comprehensive theory of green criminology, but to have a "green sensitivity" when examining green crime. Lately, Brisman (2014:

29) presumes that uncovering the aetiology of environmental crime and harm is not the *raison d'être* of green criminology. It is, however, surely the case that green criminologists would not renounce the use of theoretical foundations, but employ various existing theoretical frameworks from mainstream criminology, as well as other scholarly fields, to analyse the nature and causes of environmental crime. To name a few, these are masculinities theories (Groombridge 1998), control theory (Du Rees 2001), corporatist views (Lynch and Stretesky 2003), non-speciesist approach (Beirne 2007), environmental and social justice theories (White 2008a), deep ecology perspective (Wyatt 2012), strain theory (Agnew 2012a, b) and the ecofeminism perspective (Sollund 2013). Most extensively, Agnew (2013) brings together five major theories in mainstream criminology (e.g. strain, social control, self-control, social leaning and opportunities theories) to explain what he terms "ordinary harms" that contribute to ecocide. This diverse use of theoretical framework has played an important role in the development of green criminology.

It is now argued that the time to introduce a unified green criminological theory has arrived; such a theory would help green criminology establish its theoretical and definitional consensus and make a greater meaningful contribution to the domain of criminology (Graife and Stretesky 2013; Lynch and Stretesky 2014). In the past few years, within the field of green criminology, there have been a number of efforts made to introduce distinctive theoretical arguments on the cause of green crime. Dybing (2012: 291), for example, introduces the notion of "distance" as "a useful concept to contextualise the social cause of environmental harm". The introduction of a notion of "justice" through the frameworks of human, animal and ecological rights (White 2008a; White and Heckenberg 2014), "power and justice" (Wyatt et al. 2013), "popular green criminology" (Kohm and Greenhill 2013) and "cultural green criminology" (Brisman and South 2014) may also offer fresh approaches to an understanding of the roots of green crime.

Perhaps, the most comprehensive contribution to this topic is the work undertaken by a team of green criminologists (Graife and Stretesky 2013; Long et al. 2014; Long et al. 2012; Stretesky et al. 2014). They base their work on the theory of the "treadmill of production" originally developed by Schnaiberg (1980) to enhance the theoretical framework that can be applied to "truly understand and remedy green harms" (Lynch et al.

2013: 997). This framework, now known as the "treadmill of crime", is believed to help provide a radical explanation for the political economy of green crime with its focus on the forms of ecological disorganisation generated by the capitalist system worldwide.

More specifically, under the capitalist mechanism of the "treadmill of production", global production is constantly expanding. Consequently, this mechanism must extract natural resources and convert them into manufacturing processes, which inevitably "destroys the integrity of nature and its reproductive network" (Stretesky et al. 2014: 4). A fundamental element in the treadmill of crime theory is a dual simultaneous mechanism of "ecological withdrawal" and "ecological additions". Specifically, since the driving force of capitalism is the constant expansion of production and profit, it also consistently expands its extraction of raw materials from nature, damaging the finite natural world. Simultaneously, the procedures of both withdrawing the materials and processing them produce massive amounts of pollution and other forms of "ecological additions". These simultaneous processes create a continuous and expansionary mechanism of ecological disorganisation, which is the aetiology of green harm and crime (Stretesky et al. 2014).

In the present research, some of the conceptual elements involved in these perspectives are taken into account in the course of identifying and evaluating the main drivers of timber trafficking in Vietnam.

The Perpetrators of Green Crime

Green criminology has offered fresh and critical ways to define green crime and address its causes. The same applies to green perpetrators. This is another important dimension of green criminology, one that carries out a critical examination of the responsibility of the powerful and corporations for generating numerous environmental harms (White 2010b). Although the range of perpetrators of green crime is wide and diverse, "from solo fly-tippers to huge multinational corporations" (Bell and McGillivray 2008: 264), identifying and naming the corporation and state actors as "criminals" is one duty of green criminology (White 2010b: 6).

Green criminologists reach a high consensus on this point. They strongly challenge the continuing orthodox viewpoint that (1) crime is the conduct of the poor and the powerless (Reiman 1995) and (2) criminology is mainly a science designed for controlling and oppressing the marginalised (Lynch 2000). Instead, green criminology "tends to begin with a strong sensitivity toward crimes of the powerful" (White 2013a: 22) who are "well-connected members of society" (Spapens 2014: 224). Simon (2000) reveals two main types of powerful perpetrators against the environment: (1) governmental institutions and (2) chief executives and key shareholders of industrial corporations. Specifically, White (2010a) points out that those who control the law are very frequently those whose activities should be criminalised for the purpose of maintaining planetary well-being. To elaborate this argument, South et al. (2013) demonstrate that the legislators tend to be members of socio-economic elites whose organisations of production are culpable for many, if not most, examples of environmental harm. These law-makers then formulate laws that facilitate their business, but which also cause harm for others, for instance, pollution.[10]

There are several studies that address the responsibilities of states and corporations for different forms of environmental harm and crime. Lynch and Stretesky (2001: 165), for example, examine major manufacturers of pesticides and dioxins that "show a blatant disregard for the effects of their products and by-products on human and animal populations". Kramer (2014: 24) investigates the destructive relationship between the fossil fuel industry and nation states that "allows catastrophic climate change and its victimisation to continue unabated". In the sphere of timber trafficking, the biggest timber harvesters in many countries are state companies (Richards et al. 2003), and among the richest, most powerful men in many parts of the world are "untouchable timber barons" who have very close relationships with major political leaders (Telapak and EIA 2007: 14).

In short, the discussion of the perpetrators of green crime suggests that there are a wide array of actors who engage in green crime, of whom the powerful may be those most often overlooked by the public and academia, despite their generation of massive environmental damage. This encourages the present research to examine a variety of offenders, with close attention paid to powerful actors engaging in timber trafficking in Vietnam. Additionally, this research also needs to obtain a detailed understanding of the distinctive characteristics of each type of timber trafficker.

This offender classification is important because different criminal groups are driven by different motivations, commit different patterns of behaviour and should be tackled by different controlling measures that should "fit the circumstances of the offenders", rather than adopt a "blanket approach" (Nurse 2013b: 141).

Victims of Green Crime

Another central feature that distinguishes green criminology from other criminological perspectives is that of a radical approach in identifying "who or what precisely is being victimised" (White 2008a: 122). This identification is "crucial" to fully understand various impacts of green crime (White and Heckenberg 2014: 343). Despite this importance, literature that is concerned with victims of environmental crime is somewhat embryonic. Indeed, all scholars on this topic share a common remark that both criminologists and victimologists have generally excluded victims of environmental crime in their research, and thus "the voices and views of these victims of environmental harm have largely remained absent from the relevant literature, and indeed from policy debates" (Hall 2014a: 135).

This may be, first, because victims of environmental harm are not commonly recognised as victims of crime, but instead environmental crime is considered as "victimless", or "lacks concrete identifiable victims" (Skinnider 2011: 23), or involves "invisible victims" (Spapens 2014: 221), or in the form of wildlife crime, it is seen as "harmless" (Crow et al. 2014: 185). As a result of this underestimation, the victims are frequently overlooked or go unnoticed for a lengthy period of time (Skinnider 2011). Second, victims of green crime are often incapable of fully speaking up about their suffering. Spapens (2014: 224) notes that "flora and fauna do not report crimes. The same is often true for humans".

Third, as addressed earlier, a large quantity of environmental crime is not criminalised, and may even be considered as "legal and takes place with the consent of society" (Korsell 2001: 133), meaning that green victimisation does not fit neatly within "standard" conceptions of victims of crime that have been commonly adopted by most criminal justice systems (Hall 2013b). As a consequence of this exclusion, criminology, and particularly victimology, have excluded "those who most need to be empowered and to have their rights protected" (Lynch and Stretesky 2014: 81). This study intends to reduce in part this exclusion by investing its principal effort to systematically scrutinise the diverse impacts of timber trafficking, particularly on marginalised individuals, in the context of Vietnam.

When studying green victims, two key aspects must be examined: (1) who and/or what are victims of green crime; and (2) how the victims suffer as a result of the crime. Concerning the first aspect, the work undertaken by Williams (1996) sets out a key foundation. Williams (1996) postulates that there are perspectives in environmental victimology which intersect with critical victimology, namely those questioning who is judged as a victim and who benefits from the dominant viewpoint on crimes and their victims. Williams (1996: 204) then defines environmental victims:

> Those of past, present, or future generations who are injured as consequences of change to the chemical, physical, microbiological, or psychological environment, brought about by deliberate or reckless, individual or collective, human act or act of omission.

Yet, as a result of following a conventional human-centred conception, it can be seen that the types of victims addressed by Williams above are humans only. Contemporary green criminology recognises that non-human species and the environment are "capable of suffering such harm, and thus deserving of protection" (Wyatt 2012: 27). Accordingly, there is a significant part in the current discussion of green criminology that emphasises the importance of non-human species and the environment as real victims of green crime.[11]

After recognising who and/or what constitute the victims of green crime, it is important to examine to what extent and how these victims are harmed. Regarding this point, green criminologists are in line with key arguments found in the two closely connected concepts of environmental justice and environmental racism.[12] Concerning the environmental justice concept, as discussed later in "The Concept of Security" section, it is argued that various forms of green harm and crime such as climate change, toxic dumping, chemical spills, industrial pollution, nuclear testing and illegal fishing affect every group of victim, but more importantly, there is an endemic hierarchy of victimisation levels, in which vulnerable groups such as poor countries, lower classes, aboriginal communities, women, children and non-human species are likely to be victimised more severely (Bryant 1995; Bullard 1996; Schlosberg 2007; Walker 2012; White 2008a, 2011). Women, children and non-human species are all, for example, regarded as "natural resources", and are "significantly devalued as individuals through the objectification of their bodies" (Sollund 2013: 324). In the USA, Bullard (1996: 22) observes that "it is the poorest among the nation's inhabitants who are being poisoned at an alarming rate".

Similarly, environmental racism scholars are particularly concerned with the racial discrimination associated with the distribution of environmental harm especially toxic pollution (Chavis 1994; Holifield 2001; Pulido 2000; Stretesky and Lynch 1998). As defined by Chavis (1994: xii):

> Environmental racism is racial discrimination in environmental policy-making and enforcement of regulations and laws, the deliberate targeting of communities of color for toxic waste facilities, the official sanctioning of the presence of life threatening poisons and pollutants for communities of color, and the history of excluding people of color from leadership of the environmental movement.

For Chavis (1994), intent or deliberate discrimination is the defining element of environmental racism. Whereas, Pulido (2000: 33) argues that "multiple forms of racism exist, including less conscious forms not characterized by malicious intent and hostility", meaning that "any decision-making processes and distributive patterns that burden minority groups disproportionately provide sufficient evidence of environmental racism" (Holifield 2001: 83). Whether intent is a defining element or not, the concept of environmental racism, together with the environmental justice concept, have inspired green criminologists to pay significant attention to addressing the disproportionate levels of victimisation of the marginalised caused by green harm and crime. This aspect will also be of a substantial concern when examining the impacts of timber trafficking in Vietnam, which will be revealed in Chap. 5.

Regardless of kinds of victim and levels of victimisation, green victimisation is presumed to have several typical characteristics. Among scholarly pioneers on green victimisation, Skinnider (2011) proposes four distinctive features: (1) the unawareness or very late awareness of victims about the victimisation, (2) the uncertainty over who is the culprit of the victimisation and who is responsible, (3) the seriousness of green victimisation is due more to the numerous number of victims than the serious impacts on individual victims, and (4) the involvement of repeat offences.[13] However, Hall (2013a: 26) argues that the work by Skinnider is "largely descriptive", grounded in the Canadian perspective and "does not draw on wider criminological and victimological theory". Accordingly, at first Hall (2013a, b) presents a typology of green victimisation that is split into four categories: health impacts, economic impacts, social and cultural impacts, and security impacts. Afterwards, Hall (2013a) suggests a distinguishing attribute of green victimisation, which is subsequently substantiated by Hall and Farrall (2013) and Pemberton (2014), that there is a substantial overlap between victim and offender. It is argued that worryingly, some of those most severely impacted by green victimisation might actually find themselves recast as green offenders. In other words, some kinds of real victims "substantially overlap" with the real perpetrators of green crime (Hall 2013a: 38). These victims "are often supportive of, or participate in, the activities that harm them" (Pemberton 2014: 68). Additionally, it is posited that being a green offender increases the risk of becoming a green victim (Pemberton 2014).

To advance the discussion on victimisation from green crime, Hall (2011) initiates and then Hall (2013a, 2014a) elaborates on the term *green victimology* to study key issues and the complexities relevant to the impacts of environmental harm and crime on individuals and (sometimes large) collectives of individuals.[14] Green victimology is also independently termed, and provided with "a preliminary foray", by Lynch and Stretesky (2014: 101). To develop the domain of green victimology, these founders call for a critical approach, an interdisciplinary methodology, a careful consideration of the values of all three broad groups of victims consisting of human, non-human beings (flora, fauna, insects and microbes) and ecosystems, and their constituent parts (Hall 2011, 2013a, 2014a; Lynch and Stretesky 2014).

From the discussion above, it can be seen that the preliminary boundary and foundations of green victimology have been set out. However, as clearly stated by Hall (2013a: 145), "green victimology still has long way to develop". Some of the main areas for further development of green criminology and green victimology will be discussed in the next sub-section.

Challenges and Limitations of Green Criminology

Despite contributions from green criminology to the understanding of the essence, extent, causes, offenders and victims of green crime, it appears that green criminology faces some philosophical and analytical critiques. In a provoking publication named "Against Green Criminology", Halsey (2004) heavily criticises core arguments in green criminology, asserting that green criminology "does not capture the inter-subjective, inter-generational or inter-ecosystemic processes which combine to produce scenarios of harm"; and that it misinterprets the nature and extent of the task at hand (Halsey 2004: 835).[15] Halsey ultimately suggests that the term "green" should be eliminated from the criminological agenda.

Added to the critique by Halsey are a number of refinements that green criminology may need to undertake to become a more influential body of knowledge. To begin with, much literature on green criminology appears to neglect the indispensable connection between green crime and political economy (White 2013a, b). As environmentally harmful acts are driven regularly and persistently by economic purposes within the overarching context of a distinct global political economy, discourse on these acts must always take into account political economic aspects (White 2013b). In other words, to thoroughly understand environmental harm and crime, green criminological research should "continually return to political economic groundings" that formed the original intention of green criminology (Lynch and Stretesky 2014: 179). However, in the past two decades, the majority of the discussions under the banner of green criminology appear to overlook the connection (White 2013a: 27).

More pragmatically, while green criminology calls for minimising or even eliminating ecologically harmful practices, it appears that effective solutions for dealing with the side-effects and the costs of this elimination are not recommended in depth in the agenda of green criminology. In other words, much remains to be done to find solutions for balancing positive and negative outcomes when a green policy is implemented. It is particularly important when the negative consequences impact underprivileged communities. Research in green criminology, therefore, needs to address the question of how indigent communities that rely heavily on natural resources such as timber, minerals and fish can overcome the hardships caused by implementing green policies of, for example, banning the harvesting of such resources. For instance, after examining the significant hidden costs paid by deprived communities in the north-east of England resulting from the closure of one of the world's largest producers of aluminium on the grounds of its causing pollution, Davies (2014: 312) is concerned that

> The moral and ethical challenge for a green criminology is when, why and how should green justice override other risks and harm to communities. In this case, a green view of justice collides with other communitarian accounts of justice.

Furthermore, Davies (2014: 312) suggests a further sub-set of research around communitarian victimisation that "seeks to explore unidentified regressive impacts resulting from some environmental policies and the potentially discriminatory nature of environmental justice".

Another limitation is that geographically, literature on green criminology is dominated by very "western, transatlantic and global north" scholars (South 2012b: 16). To be a global scholarly discipline, green criminology must "capture the voices of researchers, scholars and activists from different parts of the world by considering their views and building them into our research and activities" (White and Heckenberg 2014: 188). There is no doubt that environmental problems in general, and timber trafficking in particular, are severe and omnipresent in the East, particularly in many developing countries. Yet, green criminology inspires very little interest from Eastern criminologists. Regions of the world such as India, Central Africa and Southeast Asia, while surely important to be involved in green criminology, "have often been side-lined by Western criminology" (Hall 2013a: 150).

With regard to green victimology, it has little intellectual connection with green criminology (Spencer and Fitzgerald 2013). At the same time, its arguments seem to be relatively speculative. Hall, for example, suggests four aspects (health, economic, social and cultural, and security impacts) that are gravely affected by green crime, but these observations seem to be general, lacking penetrating and multidimensional insights into the victimisation process, and are, moreover, not sufficiently underpinned by empirical evidence. Admittedly, the topic of victims of environmental crime still remains "a surprisingly under-researched and under-theorised aspect of environmental degradation" (Hall 2013a, b: 145).

Finally, environmental crime evidently cannot be separated from security because the former is one of the key, direct and intensive threats to the latter. This suggests that scholarly discussions in green criminology should have a strong connection with academic discourses in the realm of security studies. However, as observed by Hauck (2007), Elliott (2007) and South (2012a, 2015), investigations of how environmental crime is linked to security, particularly non-traditional security perspectives, have rarely been undertaken by criminologists in general and green criminologists in particular. Indeed, South (2012a: 105) observes that "there is still much to explore in the territory where matters of environmental harms, crime and conflict all meet and overlap with questions concerning security, rights and law".

To summarise the review of literature on green criminology, it can be seen that as a result of the practical complexity of environmental issues and the intersection among conceptual approaches surrounding the environment, crime and harm, there are necessarily various ways to study environmental harm and crime. Irrespective of certain criticisms, in particular ones by Halsey,[16] and some limitations, it should be acknowledged that the creation and development of green criminology is imperative and meaningful for both academic and practical purposes. Green criminology helps this research establish its conceptual approaches to define timber trafficking, to ponder some of its possible driving forces, to emphasise the role of powerful offenders, and to identify the types of victim that may bear the brunt of suffering from the crime.

However, green criminology certainly still needs both theoretical and empirical contributions, particularly ones that could lessen the above-mentioned limitations, such as the lack of practical solutions, and multi-dimensional insights into an Eastern context. One of the main contributions of this research is its integration of theoretical perspectives from two different scholarly fields of security studies and green criminology in the undertaking of an empirical investigation into timber trafficking, as one of the important forms of green crime, in the context of the Eastern society of Vietnam. The second part of the chapter will explore the convergences between security studies and green criminology, which serves as justification for combining these two fields into this research. Then it will review key perspectives in security studies in a concerted effort to selecting a relevant and applicable one to develop a conceptual framework that can be employed to systematically evaluate impacts of timber trafficking in Vietnam. It should be noted that the brief review below is not designed to include the entirety of either existing literature on security studies or any reviewed perspectives, but rather it concentrates on the key arguments from the chosen perspectives that are most pertinent to enrich the green criminological discourse.

Human Security and the Evaluation of Victimisation from Timber Trafficking

The Concept of Security

If the scholarly sphere of green criminology is relatively new, the academic realm of security studies, which is chiefly concerned with national security, has been attracting huge academic and political discussions for almost a century. Today, while the term "national security" is often used interchangeably with "security", both terms seem to have been be abused since they are utilised in various fields other than politics (Baldwin 1997). Indeed, *security* is judged as a "promiscuous" notion, "wantonly" employed in a variety of disciplines such as social security, health and safety, financial security, policing and community safety, international relations and peacekeeping (Zedner 2009: 9). Hence, Baldwin (1997: 26) concludes that "no social science concept has been more abused and misused than national security". The good news, however, is that nowadays there is already an established consensus about what security studies entails; at its most basic, it is about threats to survival (Collins 2007).

There are different definitions of security. In common usage, the word "security" refers to "freedom from various risks" (King and Murray 2001). Wolfers (1952: 485) characterises security as the "absence of threats to acquired values", which appears to capture the basic intuitive notion underlying most uses of the term. However, Baldwin (1997: 13) assumes that there is some ambiguity in Wolfers's phrase of "absence of threats", so it may be clearer and more realistic if security is defined as "a low probability of damage to acquired values". In a more comprehensive visualisation, some authors such as Zedner (2009) and Hameiri and Jones (2013) argue that the condition of being without threat is only the "objective state" of security, thus it is necessary to perceive the "subjective state" of security.[17]

Security is an important concept, and it is hazardous if employed without being clearly clarified and specified (Wolfers 1952). Clarifying the concept of security is not only conceptually and analytically important, but it also practically serves as a restraint on rampant applications of the concept (Zedner 2009). In an influential essay, Baldwin (1997) posits seven specific dimensions that need to be clarified in understanding and applying the notion of security. They are security for whom, security for which values, from what threats, how much security, by what means, at what cost, and in what time period.

Among the seven elements, Baldwin confirms that the first two elements (security for whom and security for which values) are most vital and sufficient to define the concept of security, while the last five considerations will help provide guidance for pursuing the concept in reality. However, Little and Cocklin (2007) argue that the two factors which shape definitions of security are what entities we are trying to make secure (security for whom) and what we define as threats to security (security from what threats). It is, therefore, suggested that the first three components proposed by Baldwin, consisting of security for whom, security for which values and security from what threats, are the most crucial prerequisites for the concept of security.

While the three aspects are always interrelated, different perceptions regarding these aspects will inspire different approaches in security studies that are commonly classified into two strands: traditional and non-traditional. Traditional security perspectives mostly under the banner of *state security*, consider states as the central actors of security and the most powerful international actors; security is therefore security of the states. This long-standing concept has dominated thinking in international relations because in global affairs, security of states is always the central and most vital concern (Morgan 2007). Core elements of state security are autonomy, physical safety from foreign military attacks, development and rule in which autonomy would be overriding[18] (Morgan 2007).

However, after the Cold War, there have been changes, both negative and positive, that require new approaches to effectively and thoroughly cope with security issues (Okubo 2011; Sen 2000). The negative changes mean that traditional perspectives are not able to explain a wide range of contemporary security issues, in particular the fact that secure states do not automatically mean secure people (Human Security Centre 2005; Voelkner 2012). Kaldor (2007: 196) claims that a narrow state-centred perception "would do nothing to overcome the insecurity experienced by individuals and communities in large parts of the world, especially in the developing world". Indeed, protecting citizens from foreign attacks would be a necessary condition for the security of individuals, but surely it is not a sufficient and not always unique and effective one (Newman 2010). Even worse, although the state remains the "fundamental purveyor" of security, "it often fails to fulfil its security obligations—and at times has even become a source of threat to its own people" (Commission on Human Security 2003: 2). In fact, during the past 100 years, far more people have been killed by their own governments than by foreign armies (Human Security Centre 2005). The old-fashioned analytic frameworks defined by wars between states has become largely irrelevant to violent conflicts within states that account for more than 95% of armed conflicts today[19] (Human Security Centre 2005).

What is more, in fact, is that by far the biggest killers in the world are not interstate wars anymore, but extreme poverty, preventable disease and the consequences of pollution (Human Security Centre 2005). A number of other issues, such as population pressures, increasing immigration, economic crisis, increasing inequality and international criminal activity, also have profound implications for the security scenarios at both national and global levels (Alkire 2003; Okubo 2011). Thus, any idea of security which disregards this reality is "conceptually, empirically and ethically inadequate" (Newman 2010: 80). In parallel with the negative changes, on the positive side, there have been technological advances and increased international collaborations that also require new approaches to security to be developed (Okubo 2011; Sen 2000). The meeting of these negative and positive changes brings about the momentum for the creation of non-traditional security perspectives.

Since the 1980s, several major studies (Brown 1986; Buzan 1991; Mathews 1989; Ullman 1983) have begun to embark on the work of "redefining security". All these efforts leave a clear message: the state is no longer able to monopolise the concept and practice of security, while security threats are not confined to military ones from foreign states (Lodgaard 2000). Since then, there has been a variety of approaches to non-traditional security studies such as cooperative security,[20] comprehensive security,[21] environmental security[22] and human security. Each of these approaches is further divided into varying perspectives.

The crucial question now is that among the various conceptual perspectives both conventional and non-conventional in security studies (state security, cooperative security, comprehensive security, environmental security and human security), is there any specific perspective that can be (and should be) employed (or be adjusted) in this research to identify and gauge the victimisation from green crime including timber trafficking.

Interestingly, if green crime is considered as a security threat, the referent object of security, known as security for whom, is evidently the victim of green crime including humans, non-humans and ecosystems. Also, in this sense, the acquired values to be protected from the threat of green crime could be the physical safety and well-being of the crime victims. This means that green crime, its victims and the victim values neatly fit into the security framework particularly with the three most vital dimensions: security for whom, for which values and from which threats. Accordingly, the choice of which specific approach in security studies to adopt for a green research depends on which types of green victims are particularly targeted in a given green criminological study. If a study is, for instance, particularly concerned with states as the key victim of green crime, then the paradigm of state security may be preferred.[23] Likewise, if research focuses on the natural environment as the main victim, an approach in environmental security frameworks would be productive.

While it is this research's stance that any type of victim of green crime should be recognised and deserve equal attention from green research, this study will employ a human security framework to evaluate the victimisation from timber trafficking in the context of Vietnam. This decision is rationalised in three ways. First, rather than examining all possible types of victims of timber trafficking, revolving around human victims would keep this research focused and manageable. Second, as analysed earlier, human security, as a non-traditional security perspective, strongly shares some fundamental conceptual foundations with green criminology, particularly those concerning the role of states and the current structure of power and justice.

Third, as introduced in Chap. 1 on the potential impacts of timber trafficking worldwide, among the various victims, ordinary people, particularly disadvantaged groups such as forest-based indigent dwellers, are likely to suffer some of the most severe consequences of timber trafficking. A human security perspective, which emphasises the security of ordinary people, may well be appropriate and productive to this research. In light of these three considerations, it is envisaged that inviting a human security perspective into the sphere of green criminology would possibly be a useful means to investigate impacts of timber trafficking in this research. The following will be devoted to critically discussing the literature on human security.

Human Security

Human security is relatively new in the security agenda with close attention to the field only recorded from the early 1990s in conjunction with the erosion of the narrow, state-centric, militarised paradigm of security in policy and academic circles (Hampson et al. 2002; Newman 2010). However, the historical development of the idea of human security can be traced to the rising dissatisfaction with prevailing notions of development and security in the 1960s, 1970s and 1980s (Bajpai 2000). These ideas were addressed by prominent academics, intellectuals and leaders in the reports of a number of multinational independent commissions (Bajpai 2000). These philosophical and political ideas, which prioritise ordinary people, culminated in the first introduction of the concept of "human security" in 1994 by The United Nation Development Program (UNDP 1994). In this landmark report, UNDP (1994: 22) declares that "the concept of security has for too long been interpreted narrowly", overlooking "the legitimate concerns of ordinary people who sought security for their daily lives". In brief, based on a normative humanism and ethical responsibility, human security aims to "re-orient security around the individual", not states, though the latter remains an indispensable means for ensuring individual rights, producing economic opportunities and guaranteeing good governance (Dodds and Pippard 2012: 29).

As with other concepts, the formulation of the concept of human security serves certain purposes, the uppermost of which is an attempt to raise profound attention to human security issues in an "insecure and globalising world" (Dodds and Pippard 2012: 35). Pragmatically, it attempts also to invest significant resources more in poverty relief and less in military activities (Kerr 2007). This is because as soon as a state recognises something as a "security issue", it often has the option of dealing with this issue in a manner commensurate with the way it deals with "a war"—that is, one associated with "extraordinary allocations of resources" (Barnett et al. 2010: 6).

Even though there is a long pre-establishment of, increasing policy attention to, and sizable academic investment in, human security, the methodological, definitional and conceptual debates have provided no "real consensus" on the meaning of human security and the application of the human security paradigm (Hampson 2008: 230). There are a number of individual efforts to provide definitions of human security (Alkire 2003; Commission on Human Security 2003);[24] but some scholars such as Thomas (2000), Tadjbakhsh and Chenoy (2007) and Newman (2010) consider human security as a "paradigm", while others see it as a description, a concept, a doctrine, a theory and an ideology (Fukuda-Parr and Messineo 2012).[25] All of these considerations, however, have a highly agreed starting point: that security policy and security analysis, if they are to be operative and legitimate, must revolve around the individual as the referent and primary beneficiary, and that state security is only one of the means to achieve human security (Newman 2010). This has a further implication: that the safety of the individual is the key to global security and thus, when the safety of individuals is threatened, so too is international security (Hampson 2008).

In parallel with the convergence that the primary referent of security is not the state, but individuals, in the course of determining what specific threats should be securitised, there is an analytical divergence in which two main ramifications are offered. On the one hand, the narrow approach to human security taken prominently by the Canadian government and the Human Security Report Project (HSRP), formerly known as the Human Security Centre, focuses on the consequences of armed conflict and the political violence posed to civilians by repressive governments and situations of state failure (Human Security Centre 2005; MacFarlane and Khong 2006; UNDP 2005a). This perspective reflects the negative link between human security and state security, arguing that failed states can no longer provide effective governance of, and invariably fail to obtain, human security, and that at times "outwardly aggressive and inwardly repressive regimes can be major sources of human insecurity" (Lodgaard 2000: 3).

On the other hand, the broad approach to human security is interested in not only threats from political and armed violence, but also non-armed threats to human integrity such as endemic diseases, natural disaster, starvation and severe poverty. Thus the values to be secured in this approach cover not only "freedom from fear", but also "freedom from want" (Commission on Human Security 2003). This approach is heavily influenced and inspired by the Human Development Reports, especially by UNDP (1994), in which human security, as summarised by the Commission on Human Security (2003: 4), means, first, "safety from such chronic threats as hunger, disease and repression; and second, protection from sudden and hurtful disruptions in the patterns of daily life—whether in homes, in jobs or in communities". To demonstrate the concept of security, UNDP (1994: 23) cites how individuals around the world regard security:

> A fourth-grade schoolgirl in Ghana: "I shall feel secure when I know that I can walk the streets at night without being raped". A shoe-mender in Thailand: "When we have enough for the children to eat, we are happy and we feel secure". A man in Namibia: "Robberies make me feel insecure. I sometimes feel as though even my life will be stolen"....

As explicitly clarified in UNDP (1994), human security is constituted by seven interrelated components as demonstrated in Table 2.2. UNDP suggests that there are substantial links and overlaps between these different elements of human security, which means "a threat to one element of human security is likely to travel—like an angry typhoon—to all forms of human security" (UNDP 1994: 33). This interrelatedness of different elements is also the emphasis of the broad approach of human security (Kaldor 2007).

Table 2.2 The broad paradigm of human security, adapted from (UNDP 1994)

Dimensions	Requirements
Economic security	An assured income and livelihood derived from work, the public, environmental resources, or reliable social safety nets
Food security	Having both physical and economic access to basic food
Health security	Having no infectious or parasitic diseases, access to personal healthcare and protective public health regimens
Environmental security	Having a healthy environment, being safe from natural disasters and having access to basic resources such as water and land
Personal security	Physical safety from, and no anxiety of, crime, particularly violent crime
Community security	Being a member of a community with cultural identity and values, and safety from oppressive community practices and from ethnic conflict
Political security	Freedom from state oppression and abuses of basic human rights

Similar to the broad approach of UNDP are a number of proposals including from Nef (1999) who suggests five elements of human security: (1) environmental, personal and physical security; (2) economic security; (3) social security; (4) political security; and (5) cultural security. As can generally be seen, the broad stance of human security is added with the element of human development for the purpose of paying attention to immediate concerns such as basic needs and peace of those who are most vulnerable (Gasper and Truong 2005).[26]

The concept of human security is questioned by some critics. The two biggest questions on human security are: (1) as a result of the expanded notion of security, how can human security, especially its broad version, be reliably measured; and (2) how, therefore, can human security be analytically useful (MacFarlane and Khong 2006; Newman 2010; Paris 2001)? To be more specific, Alkire (2003) suspects three weaknesses in the conceptual framework of human security: vagueness, incoherence and arbitrariness. Meanwhile, Florini and Simmons (1998) expect that theoretically, human security is not an elegant formulation based on a holistic, heterogeneous view of security, leading to confounding policy implications. The consequence would be that the human security approach does not help us understand the causes of threats, the operative mechanisms and the means to achieve (Fukuda-Parr and Messineo 2012), thus limiting the utility for policy analysis (Human Security Centre 2005).

Irrespective of these critiques, the use of the broad human security approach has been increasingly widespread in various fields worldwide, which has "chalked up significant accomplishments" (Paris 2001: 88). Specifically, thanks to this approach, key global governance institutions such as the IMF and the World Bank pay increasing attention to the field of human security with poverty and inequality increasingly seen as threats to national, regional and global security (Thomas 2001). What is more, the construction of human security helps establish an umbrella norm for a variety of international treaties and conventions (Kerr 2007). Foreign policy in Japan and Canada, the security policy in Europe, the establishments of the International Criminal Court, the International Campaign to Ban Land Mines and the International Commission on Intervention and State Sovereignty are "solid achievements", which prove that the human security approach formulated by UNDP has had a "significant role" to play in foreign policy (Fukuda-Parr and Messineo 2012: 14).

This research argues that like green criminology, the birth and growth of the discipline of human security is inevitable and imperative to deal with the changing nature of the contemporary environment of security. It is particularly so when it comes to developing countries where many contemporary issues such as widespread starvation, deadly epidemics, climate change and global warming, turbulent religious and ethnic conflicts, terrorism and ferocious criminal networks clearly and frequently threaten the very security of the majority of the population. These problems often go beyond the explanatory ability of traditional security frameworks. As rightly suggested by Okubo (2011: 11), in the current extraordinary context, "we need some basic, original idea or at least a frame of thinking to acknowledge our existence as human beings".

It is believed in this research that although the broader approach in human security may have to somewhat sacrifice its analytical strength, adopting this "emancipatory and empowering concept" (Voelkner 2012: 19) in the security agenda may well satisfy the actual aspiration of, and bring benefit for, the vast majority of humanity currently living on Earth. The justification is that if the security agenda focuses only on state security, it may only benefit a very small number of people working in governmental apparatuses. Likewise, if the narrow human security approach is prioritised, only people in certain countries with severe political and armed violence can gain advantage. Meanwhile, any individual worldwide can benefit from a security policy that is privileged for all people, and is extensive and intensive enough to address the core values of these people. For instance, rather than investing massively in costly arms races, as observed by MacFarlane and Khong (2006), securitising various issues such as health and the environment to ensure human security has resulted in more policy attention to, and resources for, these domains that are essential to all people.[27] Clearly, the attention and resources may be more essential and urgent for people in developing countries including Vietnam. This is a distinctive contribution of broad human security as justly argued by Barnett et al. (2010):

> A very important and distinctive contribution of human security is that it securitises (makes a priority of) what individuals themselves see as their paramount concerns, and so pluralises the meaning of security and opens up space for alternative security practices.

The above-mentioned strengths of the broad perspective of human security make it favourable to be employed in this research for the aim of evaluating impacts of timber trafficking in Vietnam. Nonetheless, it also needs acknowledging that the conceptual choice of a broad human security is by no means without limitations. By focusing on ordinary people as the main victims of timber trafficking, this research does not examine in detail the security implications for states that may also be negatively affected by timber trafficking.

Likewise, although the broad human security paradigm contains the element of environmental security that allows an examination of the victimisation of non-human species and calls for the sustainable use of natural resources,[28] the underlying basis for the inclusion of this element is because the security of the environment is a vital condition for the security of humans. In other words, the inclusion of environmental security may not be principally based on the intrinsic values of non-human species. It is thus anticipated that the chosen framework would not offer a significant space for a comprehensive scrutiny of the impacts of timber trafficking on non-human species.

The Convergence of Green Criminology and Security Studies

It is largely admitted that global environmental issues are closely linked to national security (Allenby 2000; Barnett 2007; Biswas 2011), and pose new and in some cases unprecedented threats to human security[29] (Lonergan 1999; Matthew et al. 2010; Sygna et al. 2013). South (2012a: 99, 104) asserts that

> Evidently, environmental change and harms pose a challenge to the security and sustainability of nations and their populations… Environmental

> insecurity defines the political and policy mood, with implications for international and national security regimes, as well as for crime and conflict.

It can be observed that "environmental insecurity" can be generated by a number of vital issues such as climate change, global warming, transnational environmental crime, deforestation, pollution and biodiversity degradation, all of these should not be beyond the core curiosity of green criminology. However, as indicated by Elliott (2007), Hauck (2007) and South (2012a), the link between environmental crime and security perspectives has attracted little attention from criminologists let alone green criminologists. Until very lately, in responding to new challenges in the epoch of Anthropocene, a number of leading criminologists have made a "wholehearted" call for connecting criminology, probably via green criminology, with security studies, which is coined as "security-ology" (Shearing 2015; South 2015). Immediately, this call is, however, firmly objected by Floyd (2015: 277) who believes that this idea is analytically weak and practically unseen, and that "not only does it necessitate the rethinking of the meaning of crime, but also criminalisation [not securitisation] can be used to achieve compliance with environmental legislation".

It is argued in the present research that there are significant theoretical and practical convergences between green criminology and security studies; and that employing a non-traditional security perspective such as human security would be relevant and fruitful to investigate the impacts of green crime. Indeed, there are probably four main pillars that underpin this argument.

The initial foundation stems from the nature of criminology and green criminology per se. Considered as a "rendezvous discipline" (Holdaway and Rock 1998), disciplinary hybridity is one of the greatest strengths of criminology (Zedner 2007). The same applies to green criminology. Green criminology is not planned to be a "unitary enterprise", but it is "an open framework" (Ruggiero and South 2013: 361), and diversity is one of its great strengths (South 2012b; South et al. 2013). Leading figures in green criminology consistently argue that in order to adequately address the present green issues, green criminology needs to open up and collaborate with ideas and materials from a wide array of disciplines within both social sciences and natural sciences. Such an interdisciplinary orientation, therefore, allows green criminology to obtain "diverse conceptual and empirical insights into the nature and dynamic of environmental wrongdoing" (White and Heckenberg 2014: 25); and creates the "scope for transposing ideas and solutions between subject areas" (Hall 2013a, b: 140). To put it differently, green criminology can (and should) be blended with other academic fields. The field of security studies, in this sense, is not an exception.

Second, the domain of security studies has strong conceptual links to criminology in general and green criminology in particular. Carrabine et al. (2009: 403) point out that green crime is a feature of a global "risk society". The "risk society" here refers precisely to an "insecure society" because, as Barnett (2007: 183) confirms, insecurity closely links to risk: "in its most basic sense, insecurity is the risk of something bad happening to a thing that is valued". Specifically, green crime leads its victims to insecure situations, so it is rational and logical to adopt relevant tools, which assess the status of (in)security of the victims, to assess the victimisation from green crime. Green criminology thus "encompasses some of the tenets" of security studies (Hauck 2007: 270), and "the theories and histories of security that we now have are certainly useful to criminology" (Valverde 2014: 383).

Zedner (2009) provides an excellent investigation into the increasing connectedness between the concept of security and criminology. She asserts that while security is "too big an idea to be constrained by the strictures of any single discipline", the very nature of criminological efforts and central criminological presumptions have now been shifted by security[30] (Zedner 2009: 10). Valverde (2011, 2014), meanwhile, proposes an alternative way of controlling crime via the provision and governance of security. Most recently, Shearing (2015: 264), a well-regarded international criminologist, posits that "pursuing lines of enquiries that challenge criminology's established boundaries is crucial if it is to continue to advance understanding of the governance of security".

There is an interesting coincidence that both human security as a nontraditional security approach and green criminology were born at a similar time, in the early 1990s. They emerged as new forces promising to efficiently deal with the changing world.

More fundamentally, both are radical schools of thought that challenge conventional thinking particularly about the state's role. Human security challenges the state as a referent object of security; while green criminology frequently questions the state as a notable perpetrator of environmental harm and crime. Likewise, both disciplines challenge the current structure of power. Green criminology criticises the injustice and unequal allocation of power in obtaining access to natural resources (Walters et al. 2013; White 2007); whereas a human security perspective criticises the existing unfair structures of power that "determine who enjoys the entitlement to security and who does not" (Thomas 2001: 160).

Third, transnational organised crime is widely proved as a direct, existential threat to national security (Castle 1997; Guymon 2000; Levitsky 2003; Okubo and Shelley 2011; Picarelli 2008; Tadros 2008). If so, there is no reason to assume that transnational and/or organised environmental crime cannot pose such similar security threats. As is evident, frequently the consequences of green crime are not less pervasive and serious than street crime (Lynch 2013); green crime is often detrimentally definitive, long-lasting, and even permanent (Wright 2011); the profits of some forms of green crime such as wildlife and timber trafficking are not much lower than the "big three" (arms, human and drug trafficking) (Interpol 2012; Khatchadourian 2008; Schneider 2008; South and Wyatt 2011); and environmental crime is closely connected to other serious crimes (Liddick 2011). In other words, green crime should be capable of being an appreciable security threat.

The final pillar is that there have already been efforts to successfully link environmental harm and crime with security issues.[31] Although these efforts do not offer significant space in which various perspectives in security studies are analysed, they nonetheless bring about a twofold message. First, they exemplify the possibility and efficiency of employing conceptual perspectives from security studies to the field of green criminology. Second, they construct a basic foundation for further analyses, reveal some gaps in knowledge, and inspire further efforts to fill these gaps. Taking this into account, the following chapters will visit a number of perspectives in security studies for the purpose of selecting a pertinent approach and then integrating it into this research's conceptual framework.

Conclusion

In this exploratory study, developing a conceptual framework plays an important role in guiding the investigation of timber trafficking in Vietnam. This research adheres to a socio-legal approach in which timber trafficking is any illegal act that breaches not only the Vietnamese criminal law, but also administrative laws and regulations. The socio-legal methodology then helps this research determine the range of activities examined, including harvesting, smuggling, trading and processing. Each of these stages may involve different actors and, at the same time, can cause different impacts. In the end, each stage needs to be tackled by different measures.

The critical thought in addressing powerful perpetrators of green crime, especially state authorities and large corporations, is also taken into consideration in the course of revealing timber traffickers and the vast environmental damage resulting from their actions. Another green criminological facet informing this research is the concern over a hierarchy of victimisation where the marginalised groups tend to suffer from the most harmful outcomes.

In the field of human security, it has been argued that the powerful actors who unjustly build their security and prosperity can be a security threat to ordinary people. This threat can be seen in various ways including their involvement in, or at least their strong support of, destructive types of environmental crime. Indeed, one of the pioneering green criminologists, White (2014: 835), points out that the pursuit of a security framework based on the injustice and wrongdoing of states, corporations and organised groups "simultaneously fosters global crimes such as ecocide, contributes to the proliferation of specific conventional environmental crimes and hampers the exercise of justice". In other words, for the sake of their own interest (e.g. security and prosperity), the powerful may encourage green crime, and accordingly, put others' security at risk. In this sense, it is suggested for this research that like the above-mentioned critical conceptualisation in green criminology of the offender of green crime, the powerful can be an actor in, or at least a supporter of, timber trafficking, and hence contribute to victimisation of the disadvantaged.

In short, this research uses both human security and green criminology to carry out a critical examination of the different roles of the powerful and the marginalised in the operation of timber trafficking as well as the varying degrees of suffering of both groups.

Furthermore, thanks to the incorporation of seven different components, the broad human security paradigm provides this research with a comprehensive approach to conceptualise the victimisation from timber trafficking. This security approach not only supports the observation by Hall (2011, 2013a, 2014a, b) that green crime generates health, economic, social and cultural impacts, but also expands Hall's observation by including environmental, personal and political elements. This expansion may help this research understand the victimisation from timber trafficking more comprehensively.

Finally, since the overarching means of achieving human security is via human emancipation and human development (Alkire 2003; Bajpai 2000; Sen 2000; Thomas 2000; UNDP 1994), the use of human security in this research suggests that solutions for curtailing human insecurity associated with timber trafficking should entail the measures that empower and bring justice for ordinary people. Again, this conceptual approach echoes the suggestions in green criminology that focus on justice and equal distribution of power as a means to effectively tackle green crime (White 2008a; White and Heckenberg 2014; Wyatt et al. 2013).

Chapters 1 and 2 have provided an overall picture of timber trafficking, defined relevant terms, identified important gaps in knowledge in the domain, and constructed a pertinent conceptual framework. The next chapter will introduce the socio-legal situation of Vietnam, which helps contextualise the findings of a detailed investigation into timber trafficking in the country.

Notes

1. There has been a terminological debate upon this. Herbig and Joubert (2006: 54), for example, reject the use of the extremely inadequate term "environmental criminology", instead recommending the use of "conservation criminology" that, they argue, is "the most representative expres-

sion" for research on natural resource crime. "Conservation criminology" is also used by Gibbs et al. (2010) who integrate criminology with natural resource disciplines and risk and decision sciences. However, White (2008a: 7) points out that conservation criminology is "ambiguous and implicitly assumes a particular narrow understanding of 'resources' as well seeming to ignore the dynamic and changing character of 'nature'". Walters (2010: 308) uses the term "eco-crime" because the term "eco" helps to comprehend the "complex networks of evolution and interaction involving species and their habitats".

2. Leading proponents of the aim of green criminology come from the UK (in Carrabine et al. 2009), the USA (in Lynch and Stretesky 2007) and Australia (in White 2008a). White (2008a), for instance, suggests that the aim of green criminology is to (1) scrutinise the nature of environmental harm, (2) examine the nature of the controlling regime of environmental harm, and (3) explore the nature of the relationship between changes in specific environments and the criminalisation process.
3. The first green criminological work examining in detail the philosophical positions that set out the foundations for defining and evaluating green harm and crime was produced by Halsey and White (1998). Halsey and White look into three philosophies—anthropocentrism, ecocentrism and biocentrism—each of which produces differing means for assessing how much harm from human activity is acceptable, and how much is unacceptable, and accordingly the subject of criminological and regulatory inquiry. Human-centred or anthropocentric perspective prioritises the conventional instrumentalist view, whereby humans are dominant over the natural world; consequently, non-human species and the natural environment are regarded as only instrumentally valuable (Halsey and White 1998). Ecocentrism is not so much concerned with whether or not human values are more worthwhile than non-human welfare, but it recognises the intrinsic value of the ecosystems, emphasising that "human and their activities are inextricably integrated with the rest of the natural world in communal or communal-like arrangements" (Steverson 1994: 71). The biocentric perspective argues that human activities need to be controlled, and that any human activities that disrupt the biotic system should be outlawed (Gibbs et al. 2010; Halsey 2004; Halsey and White 1998). It is by adopting a biocentric stance that the plight of animals is highlighted as an environmental harm (Walters 2010).

4. White (2008b, 2011) suggests three approaches for defining environmental harm: conventional criminological conceptions, ecological conceptions and green criminological conceptions, which are based on legal status, ecological sustainability and justice, respectively. Meanwhile, Gibbs et al. (2010) sum up five approaches consisting of liberal ecology or legalist perspectives, socio-legal perspectives, environmental justice perspectives, ecocentric perspectives and biocentric perspectives. Highlighting the human-centred approach, Lynch and Stretesky (2003) and Brisman (2012) offer two contrasting approaches to defining environmental harm and crime comprising the corporate perspective and the environmental justice perspective.
5. Similarly, in the domain of timber trafficking, globally Goncalves et al. (2012: vii) argue that "the criminal justice system has been used in the fight against illegal logging, but only in very sporadic instances and in limited and ineffective ways. Moreover, in those few cases, it has tended to target low-level criminals whose involvement in illegal logging is due to poverty".
6. The redefinition as discussed above can also serve as a foundation for clarifying two concepts of *environmental crime* and *environmental harm*. On the one hand, since conventional criminology is only concerned with criminalised behaviours, environmental crime or green crime refers to environmentally harmful acts that are prescribed in criminal laws. Typical forms of the green crime are air and water pollution, soil quality, deforestation, conservation of natural resources, and protection of biodiversity (Brisman 2012). Thus in mainstream criminology, ecologically harmful behaviours such as massive planned logging or cruel abuse of animals are not judged as green crimes. In stark contrast, any forms of environmental harm that may be unnoticed by criminal laws may be still conceived as environmental crime in green criminology if the acts are significantly harmful to the environment. Green criminology therefore takes "harm" as its central concept, simply summed up as "a harm-based discourse" (Beirne and South 2007: xiv).
7. In addition the acts defined in the Penal Code, the acts that violate laws and regulations concerning forest protection such as Forest Protection and Development Act (2004), the Environmental Protection Act (2008), Decree 157/2013/ND-CP, Decree 32/2006/ND-CP and Directive 35/2011/TT-BNNPTNT can also be considered as timber trafficking in this research. For further details on legal provisions relevant to timber

trafficking, see the section on "Forest Governance and Law Enforcement on Timber Trafficking" in Chap. 3 and Appendix B.

8. Article 3 of Directive 35/2011/TT-BNNPTNT in 2011 by MARD provides the principles and conditions for a logging operation, in which "the operation must (1) correctly harvest in the zones, areas and timber species that are permitted by competent authorities, (2) carry out the measures to ensure forest regeneration, minimise the detrimental impacts on the environment and the protection ability of forests…".
9. Stewart (2014: 239) observes that to avoid detection, illegal logging virtually ignores best-practice logging techniques to reduce environmental impacts, but rather employs "the most destructive forms of wood extractions".
10. Some studies look at the techniques used by the powerful perpetrators to avoid their responsibility for the massive environmental harm they generate such as investing huge resources to convince legislators to set up laws and regulations that benefit them (South et al. 2013), making green harm invisible or individualising the problem (Gaarder 2013), and taking advantage of the unevenness in legal provisions and law enforcement across national legal systems (Passas 2005).
11. By combining deep environmental ethics with an ecocentric perspective in green criminology, Wyatt (2012), for example, develops a new approach to identifying victims, namely a "deeper green" perspective, that "makes non-human animals, and potentially other species, criminologically visible as subjects whose well-being is of intrinsic significance, and whose suffering warrants remediation". In this approach, deeper green criminology "encompasses the racism, sexism, and classism that are a concern of green criminology, but also includes speciesism, which is an ecocentric stance taken from deep environmental ethics that privileges long-term environmental preservation of all species over human consumption" (Wyatt 2012: 62).
12. In general, the scope of the concept environmental justice is broader than the environmental racism concept (Holifield 2001; Schlosberg 2007). The former includes at least two overlapping parts of the grassroots environmental movement: the movement against environmental racism and the anti-toxics movement (Schlosberg 2007). The latter emerged in 1987 as part of the United Church of Christ's study: "Toxic Waste and Race in the United States", and it has served as "another powerful rhetorical tool for grassroots activists" (Holifield 2001: 83).

13. Skinnider also suggests five ways to categorise green victimisation relying on (1) the nature of wrongful acts, (2) the nature of the harm to victims, (3) the extent of damage suffered, (4) the scale of crime and (5) perpetrator identifiability.
14. The key role of green victimology, as proposed by Hall (2014a: 134), "will lie in unpicking the extent to which formal justice mechanisms are incapable of dealing with cases of environmental victimisation and the extent to which this alleged incompatibility in fact reflects cultural reticence amongst legal practitioners and others involved in delivering those justice systems".
15. Specifically Halsey presumes four problematic assumptions within green criminology comprising "(a) the general reluctance to put into critical relief the concept of environmental damage and the associated (but manifestly distinct) category of environmental crime; (b) the inability to move beyond dialectical modules of society an conflict resolution; (c) the unwillingness to do away with modernist accounts of the relationship between 'words' and 'thing'; (d) and the incapacity to develop a nuanced account of human/environment interaction" (Halsey 2004: 835).
16. Beirne (2007: 61) suggests that Halsey's argument seems to be "deliberately obscure and more nihilist than constructive", and "is hoist with its own petard". Halsey now thinks himself "less wedded to that statement" (Halsey 2013: 107).
17. Zedner (2009) argues that if security is simply seen as an objective state that is a condition of being protected from threats, the "perfect attainment" of security is therefore unachievable since we could achieve security today but we could be faced with new threats tomorrow. Thus, security should also be perceived as an "objective state" or "inherently socially constructed" where we have our own feeling of safety, meaning that security is mentally achieved by having a subjective sense of safety or feeling of being secure, stemming in part from material and social conditions (Hameiri and Jones 2013; Zedner 2009).
18. State security is closely connected with *military security* which is commonly perceived as "actual freedom from the threat or use of organised violence for political purposes" (Herring 2007: 131). It is argued that that if military security cannot be ensured, the core elements of state security, especially physical safety and autonomy, cannot be guaranteed. In other words, military security is an inevitable condition to obtain state security (Herring 2007).

19. Between 1990 and 1998, interstate violent conflicts resulted in 3.6 million people killed, 24 million internally displaced, and 18 million forced to be refugees. The conflicts killed 200,000–500,000 people in Rwanda. In 2000, ethnic conflicts in Africa internally displaced 10.6 million people. "Bad governance" led to mass killings of ordinary people that were particularly worrying in Kosovo, Liberia and Iraq (Nsiah-Gyabaah 2010: 246).
20. Cooperative security considers that the orthodox construction of security is too unilateralist in a world where interdependence is joining nations together. Therefore, the unilateralist idea of security must be replaced by cooperative security (Nolan 1994).
21. Comprehensive security argues that the threats to territorial integrity and political order must be comprehensively calculated not only from other states but also from a variety of non-state factors and even natural catastrophes. As a result, the concept of security must be greatly expanded, based on sources of threat (Alagappa 1998).
22. There are various approaches to the concept of environmental security. Biswas (2011) for example focuses on the connections between environmental issues (environmental degradation, global warming, climate change) and national security. Allenby (2000) and Detraz (2009) examine different forms of environmental security including resource security, energy security and biological security. Barnett (2007) addresses key dimensions of environmental security (entity to be secure, major sources of risk).
23. It is worth noting that state security, as a traditional security approach, may show some inconsistencies with critical green criminology in the ways in which both fields view the role of states, as discussed earlier in this chapter.
24. Alkire (2003: 2), for example, provides a working definition of human security that "the objective of human security is to safeguard the vital core of all human lives from critical pervasive threats, in a way that is consistent with long-term human fulfilment".
25. Ul Haq (1995: 16) proposes a well-known description "human security is not a concern with weapons. It is a concern with human dignity. In the last analysis, it is a child who did not die, a disease that did not spread, an ethnic tension that did not explode, a dissident who was not silenced, a human spirit that was not crushed".

26. Alkire (2003), Commission on Human Security (2003), Nsiah-Gyabaah (2010), Sen (2000), UNDP (2005b) provide excellent insights into the connectedness between human security with human development and human rights. The concept of human security, for example, "speaks of basic conditions and minimum guarantees" for human development (UNDP 2005b: 14). Nonetheless, the process of development can menace some elements of human security through the generation of conflicts and disruptions (Nsiah-Gyabaah 2010).
27. While globally up to £900 billion was spent on military goals by 1985, it would cost £30 billion to provide water for people worldwide. At the same time, the costs of re-foresting the damaged tropical forests, combating desertification, and supplying contraceptives for family planning would be £1.3, £4.5 and £2.0 billion respectively (Nsiah-Gyabaah 2010).
28. The broad human security perspective supports the process of sustainable development that "regenerates the environment rather than destroying it", and calls for the enhancement of people's capacity to use natural resources in a sustainable manner, and the reduction of pressures on ecosystems (UNDP 1994: iii).
29. The collection by Matthew et al. (2010) provides excellent insights into the close connection between human insecurity and various forms of global environmental phenomenon such as deforestation, land degradation, water pollution, biodiversity loss, and coastal and marine degradation. The book clearly shows how global environmental change is raising novel, unavoidable questions of human insecurity, conflict, collaboration and sustainable development (Matthew et al. 2010).
30. There is a wide array of practices that have currently attested to this observation. The conventional clear-cut distinctions between policing and security service, crime and terrorism, domestic and national security, community safety and international peacekeeping have been blurred (Zedner 2009).
31. The most notable efforts are probably: Elliott (2007) who warns that major forms of green crime taking place in Asia Pacific "fit" the "crime as security" framework applied to other transnational organised crimes in this region; Hauck (2007) who investigates threats to security caused by non-compliance in small-scale fisheries in South Africa; Wyatt (2012, 2013) and Wyler and Sheikh (2008) who scrutinise links between wildlife (mostly wild animal) trafficking and national security, in particular

its connection with corruption, organised crime and terrorism; Hall (2013a, 2014a) who introduces some potential major consequences of environmental degradation on the security of humans in terms of their health, safety and continued prosperity; White (2014) who "explores the political, economic and ecological context within which preoccupations with environmental insecurity emerge and how they feed back into a fortress mentality"; and Agnew (2012a, b) who initially developed a comprehensive model that addresses the connection between climate change and various criminological issues that have genuine security implications such as "the ability of individuals and governments to meet basic needs and provide security" (Agnew 2012a: 28).

References

Agnew, R. (2012a). Dire forecast: A theoretical model of the impact of climate change on crime. *Theoretical Criminology, 16*(1), 21–42.

Agnew, R. (2012b). It's the end of the world as we know it: The advance of climate change from a criminological perspective. In R. D. White (Ed.), *Climate change from a criminological perspective* (pp. 13–25). New York: Springer.

Agnew, R. (2013). The ordinary acts that contribute to ecocide—A criminological analysis. In N. South & A. Brisman (Eds.), *The Routledge international handbook of green criminology* (pp. 58–72). London: Routledge.

Akella, S., & Cannon, J. (2004). *Strengthening the weakest link—Strategies for improving the enforcement of environmental laws globally*. Arlington, VA: Center for Conservation and Government at Conservation International.

Alagappa, M. (1998). *Asian security practice: Material and ideational influences*. Stanford: Stanford University Press.

Alkire, S. (2003). *A conceptual framework for human security*. Oxford: Centre for Research on Inequality, Human Security and Ethnicit—University of Oxford.

Allenby, B. R. (2000). Environmental security: Concept and implementation. *International Political Science Review, 21*(1), 5–21.

Bajpai, K. P. (2000). *Human security: Concept and measurement*. Notre Dame, IN: University of Notre Dame Joan B. Kroc Institute for International Peace Studies.

Baldwin, D. A. (1997). The concept of security. *Review of International Studies, 23*(1), 5–26.

Barnett, J. (2007). Environmental security. In A. Collins (Ed.), *Contemporary security studies* (pp. 182–203). New York: Oxford University Press.

Barnett, J., Matthew, R. A., & O'Brien, K. L. (2010). Global environmental change and human security: An Introduction. In R. A. Matthew, J. Barnett, B. McDonald, & K. L. O'Brien (Eds.), *Global environmental change and human security*. MIT Press.

Beirne, P. (2007). Animal rights, animal abuse and green criminology. In P. Beirne & N. South (Eds.), *Issues in green criminology: Confronting harms against environments, humanity and other animals*. Cullompton, UK: Willan.

Beirne, P. (2009). *Confronting animal abuse: Law, criminology, and human-animal relationships*. Lanham, MD: Rowman & Littlefield Publishers.

Beirne, P., & South, N. (2007). Introduction: Approaching to green criminology. In P. Beirne & N. South (Eds.), *Issues in green criminology: Confronting harms against environments, humanity and other animals*. Cullompton, UK: Willan.

Bell, S., & McGillivray, D. (2008). *Environmental law*. Oxford and New York: Oxford University Press.

Biswas, N. R. (2011). Is the environment a security threat? Environmental security beyond securitization. *International Affairs Review, XX*(1).

Biswas, N. R. (2011b). Is the environment a security threat? Environmental security beyond securitization. *International Affairs Review, 20*(1), 1.

Brisman, A. (2012). Crime-environment relationships and environmental justice. *Seattle Journal for Social Justice, 6*(2), 6.

Brisman, A. (2014). Of theory and meaning in green criminology. *International Journal for Crime, Justice and Social Democracy, 3*(2), 22–35.

Brisman, A., & South, N. (2014). *Green cultural criminology: Constructions of environmental harm, consumerism, and resistance to ecocide*. London: Routledge.

Brown, L. R. (1986). Redefining national security. *Challenge, 29*(3), 25–32.

Bryant, B. I. (1995). *Environmental justice: Issues, policies, and solutions*. Washington, DC: Island Press.

Bullard, R. D. (1996). Environmental justice for all. In R. D. Bullard (Ed.), *Environmental justice and communities of color* (pp. 3–22). San Francisco: Sierra Club Books.

Buzan, B. (1991). *People, states and fear: An agenda for international security studies in the post-Cold War era*. Colchester: ECPR Press.

Cao, N. A., & Wyatt, T. (2013). A green criminological exploration of illegal wildlife trade in Vietnam. *Asian Journal of Criminology, 8*(2), 129–142.

Carrabine, E., Cox, P., Lee, M., Plummer, K., & South, N. (2009). *Criminology: A sociological introduction*. New York, NY: Routledge.

Castle, A. (1997). *Transnational organized crime and international security*. Vancouver: Institute of International Relations, University of British Columbia.

Chavis, B. F. (1994). Preface. In R. Bullard (Ed.), *Unequal protection: Environmental justice and communities of color* (pp. xi–xii). San Francisco, CA: Sierra Club Books.

Clifford, M., & Edwards, T. D. (2012). *Environmental crime*. Burlington, MA: Jones & Bartlett Publishers.

Collins, A. (2007). *Contemporary security studies*. Oxford: Oxford University Press.

Commission on Human Security. (2003). *Human security now*. New York: United Nations.

Crow, M., Shelley, T. O. C., & Stretesky, P. (2014). Camouflage-collar crime: An examination of wildlife crime and characteristics of offenders in Florida. In T. Spapens, R. White, & M. Kluin (Eds.), *Environmental crime and its victims: Perspectives within green criminology* (pp. 185–204). Surrey: Ashgate.

Davies, P. A. (2014). Green crime and victimization: Tensions between social and environmental justice. *Theoretical Criminology, 18*(3), 300–316.

Detraz, N. (2009). Environmental security and gender: Necessary shifts in an evolving debate. *Security Studies, 18*(2), 345–369.

Dodds, F., & Pippard, T. (2012). *Human and environmental security: An agenda for change*. Routledge.

Du Rees, H. (2001). Can criminal law protect the environment? *Journal of Scandinavian Studies in Criminology and Crime Prevention, 2*(2), 109–126.

Dybing, S. S. (2012). Environmental harm: Social causes and shilting dynamics. In R. Ellefsen, G. Larsen, & R. Sollund (Eds.), *Eco-global crimes: Contemporary problems and future challenges* (pp. 273–294). Farnham: Ashgate Publishing Ltd.

Ellefsen, R., Larsen, G., & Sollund, R. (2012). *Eco-global crimes: Contemporary problems and future challenges*. Farnham: Ashgate Publishing Ltd.

Elliott, L. (2007). Transnational environmental crime in the Asia Pacific: An un(der)securitized' security problem? *Pacific Review, 20*(4), 499–522.

Eman, K., Meško, G., & Fields, C. B. (2009). Crimes against the environment: Green criminology and research challenges in Slovenia. *Journal of Criminal Justice and Security, 11*(4), 574–592.

Florini, A., & Simmons, P. (1998). *The new security thinking: A review of the North American literature*. New York: Rockefeller Bros. Fund.

Floyd, R. (2015). Environmental security and the case against rethinking criminology as 'security-ology'. *Criminology and Criminal Justice, 15*(3), 277–282.

Fukuda-Parr, S., & Messineo, C. (2012). *Human security: A critical review of the literature*. Working Paper 11, Centre for Research on Peace and Development (CRPD).

Gaarder, E. (2013). Evading responsibility for green harm: State-coeporate exploitation of race, class, and gender inequiality. In N. South & A. Brisman (Eds.), *The Routledge international handbook of green criminology*. London: Routledge.
Gasper, D., & Truong, T.-D. (2005). Deepening development ethics: From economism to human development to human security. *The European Journal of Development Research, 17*(3), 372–384.
Gibbs, C., Gore, M. L., McGarrell, E. F., & Rivers, L. (2010). Introducing conservation criminology towards interdisciplinary scholarship on environmental crimes and risks. *British Journal of Criminology, 50*(1), 124–144.
Goncalves, M. P., Panjer, M., Greenberg, T. S., & Magrath, W. B. (2012). *Justice for forest: Improving criminal justice efforts to combat illegal logging*. Washington, DC: World Bank.
Graife, M. B., & Stretesky, P. B. (2013). Crude laws: Treadmill of production and state violations in civil and criminal liability for oil discharges in navigable water. In N. South & A. Brisman (Eds.), *The Routledge international handbook of green criminology* (pp. 150–166). London: Routledge.
Green, P., Ward, T., & McConnachie, K. (2007). Logging and legality: Environmental crime, civil society, and the state. *Social Justice, 34*(2), 94–110.
Groombridge, N. (1998). Masculinities and crimes against the environment. *Theoretical Criminology, 2*(2), 249–267.
Guymon, C. L. D. (2000). International legal mechanisms for combating transnational organized crime: The need for a multilateral convention. *Berkeley Journal of International Law, 18*, 53.
Hall, M. (2011). Environmental victims: Challenges for criminology and victimology in the 21st century. *Varstvoslovje: Journal of Criminal Justice & Security, 13*(4), 371–391.
Hall, M. (2013a). *Victims of environmental harm: Rights, recognition and redress under national and international law*. London: Routledge.
Hall, M. (2013b). Victims of environmental harms and their role in national and international justice. In R. Walters, D. Westerhuis, & T. Wyatt (Eds.), *Emerging issues in green criminology: Exploring power, justice and harm* (pp. 218–241). Houndmills, Basingstoke, UK: Palgrave Macmillan.
Hall, M. (2014a). Environmental harm and environmental victims: Scoping out a 'green victimology'. *International Review of Victimology, 20*(1), 129–143.
Hall, M. (2014b). Victims of environmental crime: Routes for recognition, restitution and redress. In T. Spapens, R. White, & M. Kluin (Eds.), *Environmental crime and its victims: Perspectives within green criminology* (pp. 104–118). Surrey: Ashgate.

Hall, M., & Farrall, S. (2013). The criminogenic consequences of climate change—Blurring the boundaries between offenders and victims. In N. South & A. Brisman (Eds.), *The Routledge international handbook of green criminology* (pp. 120–133). London: Routledge.

Halsey, M. (2004). Against 'green' criminology. *British Journal of Criminology, 44*(6), 833–853.

Halsey, M. (2013). Conservation criminology and the "General Accident" of climate change. In N. South & A. Brisman (Eds.), *The Routledge international handbook of green criminology*. London: Routledge.

Halsey, M., & White, R. (1998). Crime, ecophilosophy and environmental harm. *Theoretical Criminology, 2*(3), 345–371.

Hameiri, S., & Jones, L. E. E. (2013). The politics and governance of non-traditional security. *International Studies Quarterly, 57*(3), 462–473.

Hampson, F. O. (2008). Human security. In P. Williams (Ed.), *Security studies: An introduction*. London: Routledge.

Hampson, F. O., Daudelin, J., Hay, J., Reid, H., & Marting, T. (2002). *Madness in the multitude: Human security and world disorder*. Don Mills, ON: Oxford University Press.

Hauck, M. (2007). Non-compliance in small-scale fisheries: A threat to security. In P. Beirne & N. South (Eds.), *Issues in green criminology: Confronting harms against environments, humanity and other animals* (pp. 270–289). Devon: Willan.

Herbig, F., & Joubert, S. (2006). Criminological semantics: Conservation criminology—Vision or vagary? *Acta Criminologica, 19*(3), 88–103.

Herring, E. (2007). Military security. In A. Collins (Ed.), *Contemporary security studies* (pp. 129–145). Oxford: Oxford University Press.

Holdaway, S., & Rock, P. (1998). *Thinking about criminology*. London: UCL Press.

Holifield, R. (2001). Defining environmental justice and environmental racism. *Urban Geography, 22*(1), 78–90.

Human Security Centre. (2005). *Human security report 2005: War and peace in the 21st century*. New York: Oxford University Press.

Interpol. (2012). *Environmental crime, It's global theft*. Interpol—Environmental Crime Programme. Retrieved from http://www.interpol.int/Crime-areas/Environmental-crime/Environmental-crime

Kaldor, M. (2007). *Human security: Reflections on globalization and intervention*. Cambridge: Polity.

Kerr, P. (2007). Human security. In A. Collins (Ed.), *Contemporary security studies* (pp. 91–108). Oxford: Oxford University Press.

Khatchadourian, R. (2008). The stolen forests—Inside the covert war on illegal logging. *The New Yorker*. http://www.newyorker.com/magazine/2008/10/06/the-stolen-forests

King, G., & Murray, C. J. L. (2001). Rethinking human security. *Political Science Quarterly, 116*(4), 585–610.

Kohm, S., & Greenhill, P. (2013). 'This is the North, where we do what we want': Popular green criminology and 'Little Red Riding Hood' films. In N. South & A. Brisman (Eds.), *The Routledge international handbook of green criminology* (pp. 365–378). London: Routledge.

Korsell, L. E. (2001). Big sick, litle sick: Strategies for combaing environmental crime. *Journal of Scandinavian Studies in Criminology and Crime, 2*, 127.

Kramer, R. (2014). Climate change: A state-corporate crime perspective. In T. Spapens, R. White, & M. Kluin (Eds.), *Environmental crime and its victims: Perspectives within green criminology* (pp. 23–10). Surrey: Ashgate.

Levitsky, M. (2003). Transnational criminal networks and international security. *Syracuse Journal of International Law and Commerce, 30*(2), 227.

Liddick, D. (2011). *Crimes against nature: Illegal industries and the global environment*. Santa Barbara, CA: Praeger.

Little, L., & Cocklin, C. (2007). Environment and human security. In J. Pretty, A. S. Ball, T. Benton, J. S. Guivant, D. R. Lee, D. Orr, M. J. Pfeffer, & H. Ward (Eds.), *The Sage handbook of environment and society*. London: Sage.

Lodgaard, S. (2000). *Human security: Concept and operationalization*. Norwegian Institute of International Affairs.

Lonergan, S. (1999). *Global Environmental Change and Human Security (GECHS): Science plan*. International Human Dimensions Programme on Global Environmental Change.

Long, M., Stretesky, P., & Lynch, M. (2014). The treadmill of production, planetary boundaries and green criminology. In T. Spapens, R. White, & M. Kluin (Eds.), *Environmental crime and its victims: Perspectives within green criminology* (pp. 263–276). Surrey: Ashgate.

Long, M. A., Stretesky, P. B., Lynch, M. J., & Fenwick, E. (2012). Crime in the coal industry: Implications for green criminology and treadmill of production. *Organization & Environment, 25*(3), 328–346.

Lynch, M. (1990). The greening of criminology: A perspective on the 1990s. *The Critical Criminologist, 2*(3), 1–4.

Lynch, M. J. (2000). The power of oppression: Understanding the history of criminology as a science of oppression. *Critical Criminology, 9*(1-2), 144–152.

Lynch, M. J. (2013). Reflections of green criminology and its boundaries: Comparing environmental victimization and considering crime from an eco-city perspective. In N. South & A. Brisman (Eds.), *The Routledge international handbook of green criminology* (pp. 43–57). London: Routledge.

Lynch, M. J., Long, M. A., Barrett, K. L., & Stretesky, P. B. (2013). Is it a crime to produce ecological disorganization? Why green criminology and political economy matter in the analysis of global ecological harms. *British Journal of Criminology, 53*(6), 997–1016.

Lynch, M. J., & Stretesky, P. (2001). Toxic crimes: Examining corporate victimization of the general public employing medical and epidemiological evidence. *Critical Criminology, 10*(3), 153–172.

Lynch, M. J., & Stretesky, P. (2007). Green criminology in the United States. In P. Beirne & N. South (Eds.), *Issues in green criminology: Confronting harms against environments, humanity and other animals* (pp. 248–269). Cullompton, UK: Willan.

Lynch, M. J., & Stretesky, P. (2014). *Exploring green criminology: Toward a green criminological revolution.* Farnham: Ashgate Publishing Ltd.

Lynch, M. J., & Stretesky, P. B. (2003). The meaning of green: Contrasting criminological perspectives. *Theoretical Criminology, 7*(2), 217–238.

MacFarlane, S. N., & Khong, Y. F. (2006). *Human security and the United Nations.* Indiana University Press.

Mares, D. (2010). Criminalizing ecological harm: Crimes against carrying capacity and the criminalization of eco-sinners. *Critical Criminology, 18*(4), 279–293.

Mathews, J. T. (1989). Redefining security. *Foreign Affairs, 68*(2), 162–177.

Matthew, R. A., Barnett, J., McDonald, B., & O'Brien, K. L. (2010). *Global environmental change and human security.* MIT Press.

Morgan, P. (2007). Security in international politics: Traditional approaches. In A. Collins (Ed.), *Contemporary security studies* (pp. 13–34). Oxford: Oxford University Press.

Natali, L. (2013). The contemporary horizon of green criminology. In N. South & A. Brisman (Eds.), *The Routledge international handbook of green criminology* (pp. 73–84). London: Routledge.

Nef, J. (1999). *Human security and mutual vulnerability—An exploration into the global political economy of development and underdevelopment.* Ottawa: International Development Research Centre.

Newman, E. (2010). Critical human security studies. *Review of International Studies, 36*(1), 77–94.

Nguyen, C. C. (2010). *The theoretical and practical issues on environmental crimes in the Vietnamese criminal law.* Master degree (in Vietnamese).

Nolan, J. E. (1994). *Global engagement: Cooperation and security in the 21st century*. Washington, DC: Brookings Institution Press.

Nsiah-Gyabaah, K. (2010). Human security as a prerequisite for development. In R. A. Matthew, J. Barnett, B. McDonald, & K. L. O'Brien (Eds.), *Global environmental change and human security* (pp. 237–259). Cambridge, MA: MIT Press.

Nurse, A. (2013a). *Animal harm: Perspectives on why people harm and kill animals*. Farnham, Surrey, UK; Burlington, VT: Ashgate.

Nurse, A. (2013b). Perspectives on criminality in wildlife. In R. Walters, D. Westerhuis, & T. Wyatt (Eds.), *Emerging issues in green criminology: Exploring power, justice and harm* (pp. 127–144). Houndmills, Basingstoke, UK: Palgrave Macmillan.

Okubo, S. (2011). Glogalisation, human security and the right to live in peace. In S. Okubo & L. Shelley (Eds.), *Human security, transnational crime and human trafficking: Asian and Western perspectives* (pp. 9–35). Hoboken: Taylor and Francis.

Okubo, S., & Shelley, L. (2011). *Human security, transnational crime and human trafficking: Asian and Western perspectives*. Hoboken: Taylor and Francis.

Paris, R. (2001). Human security: Paradigm shift or hot air? *International Security, 26*(2), 87–102.

Passas, N. (2005). Lawful but awful: 'Legal corporate crimes'. *The Journal of Socio-Economics, 34*(6), 771–786.

Pemberton, A. (2014). Environmental victims and criminal justice: Proceed with caution. In T. Spapens, R. White, & M. Kluin (Eds.), *Environmental crime and its victims: Perspectives within green criminology* (pp. 63–86). Surrey: Ashgate.

Picarelli, J. T. (2008). Transnational organized crime. In P. Williams (Ed.), *Security studies: An introduction* (pp. 453–467). London: Routledge.

Potter, G. (2010). What is green criminology? *Sociology Review, 20*, 8–12.

Potter, G. (2014). The criminogenic effects of environmental harm: Bringing a 'green' perspective in mainstream criminology. In T. Spapens, R. White, & M. Kluin (Eds.), *Environmental crime and its victims: Perspectives within green criminology* (pp. 7–21). Surrey: Ashgate.

Pulido, L. (2000). Rethinking environmental racism: White privilege and urban development in southern California. *Annals of the Association of American Geographers, 90*(1), 12–40.

Reiman, J. (1995). *The rich get richer and the poor get prison: Ideology, crime, and criminal justice* (pp. 237). ISBN 0-02-439250-2.

Richards, M., Wells, A., Gatto, F. D., Contreras-Hermosilla, A., & Pommier, D. (2003). Impacts of illegal ity and barriers to legality: A diagnostic analysis of illegal logging in Honduras and Nicaragua. *International Forestry Review, 5*(3), 282–292.

Ruggiero, V., & South, N. (2013). Green criminology and crimes of the economy: Theory, research and praxis. *Critical Criminology, 21*(3), 359–373.

Schlosberg, D. (2007). *Defining environmental justice: Theories, movements, and nature*. Oxford: Oxford University Press.

Schnaiberg, A. (1980). *The environment: From surplus to scarcity*. New York: Oxford University Press.

Schneider, J. L. (2008). Reducing the illicit trade in endangered wildlife. The market reduction approach. *Journal of Contemporary Criminal Justice, 24*(3), 274–295.

Sen, A. (2000). Conference paper: Why human security. In *International Symposium on Human Security*, Tokyo.

Shearing, C. (2015). Criminology and the Anthropocene. *Criminology and Criminal Justice, 15*(3), 255–269.

Simon, D. R. (2000). Corporate environmental crimes and social inequality: New directions for environmental justice research. *American Behavioral Scientist, 43*(4), 633–645.

Skinnider, E. (2011). *Victims of environmental crime: Mapping the issues*. International Centre for Criminal Law Reform and Criminal Justice Policy.

Sollund, R. (2013). The victimization of women, children and non-human species through traffcking and trade—Crime understood through an ecofeminist perspective. In N. South & A. Brisman (Eds.), *The Routledge international handbook of green criminology* (pp. 317–330). London: Routledge.

South, N. (1998). A green field for criminology? A proposal for a perspective. *Theoretical Criminology, 2*(2), 211–233.

South, N. (2012a). Climate change, environmental (in)security, conflict and crime. In S. Farrall, T. Ahmed, & D. French (Eds.), *Criminological and legal consequences of climate change* (pp. 97–111). Oxford: Hart.

South, N. (2012b). Conference paper: Green criminology: Theories, concepts and connected meanderings. In *ESRC Green Criminology Seminar*, Northumbria University, Newcastle Upon Tyne.

South, N. (2015). Anticipating the Anthropocene and greening criminology. *Criminology and Criminal Justice, 15*(3), 270–276.

South, N., & Beirne, P. (2006). Introduction. In N. South & P. Beirne (Eds.), *Green criminology* (pp. xxvii–xxvii). Aldershot: Ashgate.

South, N., Brisman, A., & Beirne, P. (2013). A guide to a green criminology. In N. South & A. Brisman (Eds.), *The Routledge international handbook of green criminology* (pp. 27–42). London: Routledge.

South, N., & Wyatt, T. (2011). Comparing illicit trades in wildlife and drugs: An exploratory study. *Deviant Behavior, 32*(6), 538–561.

Spapens, T. (2014). Invisible victimes: The problems of policing environmental crime. In T. Spapens, R. White, & M. Kluin (Eds.), *Environmental crime and its victims: Perspectives within green criminology* (pp. 221–236). Surrey: Ashgate.

Spencer, D. C., & Fitzgerald, A. (2013). Three ecologies, transversality and victimization: The case of the British Petroleum oil spill. *Crime, Law and Social Change, 59*(2), 209–223.

Steverson, B. K. (1994). Ecocentrism and ecological modeling. *Environmental Ethics, 16*(1), 71–88.

Stewart, D. (2014). Project leaf, and interpol's work on illegal logging and forest crime. In T. Spapens, R. White, & M. Kluin (Eds.), *Environmental crime and its victims: Perspectives within green criminology* (pp. 236–247). Surrey: Ashgate.

Stretesky, P., & Lynch, M. J. (1998). Corporate environmental violence and racism. *Crime, Law and Social Change, 30*(2), 163–184.

Stretesky, P. B., Long, M. A., & Lynch, M. J. (2014). *The treadmill of crime: Political economy and green criminology*. Oxon and New York: Taylor and Francis.

Sygna, L., O'Brien, K., & Wolf, J. (2013). *A changing environment for human security: Transformative approaches to research, policy and action*. Routledge.

Tadjbakhsh, S., & Chenoy, A. M. (2007). *Human security: Concepts and implications*. Oxon: Routledge.

Tadros, V. (2008). Crimes and security. *The Modern Law Review, 71*(6), 940–970.

Telapak & EIA. (2007). *The thousand-headed snake: Forest crimes, corruption and injustice in Indonesia*. Bogor, Indonesia.

Thomas, C. (2000). *Global governance, development and human security: The challenge of poverty and inequality*. London: Pluto.

Thomas, C. (2001). Global governance, development and human security: Exploring the links. *Third World Quarterly, 22*(2), 159–175.

Ul Haq, M. (1995). *Reflections on human development*. New York, NY: Oxford University Press.

Ullman, R. H. (1983). Redefining security. *International Security, 8*(1), 129–153.

UNDP. (1994). *Human Development Report 1994*. New York: United Nations Development Program.

UNDP. (2005a). *Human Development Report 2005; International cooperation at a crossroads: Aid, trade and security in an unequal world.* New York: United Nations Development Programme.

UNDP. (2005b). *Overcoming fear: Citizen (in)security and human development in Costa Rica (Summary).* New York: United Nations Development Programme.

UNODC. (2012). *Wildlife and forest crime analytic toolkit—Revised edition.* New York: United Nations Office for Drugs and Crime. Retrieved from http://www.unodc.org/documents/southeastasiaandpacific//Publications/wildlife/CJS_Response_-_VIETNAM_01_13_Dec_201.pdf

Valverde, M. (2011). Questions of security: A framework for research. *Theoretical Criminology, 15*(1), 3–22.

Valverde, M. (2014). Studying the governance of crime and security: Space, time and jurisdiction. *Criminology and Criminal Justice, 14*(4), 379–391.

Voelkner, N. (2012). *Human security assemblages in global politics: The materiality and instability of biopolitical governmentality in Thailand and Vietnam.* Submitted for the degree of DPhil, Department of International Relations, University of Sussex.

Walker, G. P. (2012). *Environmental justice: Concepts, evidence and politics.* London: Routledge.

Walters, R. (2010). Toxic atmospheres air pollution, trade and the politics of regulation. *Critical Criminology, 18*(4), 307–323.

Walters, R., Wyatt, T., & Westerhuis, D. (2013). *Emerging issues in green criminology: Exploring power, justice and harm.* Houndmills, Basingstoke, UK: Palgrave Macmillan.

Westerhuis, D. S. (2013). A Harm analysis of environmental crime. In R. Walters, D. Westerhuis, & T. Wyatt (Eds.), *Emerging issues in green criminology: Exploring power, justice and harm* (pp. 197–217). Houndmills, Basingstoke, UK: Palgrave Macmillan.

White, R. (2007). Green criminology and the pursuit of social and ecological justice. In P. Beirne & N. South (Eds.), *Issues in green criminology: Confronting harms against environments, humanity and other animals.* Cullompton, UK: Willan.

White, R. (2008a). *Crimes against nature: Environmental criminolgy and ecological justice.* Cullompton, UK: Willan.

White, R. (2008b). The criminalisation of environmental harm. *Criminal Justice Matters, 74*(1), 35–37.

White, R. (2009). Introduction: Environmental crime and eco-global criminology. In R. White (Ed.), *Environmental crime: A reader* (pp. 1–8). Cullompton, UK: Willan.

White, R. (Ed.). (2010a). *Global environmental harm—Criminological perspectives*. Cullompton, UK: Willan.
White, R. (2010b). Globalisation and environmental harm. In R. White (Ed.), *Global environmental harm: Criminological perspective*. Cullompton, UK: Willan.
White, R. (2011). *Transnational environmental crime: Toward an eco-global criminology*. London: Routledge.
White, R. (2013a). The conceptual contours of green criminology. In R. Walters, D. Westerhuis, & T. Wyatt (Eds.), *Emerging issues in green criminology: Exploring power, justice and harm* (pp. 17–33). Houndmills, Basingstoke, UK: Palgrave Macmillan.
White, R. (2013b). Eco-global criminology and the political economy of environmental harm. In N. South & A. Brisman (Eds.), *The Routledge international handbook of green criminology*. London: Routledge.
White, R. (2013c). Guest editor's introduction. *Journal of Social Criminology*, Green Criminology Issue.
White, R. (2014). Environmental insecurity and fortress mentality. *International Affairs, 90*(4), 835–851.
White, R., & Heckenberg, D. (2014). *Green criminology: An introduction to the study of environmental harm*. Oxon, UK; New York: Routledge.
Williams, C. (1996). An environmental victimology. *Social Justice, 23*(4), 16–40.
Wolfers, A. (1952). 'National security' as an ambiguous symbol. *Political Science Quarterly, 67*(4), 481–502.
Wright, G. (2011). Conceptualising and combating transnational environmental crime. *Trends in Organized Crime, 14*(4), 332–346.
Wyatt, T. (2012). *Green criminology & wildlife trafficking: The illegal fur and falcon trades in Russia Far East*. Saarbrücken Lap Lambert Academic Publishing.
Wyatt, T. (2013). Uncovering the significance of and motivation for wildlife trafficking. In N. South & A. Brisman (Eds.), *The Routledge international handbook of green criminology*. London: Routledge.
Wyatt, T., Westerhuis, D., & Walters, R. (2013). Introduction. In R. Walters, D. Westerhuis, & T. Wyatt (Eds.), *Emerging issues in green criminology: Exploring power, justice and harm* (pp. 1–13). Houndmills, Basingstoke, UK: Palgrave Macmillan.
Wyler, L. S., & Sheikh, P. A. (2008). *International illegal trade in wildlife: Threats and US policy*. DTIC Document.
Zedner, L. (2007). Pre-crime and post-criminology? *Theoretical Criminology, 11*(2), 261–281.
Zedner, L. (2009). *Security*. London: Routledge-Cavendish.

3

The Environment and Forests in Vietnam

This chapter is designed to provide basic information on Vietnam's environment and forests, which is essential for understanding timber trafficking, its impacts on human security and for recommending appropriate solutions to tackle the crime in the country. To achieve this, the chapter will be sub-divided into three sections. The first section will look at the main environmental issues in Vietnam including the problem of environmental crime and a number of drivers of these problems. The information in this section will help readers contextualise its findings on the typology and drivers of timber trafficking as a major form of environmental crime in Vietnam. The second section will introduce the country's forest resources and then evaluate their significance to the Vietnamese, particularly for forest-based inhabitants, as well as for the booming industry of timber processing in Vietnam. An understanding of the forest's significance is essential to appreciate various detrimental impacts of timber trafficking and its inescapable outcome—forest loss. This section will also examine the long-lasting tradition of consuming endangered timber, which plays a key role in facilitating the illicit market of timber in the country. The final section of the chapter will introduce Vietnamese legislation and law enforcement concerning timber trafficking.

N.A. Cao, *Timber Trafficking in Vietnam*, Palgrave Studies in Green Criminology,
DOI 10.1007/978-3-319-64280-2_3

Environmental Issues in Vietnam

Since the late 1980s, Vietnam has launched an exhaustive reform of economic innovation, adopting a "market economy with socialist direction", which offered economic freedom for a number of private sectors (Hayton 2010; Kokko 2004). This was, therefore, considered as "one of the most dramatic turnarounds in economic history" (Dollar and Litvack 1998: 1), fundamentally changing the previous fully subsidised and centrally planned economic structure (Drury 2009). On the one hand, the economic achievements after more than 20 years of undertaking this innovative shake-up have been truly "remarkable" (World Bank 2012: 1), resulting in the Vietnamese economy experiencing significant growth, witnessing an overall annual rise of 7.0% since 1987, and rendering it the third-fastest growing developing economy in the world (SRV 2012).[1] On the other hand, while economic development with "the growth-at-all-costs strategy" is prioritised (Brunner 2012: 1), the lack of effective implementation of environmental protection measures has left Vietnam with a range of environmental concerns (ADB 2005; Ministry of Planning and Investment 2008; MNRE 2010; USAid 2013). A study by Pham (2011: 3) stresses that "in many areas, due to the priority for economic growth, business investments are excessively encouraged and abundantly permitted without sufficient assessments of environmental impacts. It is particularly so in the projects either emphasised as key economic industries or delivered by state-owned corporations".

The latest global Environmental Performance Index[2] indicates that Vietnam earns 38.17 out of a possible score of 100 and ranks 136 out of 178 countries surveyed worldwide (Environmental Performance Index 2014). In an effort to broadly assess the prospects of environmental security in Vietnam, Nguyen and Nguyen (2010) emphasise five major environmental threats currently challenging the country. They are (1) disputes over resources and ecological services such as water, land, forest and fishing resources; (2) natural disasters and environmental incidents especially climate change; (3) pollution; (4) ecological disorder; and (5) environmental refugees.

The impacts of climate change are a prominent example of how environmental problems can severely affect the country. Research by the

World Bank points out that Vietnam is one of the countries most severely affected by climate change (Dasgupta et al. 2007). Among 84 coastal developing countries examined for impacts of sea level rise, Vietnam ranks first regarding the severest impacts on population, GDP, urban extent and wetland areas; and ranks second in terms of impacts on land area (behind the Bahamas) and agriculture (behind Egypt). The study also warns that the outcomes of sea level rise for Vietnam are "potentially catastrophic"; therefore "intermediate planning for adaptation" is imperative (Dasgupta et al.: 2007: 2). Notwithstanding the urgent situation, some commentators such as Pham et al. (2012: 39) note that "rapid economic growth continues to be a higher priority than environmental protection".

Among the major crimes in Vietnam,[3] the offences against environmental laws and regulations are complex and serious, particularly in the sectors of industrial waste treatment, importation of outdated technologies and toxic equipment, forest protection and food safety (Nguyen 2010b; Pham 2011; Viet Ba 2011). In recent years, various forms of environmental crime in Vietnam, as pointed out by Pham (2011: 1), "bring about profound impacts on the country's sustainable development and on the people's life such as causing water and air pollution and unsafe foods. In some localities, environmental crime is a potential trigger of insecurity and disorder". A report by the national Department of Environmental Police summarises that in the past four years, this force alone detects and investigates around 7200 violations of environmental laws and regulations, in which nearly 200 cases are charged criminally and the fines issued were worth approximately £6 million (Pham 2011).

However, while it is estimated that the detected number of violations of environmental law represents only 10–12% of the actual number of violations (Ha An 2012), the rate of violations criminally charged, as mentioned in Chap. 2, is exceedingly low, representing only 1.1–3% of the total number of violations detected in Vietnam (Nguyen 2010b). The criminally charged cases are from only two environmental offences: forest destruction (Article 189) and wildlife trafficking (Article 190), whereas the nine other environmental offences defined in Chapter XVII of the Penal Code have never been prosecuted for, despite the fact that there are many major incidents, disclosed by the media and confirmed by enforcement agencies, as serious violation of laws and regulations on environmental protection (Ha An 2012; Huong Nguyen 2010a). A large array of pollution violations, for instance, committed by industrial and manufacturing companies has not been adequately addressed, with no violators ever brought to a criminal court (Pham 2011; Viet Ba 2011).[4]

One of the explanations for this stems from the drawbacks of the current Vietnamese criminal law. Indeed, the Penal Code of Vietnam only offers general definitions of environmental offences, in which causing "serious consequences" is a compulsory requirement in order to charge environmental offenders.[5] However, in practice, it is highly challenging for law enforcement agencies to gather sufficient scientific evidence to prove these "serious consequences", and even more difficult to prove the actual causal relationship between the suspect's actions and these consequences (Ha An 2012; Nguyen 2010b; Tran 2001). Furthermore, while many environmental offences are collectively committed by corporations, the current legislative regime has not yet recognised organisations as a subject of criminal sanctions. This is considered as a major drawback of the Vietnamese criminal law (Binh An 2014), which reveals that Vietnam is no exception to the general trend that, irrespective of the pervasiveness and severity of green crime, the criminal law is limited in its effectiveness in tackling the crime. The ineffectiveness of the control of environmental crime in Vietnam is further explained by its perpetrators; many of these are often highly knowledgeable and capable of taking advantage of legal loopholes and rampant corruption in environmental inspection agencies to conceal their environmentally destructive practices (Pham 2011). In short, environmental crime has been, and continues to be, a major challenge to the Vietnamese criminal justice system.

Forest Resource and its Importance

Forest Resource

According to the National Forest Declaration published by the MARD (2014), as of 31 December 2013, Vietnam has almost 14 million hectares of forest with a forest cover of 41%. From 2000 to 2010, while almost all

Southeast Asian countries suffered from severe deforestation, the rate of forest cover in Vietnam increased by 15% during this decade (UNODC 2013b).

Based on function, forests in Vietnam are classified into three types: (1) *special-use forests* primarily used for natural preservation, scientific research, protection of cultural places and tourism; (2) *protected forests* principally protected for environmental reasons; and *production forests* mainly used for timber and non-timber forest product (NTFP) manufacturing (MARD 2009). By the end of 2013, the areas of these forest types were 2.1 (15%), 4.7 (35%) and 7.0 (50%) million hectares, respectively (MARD 2014). Meanwhile, based on the originating formation of forests, Vietnamese forests are categorised into two types, natural forests and plantation forests (MARD 2009), which by 2013 occupied 10.4 (74%) and 3.6 (26%) million hectares, respectively (MARD 2014). Figure 3.1 shows the proportions of these differing forest types in Vietnam.

Given the quality of forest and the timber value in Vietnam, it is widely observed that although, during the past two decades, the forest cover rate has been consistently increasing, forest quality is progressively falling (Pham et al. 2012). The average timber volume of natural forests is currently around 76.5 m^3 per hectare (FSIV and FAO 2009). While the total

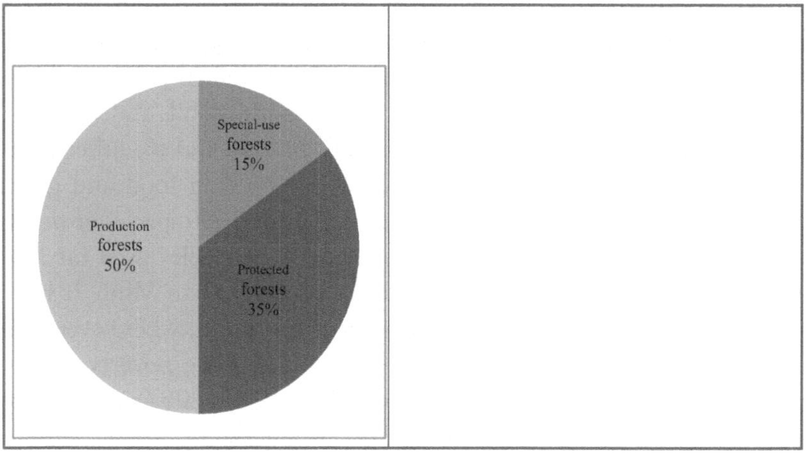

Fig. 3.1 Proportions of different types of forest. Adapted from MARD (2014: 1)

area of rich forests has declined continuously, the area of poor and secondary forests has increased rapidly from 7 million hectares in 1990 to 10.2 million hectares in 2005, constituting over 80% of total forest area in Vietnam (FSIV and FAO 2009). The apparent reason for this is that the majority of new forests are plantation, artificial and monolayer types, while the areas of intact and multilayer forests are substantially shrinking (MNRE 2010).

The National Environmental Report 2010 by MNRE (2010) indicates that Vietnam now has 570,000 hectares of old-growth forests, accounting for 8% of forested areas, whereas the Centre for People and Forest (RECOFTC 2014) notes that Vietnam has only 384,000 hectares of old-growth forest remaining, covering just over 1% of the land. At the same time, forest area per capita in Vietnam declined from 0.7 hectares in 1943 to 0.15 hectares in 2004, much lower than the average forest area per capita in Southeast Asian countries at 0.42 hectares (Vo and Vo 2009: 3).

Forests and Local Residents

Even with the decline in timber capacity, forests in Vietnam still have a variety of NTFPs that are important for the Vietnamese in general and mountainous inhabitants in particular. These products can be categorised into six groups (1) fibre products such as bamboo, rattan, and leaves; (2) foodstuffs such as bamboo shoots, vegetation, leaves, fruits, grains, spices, honey, swallows' nests and edible insects; (3) medicinal plants and aromatic substances; (4) extracted products such as resins, oils and stains; (5) forest animals and their products, birds and insects; and (6) other products for example decorative plants, leaves for packing food and goods (FSIV and FAO 2009). Vietnamese forests provide 2000 species of woody plants, over 3000 species used for medicine, 400 species used for food and fodder, and 500 species used for oil extraction (Xuan Minh 2014).

Similar to what is found in other parts of the world, where over 90% of the 1.2 billion people worldwide living in extreme poverty rely on forests for some part of their livelihood (World Bank 2007), forests play an important role in maintaining the livelihood of 24–30 million Vietnamese people residing inside or near forests (Nguyen et al. 2007;

RECOFTC 2014; Xuan Minh 2014). The importance of forest resources is particularly intense among ethnic minorities living in mountainous areas where forest resources often provide up to 50% of their livelihoods (RECOFTC 2014). Even in some highland areas, most households make a living from forest and forest-related activities (FSIV and FAO 2009), and for many locals, "forests are the unique source of livelihood" (Pham 2008: 108).[6]

Several reasons account for forest dependence among forest dwellers. First, the long-lasting tradition of relying on forest resources has existed and been preferred "for thousands of years" by many poor communities residing in forested areas in Vietnam (Nguyen 2013: 103). This preference is due not only to the geographical linkage, but also to the attributes of natural forest resources that "lend themselves well to exploitation by the poor" (Sunderlin and Huynh 2005: 6).

Second, the forest-dwelling people have very limited choices for alternative work, so are heavily reliant on either farming or forest-based employment (World Bank 2009). Meanwhile, they have little cultivable land for farming. In the Na Mau commune, Bac Kan province, agricultural land only makes up 1.6% of the total land (Tran et al. 2010). Similarly, Xuan Trach Commune, located next to the Phong Nha Ke Bang National Park, has some 17,700 hectare of natural land, but only 900 hectares (5%) is agricultural land (Xuan Trach Comnune 2013). The limited land for farming means that forest-based work plays an important role in the local people's livelihood.

Moreover, since mono-agricultures, such as growing rice, are dependent on the weather and have tremendously low productivity, they cannot entirely meet the basic income and even food security of the local people. Working inside forests thus becomes a common solution to minimise economic and food insecurity. Tran et al. (2010: 208) indicate that the average annual income among the Na Mau commune residents is only about £23–27 per person, and that 50% of the commune households lack food for at least one month each year. To cope with the shortage, while some people visit the forest to collect NTFPs such as bamboo, fungi, cucumber and forest fruits, others take part in illegal logging and wild animal poaching (Tran et al. 2010). There are other coexisting factors that exacerbate the forest dependence of people living in forested areas such as isolated geographical location, poor infrastructure, absence of a national electricity grid and poor education (ICEM 2003; Sunderlin and Huynh 2005; Tran et al. 2010; World Bank 2009).

Examining the significance of forests for forest-dwelling inhabitants reveals two main implications for this research. First, since forests play an important role in creating livelihoods for the inhabitants, timber trafficking with its inevitable outcome—forest loss—will threaten their livelihood. The more reliant on forests the residents are, the more severe the impacts of timber trafficking will be on their employment. The ways in which timber trafficking affects the livelihood and income of forest dwellers will be systematically examined in Chap. 5. Second, since the harvesting of forest resources including timber is an important part of many forest-based communities, in order for any solutions aimed at timber trafficking to succeed, the importance of this harvesting needs to be recognised; such solutions will be proposed in Chap. 6.

Forests and the National Economy

Forest resources are important not only for forest-dwelling communities, but also for the national economy of Vietnam. By 2012, logging from the forests in Vietnam produced a total of over 5 million m^3 of logs (Nguyen and Phan 2014; To et al. 2014). This timber supply plays an important role in meeting various timber demands, notably the timber processing industry.

Stimulated by rapid economic growth, in the past decade, the timber processing industry in Vietnam has seen "spectacular progress" with an annual growth of 15%, making the country one of the world's largest manufacturing centres for wood products especially furniture products (Forest Trends 2010; Nguyen and Phan 2014; To et al. 2014). Vietnam has now become the biggest exporter of furniture and other timber products in Southeast Asia (FSIV and FAO 2009). In 2013, the export turnover peaked at £3.7 billion, jumping up 19% from the last year (Le Chi 2013; To et al. 2014). While the nominal GDP of Vietnam in 2013 was £104 billion (International Monetary Fund 2013), timber exports account for 3.56% of the country's GDP, and timber is the fifth most

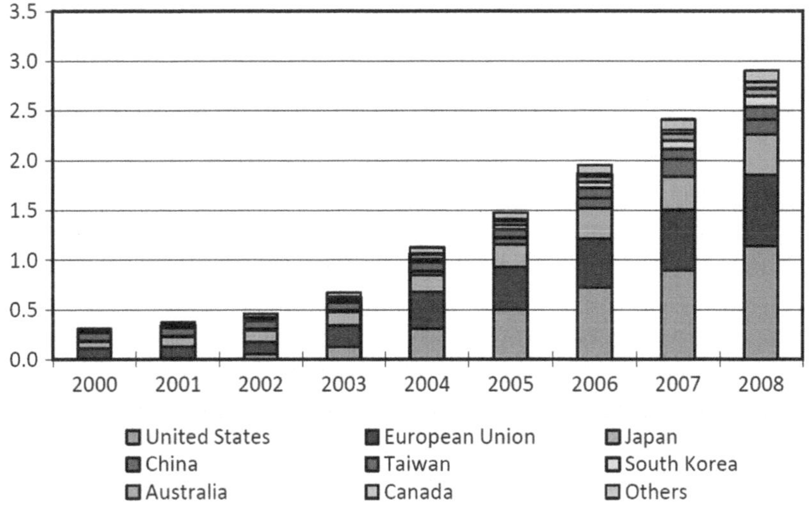

Fig. 3.2 Vietnam's export of timber products to main countries (by $ billion). Source: To and Canby (2011: 23)

valuable export commodity (To et al. 2014). In Binh Dinh province, by 2010, income derived from the export of timber products accounted for 60% of the total revenue of the local government (To and Canby 2011). Figure 3.2 shows the growth in exports of forest products by the Vietnamese firms.

Accompanied by overseas markets is domestic timber demand that is also very high. In 2012, while the total export volume of timber and its products was 19.6 million m^3 of log, the domestic market consumed 11.2 million m^3, worth £2 billion (Le Chi 2013; To et al. 2014). In addition to monetary benefits, it is important to note that the timber industry provides considerable employment for the Vietnamese. Currently, there are more than 3000 wood-processing companies in Vietnam, with 95% being privately owned, employing more than 300,000 people (To and Canby 2011). Since the industry is benefiting all three groups—the state, the timber companies and the workers—it is expected that it is going to constantly expand in future (Nguyen and Phan 2014).

In tandem with the apparent significant rewards, the rapid growth of the timber processing industry also generates increasing pressure on the forests not only in Vietnam, but in other countries. Additionally, there are many cases where timber processing companies have seriously polluted the environment (Duc Trung and Tan Phat 2014). In this scenario, the timber processing industry in Vietnam would be a sound demonstration of the theory of "treadmill of crime" by Stretesky et al. (2014) as discussed in Chap. 2.

Another worrying feature of this industry is that despite the growing requirements for proof of legal sourcing from Vietnam's major export markets, such as the USA, the European Union (EU) and Japan, so far the Vietnamese wood companies have been slow to respond. Certified timber from either domestic or imported sources is only a small portion of the total timber volume used in this industry. As of 2011, only 233 (less than 8%) out of the 3000 wood companies in Vietnam have obtained certification from the Forest Stewardship Council (FSC) Chain of Custody (CoC) (To and Canby 2011). Compared to other countries such as China, Thailand and Malaysia, the pace of FSC CoC certification in Vietnam is slow. This is mainly because the majority of the Vietnamese companies are small or medium sized and cannot afford FSC CoC. Although the Forest Sector Development Strategy of the country aims that by 2020, the sustainability of 30% of the forest areas in the country will be certified by international competent organisations, as of 2011, only about 15,000 hectare of plantation forest (1.1%), and no natural forests, are certified (To and Canby 2011).

The result is that according to the market standard, the majority of timber used by the Vietnamese wood firms is uncertified, and much of the suspicious timber is illegal. It is argued that the high demand for timber, coupled with limited legal supplies, leads to some timber firms being willing to use suspicious timber, creating a market for illegal logging (EIA 2011; To 2012; USAid 2013). Research by Meyfroidt and Lambin (2009) observes that on many occasions, the Vietnamese manufacturers of outdoor furniture are accused of using illegally sourced materials, and that 48.1% of timber imports to Vietnam are estimated to be illegal. This figure varies considerably from one by Lawson and MacFaul (2010) who estimate that between 2000 and 2008, only 17% of timber imports into Vietnam may have been illegally sourced. This rate remained almost unchanged by 2013 at 18% (Saunders 2014).[7]

In addition to their role in the timber processing industry, the Vietnamese forests also have another important role to play in providing capital and locations for various major projects of forestland conversion into different schemes like plantations, construction of irrigation systems, road building, hydropower plants, mineral exploitation and spiritual site building. From 2006 to 2013, the Vietnamese authorities permitted some 363,500 hectares of forest to be used for almost 2400 forestland conversion projects (FPD 2014; MARD and FSSP 2014).

Large areas of Vietnamese forests have, for example, been used to build hydropower plants to contribute to the rapidly growing demand for electricity in Vietnam.[8] The MARD reports that from 2006 to 2012, some 20,000 hectares of forest have been used to construct 160 hydropower projects in Vietnam.[9] Additionally, the forests account for the majority of about 910,500 hectares of rubber plantation area, which helped Vietnam become one of five countries having the largest area of rubber plantation as well as the highest total production of rubber worldwide, with the turnover from exports of rubber latex reaching over $2 billion by 2012 (To and Tran 2014b).

These forest-based projects are believed to bring about substantial economic and social benefits for many parties, primarily the creation of employment opportunities and infrastructure improvements for local communities and providing important goods and services for the national economy (Binh Duong Company 2010; Hydropower No.7 Management Board 2006). For example, according to the assessment by the An Khe—Ka Nak Hydropower Project, which makes use of 344 hectares of forest, every year it contributes 700 million kWh to the national electricity grid and provides irrigation water for nearly 5000 hectares of lowland farming. It adds:

> Thanks to beneficial programmes such as compensation and resettlement, the local people will have better roads, more employment opportunities, receive better social welfare, improve production methods, become familiarised with modern methods of market economy and reduce poverty. (Hydropower No.7 Management Board 2006: 107)

Nonetheless, the downside is that forest-based projects have accounted for 43% of the total forest loss in Vietnam in recent years (Environmental

Police Department 2012). They also serve as a legitimate platform for the illicit harvesting of large volumes of timber and bring about various negative social and environmental consequences, which will be carefully examined in Chaps. 4 and 5, respectively. The previous sections reviewed the importance of forest resources including timber and NTFPs for the Vietnamese forest-based residents and the national economy. The next section will discuss another important role of forests; that is, to provide timber to satisfy a popular tradition of timber consumption in Vietnam.

Forests and the Traditional Consumption of Rare Timber

Cultural elements have been seen as a driver of wild animal trafficking. The traditional perception of using rare and precious wildlife and its products for food and medicine has an important role to play in boosting the demand for wildlife (Blevins and Edwards 2009; Cao and Wyatt 2013; Drury 2009; Lin 2005; Wyatt 2013). Literature on timber harvesting, however, has paid very little, if any, attention to this element. This research is thus interested in how some cultural factors shape the process of timber trafficking in Vietnam. In this sense, this research finds that there is in Vietnam a cultural passion for using timber products made from the rare, precious and endangered timber species logged in natural forests (hereafter termed "endangered timber"). The Vietnamese have a long-standing culture of using these products for various purposes: decoration, material for house building, spirituality and traditional medicine.

Indeed, a common pastime of many wealthy Vietnamese is purchasing decorative products and wood furniture made from endangered timber to display in their living rooms. Displaying a set of table and chairs made from endangered timber is believed to make the house more sophisticated and elegant. By doing so, the owners can boast about their prosperity to others.[10] They also believe that it can prevent "evil air" from entering their homes, and that they will receive godsends and fortune in their lives and careers. In the past, Sủa timber, for example, was used to design luxurious furniture in royal places and to make caskets for deceased kings and queens. It was also ground into powder to sprinkle over the coffin in the belief that the soul in the coffin would soon be released (Le Trang 2011).

In addition to furniture products, there are several other such decorative items such as *Lộc Bình* (Bottle for Fortune), *Thần Tài* (Statue for Intelligence), *Tượng Phật* (Buddha Statue), and *Phúc Lộc Thọ* (Happiness-Fortune-Longevity Statue) that are popularly used by many Vietnamese. It has become a prevalent trend in many rich families to spend tens of thousands to possess such exceptionally expensive products made from endangered timber such as *Thủy tùng* (*Glyptostrobus pensilis*), *Cẩm lai* (*Dalbergia oliverii Gamble*), *Pơ mu* (*Fokienia hodginsii A. Henry et Thomas*) and *Giáng hương* (*Pterocarpus Pedatus Pierre*) (SGTT 2011).

Additionally, the preference for long and large timber planks made from natural forests is also increasingly popular in house building. Stories about the wealthy Vietnamese who spend tens of thousands on buying beds made from hundred-year-old Sua, Lim or Trac timber are at times reported in the media (Hanh Nguyen 2014; Viet Huong 2014). There was a notable case in Dien Bien province where a construction magnate built a 500 m^2 stilt house entirely made from Lim timber—*Erythrophloeum fordii*—an endangered timber species in Group IIA.[11] This £6.7 million house was constructed by over 10,000 skilled craftsmen over two years (Viet Huong 2014). Despite the exorbitant costs, this pastime has already been "absorbed into the blood" of consumers who enjoy the mysterious beauty of the items and the sense of meditation, peacefulness and profoundness inspired by these items (Ba Thang 2014).

In addition to the consumption of timber for decorative and spiritual purposes, some Vietnamese also use timber for medicinal purposes although this practice is possibly much less prevalent than those previously mentioned. A study by Nguyen and Nguyen (2008) estimates that three-quarters of Vietnamese adopt traditional remedies for general health problems; meanwhile, 3500 species of fauna and flora and about 20,000 tons of other flora have been used as medicine; and 95% of the traditional remedies are plant-based.[12]

In addition to the traditional use of timber for aromatherapy and pharmaceuticals, highly endangered timber species such as Ngọc Am, Lim, Ky Nam and Tram Huong can be used to treat diseases. Purchasers use Ky Nam and Tram Huong, for example, to cure urinary tract disease, malaria and abdominal pain (Ton 2006), and mushrooms grown on Lim trees in Quang Ngai province are used to cure cancers, particularly liver cancer (VTC News 2012). Traditionally seen as the "pearl of forests", Ngoc Am timber is used to make bathtubs. It was believed that bathing in these bathtubs could facilitate the emission of body toxins, increase blood circulation, prevent skin diseases and improve mental health (Le Trang 2011). Much information on the "magic" effects of these endangered timber species can be easily found online, with many websites exclusively designed to promote consumption of the timber.[13] Further details of this consumption will be provided in Chap. 6. The preceding two sections of this chapter have looked at environmental issues and the role of forests and timber for the Vietnamese people and economy. The final section will introduce the legal context with a focus on legislation and its implementation concerning the control of timber trafficking in Vietnam.

Forest Governance and Law Enforcement on Timber Trafficking

Legislatory Efforts

Although some commentators observe that economic growth may still be a higher priority than environmental protection (Pham et al. 2012), there seems to be a substantially increasing political determination to deal with environmental problems, including timber trafficking in Vietnam. There is firm evidence to underpin this observation. First, it is a UNODC evaluation that "perhaps more than any other country in the region, there is a high level of stated political commitment to counter the illegal timber trade in Viet Nam" (UNODC 2013a: 20). Vietnam is internationally recognised for its efforts in participating in a wide range of international and regional treaties, as well as enacting a large number of laws, regulations and national plans, with a view to effectively handling environmental issues in general and environmental crime and timber trafficking in particular.[14] In the past decade, the Communist Party of Vietnam, for example, has issued two resolutions (41/NQ/TW in 2004 and 24-NQ/TW in 2013) to demand that serious attention needs to be given to the mission of environmental protection in the country. This "indicates a healthy preoccupation with the environment by the national government" (USAid 2013: 18).

The efforts are not confined to joining the treaties concerning environmental protection, but actively implementing them. For example, since Vietnam joined CITES in 1994, in order to facilitate the implementation of the Convention, the Vietnamese legislative bodies have issued a number of laws and regulations on the harvesting of, and the trade in, wild fauna and flora. Vietnam has established the CITES Scientific Authority affiliated to the MARD, and it is one of the first four countries to voluntarily undertake a wildlife trade policy review (Nguyen et al. 2007). The contents of CITES have also been fully implemented into domestic laws.[15] In addition to CITES, after joining the Forest Law Enforcement, Governance and Trade (FLEGT) in 2010, Vietnam established the FLEGT standing office and three sub-technical working groups (Pham et al. 2012).

In the current Penal Code of Vietnam, promulgated in 1999 and amended in 2009, which is the country's foremost criminal legislation, all of Chapter XVII is devoted to the 11 articles defining 11 environmental offences. Key illicit activities in timber trafficking have been defined in Article 175 in Chapter XVI, namely "Breaching regulations on forest exploitation and protection". The clarification of Article 175 is provided in the Joint Circular 19/2007/TTLT/BNN&PTNT-BTP-BCA-VKSNDTC-TANDTC. In parallel to the criminal provisions, there are a number of laws and regulations that can also be applied to timber trafficking in Vietnam. The most relevant legal documents include the Forest Protection and Development Act (2004), the Environmental Protection Act (2008), and Decree 157/2013/ND-CP on Punishing Administrative Violations in the Field of Forest Management, Protection and Forest Product Management. It is calculated that, in Vietnam, there are a total of about 150 legal documents related to the undertaking of forest protection and development, most of which are currently in effect (MARD 2012).[16]

In tandem with the legislative work, a number of national programmes have been initiated and implemented to improve forest quality and forest management. The two most notable projects are Programme 377 (1992–1997) and Programme 661 (1998–2010), which received a total of almost $1 billion from both the Vietnamese government and international communities (MARD 2001; Morris et al. 2004; Sunderlin and Huynh 2005). These projects played a crucial role in reaching the 14 million hectares of forest and 41% forest cover rate in 2013, compared to 9.17 million hectares and 27% in 1990 (FSIV and FAO 2009; MARD 2014; To and Canby 2011).

Policy Framework

Globally, it is observed that an inappropriate policy framework in forest governance is a fundamental explanation for illegal logging. Under this framework, the majority of poor forest-dependent households are not allowed access to forest resources, which leads them to operate illegally to meet their basic needs for food, employment and income (Kaimowits 2007; Wells et al. 2007). Looking into the policy approach in forest governance in Vietnam, it is argued in this research that one of the most encouraging and fundamental changes in the history of forest governance in the country could be the deployment of a decentralised approach to forest ownership. Since the establishment of the Vietnamese government in 1945 until the early 1990s, all natural forests, accounting for the majority of Vietnamese forests, were nationalised, meaning that the forests were owned, managed and exploited only by state institutions, chiefly state-owned forestry enterprises (SFEs) (FSIV and FAO 2009: 17).

However, this nationalised mechanism revealed profound problems. Among the downsides were the notoriously ineffective management and excessive logging among the 420 SFEs nationwide, which resulted in extremely severe degradation of the forests (de Jong et al. 2006; McElwee 2004).[17] From the early 1990s, Vietnam has gradually recognised these fundamental shortcomings, leading to the Forest Protection and Development Act in 1991 that signalled a significant departure from a fully centralised framework to the involvement of non-state actors in forest administration (Sikor and To 2011). Overall, the current legal system defines that forestland is under public ownership with the state acting as the representative.[18] Following the implementation of this legal mechanism, forest ownership has recently shifted towards households, individuals, village communities and the private sector as they are increasingly being allocated forests (FSIV and FAO 2009).

To date, almost 4.5 million hectares of forestland have been allocated to 1.2 million households across the country (To and Tran 2014a). At the same time, the previous SFEs have been either dissolved, or shifted to mostly state forest companies (FC) and forest management boards (MB).[19] Figure 3.3 presents details of the forest areas managed by different users. It shows that the three biggest user groups in terms of forest area are management boards, households and people's committees at commune level. Overall, it is evaluated that the shift in the policy framework has proved to be effective for both forest protection and the improvement of livelihoods in some forest communities (To and Tran 2014a).

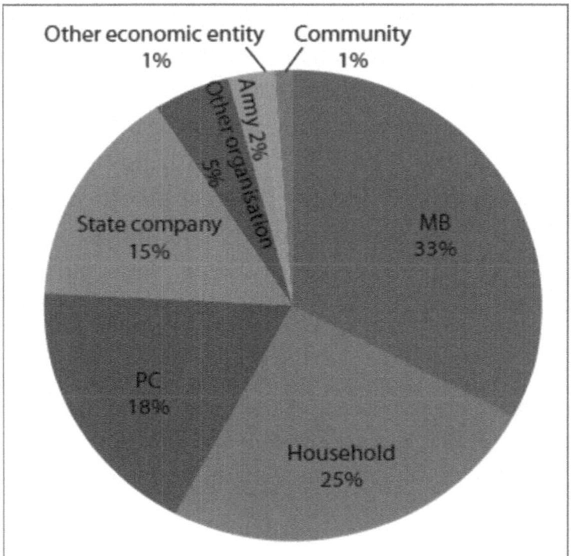

Fig. 3.3 Proportion of forest areas managed by forest user groups in Vietnam by the end of 2009. Source: Nguyen and Tran (2011: 5)

In addition to the transferral of forest ownership towards non-state actors, the Vietnamese government has made positive attempts to legalise some of the logging activities of poor forest-based households with the intention of helping them meet their basic subsistence needs, particularly cultivable forestland and residential housing. Specifically, to support poor people in general and poor ethnic minorities in particular, the Vietnamese government has launched a number of significant nationwide initiatives with the most well-known being Project 134 in 2004 and Project 167 in 2008.[20] In these programmes, backed up by a number of solutions to support poor ethnic minorities, is an approach that allows some of them to harvest a certain amount of non-endangered timber in specific forests. Point b, Clause 3, Article 2 of Decision 134/2004 states that:

> In forest localities that have approved annual logging plans, the provincial People's Committees are allowed to harvest timber to help ethnic minorities build their houses. The specific volume for each household is determined by the provincial People's Committees. (PMV 2004)

The policy also prescribes that in situations where households that qualify for the programme are unable to conduct the harvesting, it is the responsibility of the local authorities to organise logging operations for those households. To prevent any transgressions against the policy, Article 2 of Decision 134/2004 declares that "it is strictly prohibited to unscrupulously take advantage of the logging policy in destroying forests". More specifically, Article 7 in Joint Circular No.08/2009/TTLT-BXD-BTC-BKHĐT-BNNPTNT-NHNN on Guiding the Implementation of Decision No. 167/2008/QĐ-TTg clearly states:

> Timber is harvested for the only purpose of building the residential houses of the poor indigenous households that are in the list of the district people's committees. The timber is not allowed to exchange or trade in any forms. The maximum timber volume is no more than 10 m^3 of round timber for each household. (Joint Circular 2009: 5)

These projects have actually developed a general mechanism as well as specific support for poor people to have residential timber houses. After three years, in effect 520,000 poor households have had houses constructed by Project 167 (Kieu Thang 2013). However, this policy has also been abused by professional timber traffickers to illegally harvest relatively large amounts of timber, which will be scrutinised in Chap. 4.

In addition to evaluating the relevance of policy approaches, this research is also concerned about the severity of penalties imposed on timber trafficking offenders in Vietnam. This is because, in many countries, as with some other forms of green crime, punishments for timber trafficking seem to be relatively lenient, particularly when compared to the illicit gains that can be made and to the harsh sanctions targeting other "high-profile" crimes. Contreras-Hermosilla (2001), for instance, posits that "it is common for penalties for forest-related illegal acts to be minimal in comparison with the rewards of forest crime. Often, penalties are so light that they do not translate into a significant deterrent". Similarly, Goncalves et al. (2012: 10) observes that illegal logging is a perceived as "quasi-criminal" administrative offence, resulting in low fines and minimal criminal sanctions.

In the context of Vietnam, although some previous studies such as Pham (2008) and Tran (2013) argue that criminal punishment for timber offenders is too lenient, it is observed that overall the current system of sanctions in both criminal and administrative forms is not excessively lenient. The highest level of penalty imposed on timber trafficking by Clause 2 of Article 175 in the Penal Code is 10 years' imprisonment, which means the crime is categorised as a "very serious offence".[21] The punishment for cross-border timber smuggling defined by Clause 4 of Article 153 of the Penal Code is from 12 to 20 years or even life imprisonment. This level of criminal punishment is well beyond the scale of penalties for "serious crime" as defined in the United Nations Convention against Transnational Organised Crime,[22] and it is relatively high compared to some other jurisdictions. The maximum criminal punishment for illegal logging in China, for example, is seven years' imprisonment.[23] Likewise, the UK Illegally Logged Timber (Prohibition of Import, Sale or Distribution) Bill offers a maximum imprisonment of five years for illegal timber traders in England and Wales[24] (Bennett and Ares 2011).

As for monetary sanctions, the maximum monetary fine, clarified in Decree 157/2013, for an incident of timber trafficking (e.g. Point c, Clause 10, Article 22 on illegal timber transportation) is VND500 million (£17,000). This amount is perhaps not very high for a few of the large-scale timber barons, but for the vast majority of timber offenders in Vietnam, especially poor loggers and carriers, this is far too high, and thus possibly ineffective. Indeed, it is common that extremely high fines, which clearly exceed the value of the offenders' property, mean that the authorities are unable to seize any of their possessions (N. Hung 2014). One example is that from 2008 to 2012, the total administrative monetary fines issued by the Dak Glei district authorities in Kon Tum province for violations of forest protection laws were worth over £33,000. However, no considerable payment from the violators has been made yet simply because the violators have no sizeable assets to be seized (Hoai Nhon 2012). Consequently, the system of draconian fines becomes a legal loophole repeatedly abused by unlawful loggers as well as transporters to avoid legal liability. This technique will be examined in Chap. 4.

Institutions Involved in the Control of Timber Trafficking

The positive news is that many law enforcement agencies tasked with policing timber trafficking in Vietnam have been established. The first and most specialised of these agencies is the Forest Protection force, known as Kiem Lam. Established in 1946, the force is regularly deployed at central, regional, provincial, and district levels and in protected and reserve forests. It was created to implement the administrative management of forest protection, ensure law enforcement of forest protection and development, and monitor forest products (FPD 2012). At the central level, the Forest Protection Department (FPD) belongs to the MARD,[25] but at the provincial level, Forest Protection Sub-departments are under the direct leadership of the Departments of Agriculture and Rural Development in the Provincial People's Committees. These sub-departments take control of the Forest Protection Branches at district level and in special-use and protected forests (FPD 2012).

The second important body involved in combatting timber trafficking is the Environmental Crime Prevention and Combat Police forces (henceforward called the Environmental Police) that were established in 2006 and exist at three levels: central, provincial and district. Similar to other police forces, the environmental police in Vietnam are an armed force, under the absolute and direct leadership in all aspects of the Communist Party of Vietnam, the command of the State President, the unified management of the Government, and the direct governance of the Minister of Public Security (NAV 2005). According to Decision 449/QĐ-BCA, dated 4 February 2010, issued by the Minister of Public Security, the power of the Environmental Police entails directing, guiding, checking and conducting preliminary investigations into environmental crime as well as handling administrative violations in accordance with the provisions of environmental protection laws and regulations (Environmental Police Department 2010).[26]

In addition to the Forest Protection and Environmental Police forces, there are five other forces that also police timber trafficking activities. They are the Investigative Police, Customs of Vietnam, Border Guards (Border Army), Market Control and Forest Rangers. The organisational structures and responsibilities relevant to timber trafficking of these forces are summarised in Appendix B.

It is generally observed that in the past decade, these institutions' operational capacity has been constantly and considerably improved. With reference to the most specialised force, Kiem Lam, for example, 63 out of the63 provinces in Vietnam have already established the Kiem Lam Sub-departments with nearly 450 district branches formed. In addition, over 200 checkpoints have been newly created at the forest gates, while almost half of Kiem Lam officers now have university degrees (FPD 2013).

Furthermore, on the frontline, there have been promising efforts from anti-trafficking officers to curtail timber trafficking in Vietnam over the past years. Quite often the Vietnamese media present officers who are bravely committed to the fight against Lam Tac despite being severely threatened and attacked. In Gia Lai in 2014, for example, irrespective of being sprayed with pepper gas and having a hand almost cut off, two Kiem Lam officers still kept fighting against the Lam Tacs until their colleagues came to assist (Phap Luat Viet Nam 2014). In an incident when trying to prevent illegal timber transportation, one officer was attacked and severely wounded. The officer still pursued the criminals until he collapsed (Tien Thanh 2014). There are stories of Kiem Lam officers who continually stayed in makeshift camps next to endangered trees despite facing harsh risks (Minh Thi—Minh Trieu 2014).

The report that recaps the work of the Kiem Lam forces over the past 40 years confirms that "forest protection forces in the entire country have constantly grown and developed, overcoming numerous difficulties and challenges with the aim of keeping the green of the homeland" (FPD 2013: 29). Although there is not sufficient evidence to confirm this, the efforts by the forces entrusted to combat timber trafficking in Vietnam may be one of the contributors to the annual detection of over 35,000 violations of forest protection law, and to the gradual reduction in the number of violations detected over the past five years (UNODC 2013a). Nonetheless, in a study aimed at measuring the impact of law enforcement activities in tackling illegal timber trade in Vietnam, UNODC (2013a: 22) concludes:

> While there are some higher level prosecutions which have been undertaken, they are by far the exception than the rule, and have only targeted middle level players. In short, the criminal justice system is not being used optimally to counter the illicit timber trade.

Conclusion

This chapter offers background information about Vietnam which is required prior to carrying out a detailed investigation of how timber trafficking is taking place and affecting human security, and how to better control the crime in Vietnam. It is first observed that in parallel with increasing noteworthy economic improvements, Vietnam is today facing many significant environmental challenges, including environmental crime. There is a large number of serious violations against laws and regulations on environmental protection that have been detected, of which very few cases have been charged with a crime. This is, in part, due to major drawbacks in current Vietnamese criminal law regarding tackling environmental crime.

The second section of this chapter introduces timber and other forest resources and their important role for forest-dwelling people, the national economy and the cultural consumption of endangered timber in Vietnam. This section then raises two main concerns. First, there are still tens of millions of Vietnamese whose livelihood is partly or entirely reliant upon forest resources, which means that forest loss would be a real threat to their economic security. Second, there is a high and diverse demand for timber in the country. Possibly the highest demand for timber comes from the country's booming timber processing industry that consumes millions of m^3 of timber annually, with a considerable part of the processed timber thought to be illegally harvested. Furthermore, the high demand for the traditional consumption of endangered timber species is also of particular concern because the consumption is a direct and key threat to the survival of these valuable species.

The final section of the chapter looks at the work by legislative and law enforcement agencies to confront timber trafficking in Vietnam. There seems to be a strong commitment among politicians and policy makers to join a variety of international green conventions, as well as enact a relatively comprehensive system of laws, regulations and programmes, including those that directly aim to better control timber trafficking. Generally, the legal provisions with relatively heavy sanctions have provided a vital legal backbone for the fight against timber trafficking. Additionally, at the frontline of this fight, there has been an establishment of, and some effective work delivered by, various law enforcement agencies. The efforts of both legislators and law enforcers might be one of the contributors to the detection of tens of thousands of violations of forest protection law every year. However, it seems that these efforts have not translated into fundamental success in the prevention of forest crimes.

After this research context has been provided, the next chapter will detail the methodological approach and specific methods used to collect and analyse data, which helps answer the research questions.

Notes

1. From being one of the poorest countries in the world, Vietnam has, since 2010, officially moved from the group of poor countries to join the middle-income group with GDP per capita being almost $1700 in 2013

(International Monetary Fund 2013; UNDP 2013). It is forecasted that Vietnam will be "a New Tiger Economy" (Curran 2015), remain one of the most rapidly growing emerging economies by 2025, and become one of the world's top 20 economies by 2050 (Hawksworth and Tiwari 2011).

2. The Index is conducted by the Yale Center for Environmental Law & Policy and its partner. It ranks how well countries perform on high-priority environmental issues in two broad policy areas: protection of human health from environmental harm and protection of ecosystems. According to this Index, the worst aspect of Vietnam's performance is shown to be air pollution (Environmental Performance Index 2014). The Vietnam Ministry of Natural Resources and Environment estimates that environmental pollution costs 5.5% of the country's GDP annually (MNRE 2010).
3. In the entirety of Vietnam, the average number of criminal cases discovered each year has climbed from about 60,000 in the 1990s to 83,500 in the 2000s (United Nations 2007). In the past few years, while the number of criminal cases in Vietnam increases about 1–2% yearly, reaching almost 80,000 cases in 2014, the criminal situation has remained increasingly complex (Nguyen Le 2014; Quang Vu 2013). Serious criminal cases are reported as noticeably increasing in terms of the quantity, transnationality and sophistication (particularly corruption), environmental crime, trafficking of women and children, money laundering and fraud (Le 2009; UNODC 2012).
4. Currently, Vietnam has about 200 industrial zones of which, as reported by the Vietnamese National Assembly, 90% of these zones commit violations against environmental protection law. This figure is 70% according to an estimate by Environmental Police (Viet Ba 2011).
5. Article 182 in the Penal Code 2009, for example, defines the offence of "Polluting the Environment":

    1. Those who discharge pollutants into the air, water sources or soil, emit radioactivity or radiation in excess of waste limits prescribed by national technical regulations to a *serious extent* or *seriously* polluting the environment or causing other *serious consequences*, shall be imposed a fine of between fifty million and five hundred million VND, subject to non-custodial reform of up to three years or a prison term of between six months and five years.

2. Committing the crime in either of the following circumstances, offenders shall be sentenced to between three and ten years of imprisonment:
   (a) In an organised manner;
   (b) Causing *particularly serious* environmental pollution or other *very serious or particularly serious consequences*" (NAV 1999).

6. By way of example, an ethnographic study by Nguyen (2013: 84) in a village of Muong ethnic minorities in Thanh Hoa province shows that the villagers "rely very much" on forest resources through hunting wild animals, collecting firewood, medicinal herbs, quick growing species such as bamboo, wild vegetables, and low-quality timber such as eucalyptus.
7. In addition to the use of illegal timber, some other major challenges facing the timber industry in Vietnam include limited competitiveness, low added value and low labour productivity (the productivity of the Vietnam timber processing industry is around 50%, 40% and 20% of that of Philippines, China and the EU, respectively (Nguyen and Phan 2014).
8. The conditions coexisting in Vietnam such as extensive poverty reduction, urbanisation and industrialisation have resulted in a substantial and consistent growth in electricity demand in the country with increases in power demands of 15%, 17% and 20.2% in 2008, 2009 and the first quarter of 2010, respectively (Carew-Reid et al. 2010). To satisfy the increasing demand, the Government's 6th Power Development Plan for 2006–2015 intends that hydropower plants constructed in forested areas will provide one of the main sources of national electricity (Carew-Reid et al. 2010).
9. The Ministry of Industry and Trade (MIT) provides a different number from MARD. According to MIT, since the implementation of Decree 23/2006/ND-CP, to date there have been 50,930 hectares of forest used for the purpose of building hydropower plants (Bich Ngoc 2013).
10. There are some publications showing that one of the major characteristics of the Vietnamese is that "they are thrifty, but many times squander money for meaningless reasons (to save face or to show off)" (see Ha and Nguyen 2014) or "the Vietnamese are not yet rich but 'posh' because of exhibitionism" (Thanh Huyen 2014).
11. The classification of groups of timber in Vietnam is specified in Decision 2198/CNR dated 26 November 1977 by the Ministry of Forestry. In

this decision, based on physical properties, durability, practical purposes and economic values, timber species in Vietnam are categorised into eight groups: Group I—Group VIII (Do et al. 2007). The classification, however, is relatively outdated and inconsistent with the international classifications (Do et al. 2007). It is also easy to be confused because a number of timber species in Group I and Group II in this Decision are also listed in Group IA (endangered species of forest plants of which the harvest and use for commercial purposes are strictly prohibited) and Group IIA (endangered species of forest plants of which the harvest and use for commercial purposes are restricted) of Decree 32/2006/ND-CP.

12. There are a large number of traditional medicine institutions in Vietnam, including 48 hospitals, over 240 traditional medicine departments in central and provincial hospitals, over 9000 health centres reportedly licensed to practice traditional medicine, and a large number of unregistered herbalists and healers providing traditional medicine treatments to patients (Nguyen and Nguyen 2008).
13. For the effects of Tram Huong and Ky Nam see http://tramhuongkyanh.vn/, of Lim mushroom, see http://www.namlimxanhtienphuoc.vn/. It is worth noting that the traditional consumption of endangered timber is not confined to Vietnam but exist in many other Oriental societies. Tram Huong (Agarwood), for instance, which has been listed in Appendix II of CITES since 1995, is commonly used in Asia. The timber is used in Japan where it is worth $10,000 per kilogramme (UPM 2013). Some of this timber's effects have been scientifically examined and confirmed (Takemoto et al. 2008). The first international conference on timber effects was held in Malaysia in 2013 where 104 participants from 16 countries attended the conference to find out about the "magic" effects of the timber (UPM 2013). A study shows some values of Agawood: "It has a fragrance so sublime that its aficionados dubbed it the 'incense of the gods', while the ancient mariners braved the high seas in search of it. For sure, the charm and allure of gaharu or agarwood has not diminished over the ages" (UPM 2013: 1).
14. The main agreements include the Ramsar Convention on Wetlands of International Importance, the United Nations Framework Convention on Climate Change and the Kyoto Protocol, the United Nations Convention to Combat Desertification, the Vienna Convention on Protection of the Ozone Layer and the Montreal Protocol on Substances that Deplete the Ozone Layer, CITES, FLEGT, the Basel Convention on the Control of Trans-boundary Movements of Hazardous Wastes and

Their Disposal, and the Stockholm Convention on Persistent Organic Pollutants (SRV 2012; Pham et al. 2012).

15. Section II of the Joint Circular 19/2007/TTLT/BNN&PTNT-BTP-BCA-VKSNDTC-TANDTC and Article 7 in the Decree 157/2013/ND-CP specify methods of handling species listed in Appendices I and II of CITES in accordance with the current Vietnamese legislation.
16. See Appendix B for more details on policies, laws and regulations on forest protection in Vietnam.
17. Some commentators even suggest that irrespective of their legal logging status, the detrimental impacts on the forests caused by the SFE's logging were even more severe than the consequences of the Vietnam War (McElwee 2004).
18. Currently, forestland ownership is shaped by two key laws: the Land Act of 1993, amended in 1999 and 2003, and the Forest Protection and Development Act 2004, which succeeded that of 1991.
19. The shift from SFEs is regulated by Resolution 28/NQ-TW of the Communist Party of Vietnam's Politburo in 2003, and Decree 200/2004/ND-CP by Vietnam's Government in 2004. Consequently, 256 SFEs have been shifted into 148 FCs, three joint-stock companies, and 91 forest MBs; 14 ineffective SFEs were dissolved. Ten of the 148 FCs are under central management, while the 138 remaining SFEs are managed by their respective Provincial People's Committees (To and Tran 2014a).
20. Project 134 was established by Decision 134/2004/QĐ-TTg of the Prime Minister namely "A number of policies to support the households of poor ethnic minorities with production lands, residential lands, residential houses and clean water". Project 167 was created from Decision 167/2008/QĐ-TTg by the Prime Minister namely "Housing policy for the poor households".
21. According to the present criminal law in Vietnam, criminal offences are categorised into four groups: "less serious offences", "serious offences", "very serious offences" and "particularly serious offences" with the maximum penalty brackets for these offences being three years' imprisonment; seven years' imprisonment; 15 years' imprisonment; and over 15 years' imprisonment, life imprisonment or capital punishment, respectively (NAV 1999). The maximum punishment for timber trafficking is 10 years' imprisonment; accordingly, it is considered a very serious offence.
22. According to the UN Convention against Transnational Organised Crime, "serious crime shall mean conduct constituting an offence

punishable by a maximum deprivation of liberty of at least *four years* or a more serious penalty" (UNODC 2004: 5).

23. Article 344 of The Criminal Law of the People's Republic of China (1997) prescribes: "Whoever, in violation of the provisions of the Forestry Law, illegally fells or destroys precious trees shall be sentenced to fixed-term imprisonment of not more than three years, criminal detention or public surveillance and shall also be fined; if the circumstances are serious, he shall be sentenced to fixed-term imprisonment of not less than three years but not more than seven years and shall also be fined" (China National People's Congress 1997).
24. Similarly, as prescribed by the Customs and Excise Management Act (CEMA) 1979 of England and Wales, the maximum period of imprisonment for offences that are committed in connection with a breach of the Forest Law Enforcement, Governance and Trade (FLEGT) Regulation is three years only (UK Customs 2012).
25. FPD contains five divisions: (a) Forest Management and Protection; (b) Inspection and Legislation; (c) Propaganda and Force Development; (d) Special Team; and (e) Office of the Department. It takes the leadership role of three regional agencies consisting of the Forest Protection Zone I, II, and II located in North, Centre and South Vietnam respectively.
26. Apart from their role in combating wildlife trafficking, the environmental police have become a specialised body taking responsibility for dealing with other forms of environmental crime primarily air pollution, soil and water contamination, and the destruction of aquaculture resources (Do 2010). They also coordinate evaluations of the impacts of business investment projects on the environment and test environmental standards. The Divisions of Environmental Protection Police in the 63 provinces have tasks similar to these, but are in charge of only their own provinces (Environmental Police Department 2010).

References

ADB. (2005). *Vietnam: Country environmental analysis*. Hanoi, Vietnam: Asia Development Bank.

Ba Thang. (2014). Gốc cây gây sốt [Tree stumps become best sellers]. *Cong an Da Nang*, 1/11/2014. Retrieved March 25, 2015, from http://cadn.com.vn/news/65_122617_go-c-cay-gay-so-t.aspx

Bennett, O., & Ares, E. (2011). *Illegally logged timber: EU and UK legislation, Standard Note: SN/SC/5858*. Retrieved from file:///C:/Users/w10036832/Downloads/sn05858.pdf

Bich Ngoc. (2013). GS.TSKH Nguyễn Ngọc Lung: Cao su và thủy điện tận lực phá rừng! [Prof. Dr. Nguyen Ngoc Lung: Rubber and hydropower deforestate by their best!]. *Đất Việt*, 02/10/2013. Retrieved March 25, 2015, from http://www.baodatviet.vn/chinh-tri-xa-hoi/binh-luan/gstskh-nguyen-ngoc-lungcao-su-va-thuy-dien-tan-luc-pha-rung-2356287/

Binh An. (2014). Thi hành Bộ luật Hình sự: Bất lực với tội phạm môi trừờng? [Enforcing the Penal Code: Being powerless over green crime]. *Ministry of Justice*. Retrieved June 2, 2015, from http://www.moj.gov.vn/ct/tintuc/Pages/thong-tin-khac.aspx?ItemID=5291

Binh Duong Company. (2010). *Environmental impact evaluation report of the project in rubber development in 935, 936 Zones Ia Puch commune, Chu Prong district, Gia Lai province*. Gia Lai: Binh Duong Company—Binh Doan 15 (in Vietnamese).

Blevins, K. R., & Edwards, T. D. (2009). Wildlife crime. In J. M. Miller (Ed.), *21st century criminology: A reference handbook* (pp. 557–563). Thousand Oaks, CA: Sage.

Brunner, J. (2012). *Biodiversity conservation in Vietnam: A perfect storm*. IUCN Vietnam. Retrieved from http://cmsdata.iucn.org/downloads/iucn_nbsap_workshop_march_2012_paper_final.pdf

Cao, N. A., & Wyatt, T. (2013). A green criminological exploration of illegal wildlife trade in Vietnam. *Asian Journal of Criminology, 8*(2), 129–142.

Carew-Reid, J., Kempinski, J., & Clausen, A. (2010). *Biodiversity and development of the hydropower sector: Lessons from the Vietnamese experience—Volume I—Review of the effects of hydropower development on biodiversity in Vietnam*. Hanoi: ICEM—International Centre for Environmental Management.

China National People's Congress. (1997). *Criminal Law of the People's Republic of China*. Adopted at the Second Session of the Fifth National People's Congress on July 1, 1979. Revised at the Fifth Session of the Eighth National People's Congress on March 14, 1997.

Contreras-Hermosilla, A. (2001). *Law compliance in the forestry sector: An overview*. Washington, DC: World Bank.

Curran, E. (2015). Asia's about to spawn a new tiger economy: Good morning, Vietnam. *Bloomberg*, 22/3/2015. Retrieved May 1, 2015, from http://www.bloomberg.com/news/articles/2015-03-22/asia-s-about-to-spawn-a-new-tiger-economy-good-morning-vietnam

Dasgupta, S., Laplante, B., Meisner, C., Wheeler, D., & Jianping Yan, D. (2007). *The impact of sea level rise on developing countries: A comparative analysis*. World Bank Policy Research Working Paper, 4136.

de Jong, W., Do, D. S., & Trieu, V. H. (2006). *Forest rehabilitation in Vietnam: Histories, realities, and future*. Bogor, Indonesia: Center for International Forestry Research—CIFOR.

Do, A. T. (2010). *Measuring the effectiveness of the enforcement of organised wildlife trade crime: A comparative assessment between Vietnam and England and Weles*. A thesis submitted in partial fulfilment of the requirements of the University of the West of England, Bristol for the degree of Doctor of Philosophy, University of the West of England.

Do, V. B., Le, T. H., & Hoang, T. S. (2007). *Studying scientific foundations for timber classification [Nghiên cứu cỏ sở khoa học phân loại gỗ]*. Ha Noi: The Institute of Forestry Science of Vietnam.

Dollar, D., & Litvack, J. (1998). Macroeconomic reform and poverty reduction in Vietnam. In D. Dollar, P. Glewwe, & J. Litvack (Eds.), *Household welfare and Vietnam's transition* (pp. 1–28). Washington, DC: The World Bank.

Drury, R. C. (2009). *Identifying and understanding consumers of wild animal products in Hanoi, Vietnam: Implication for conservation management*. The Degree of Doctor of Philosophy, University College London.

Duc Trung & Tan Phat. (2014). Những công ty "đầu độc" môi trường [The companies that "poison" the environment]. *Sai Gon Giai Phong*, 9/11/2014. Retrieved January 8, 2015, from http://www.sggp.org.vn/moitruongdothi/2014/11/366461/

EIA. (2011). *Crossroads: The illicit timber trade between laos and Vietnam*. London: Environmental Invetsigation Agency.

Environmental Performance Index. (2014). *A global ranking for the environment: 2014 EPI results*. Retrieved June 18, 2015, from http://epi.yale.edu/

Environmental Police Department. (2010). *Ministry of Public Security issued Decision No.449 regulating the functions, duties, powers and organizational structure of the Department of Environmental Crime Prevention and Combat Police*. Department of Environmental Crime Prevention and Combat Police. Retrieved June 2, 2015, from http://www.canhsatmoitruong.gov.vn/default.aspx?tabid=441&ID=3029&CateID=479

Environmental Police Department. (2012). Conference paper: Opening paper—The work of preventing and combating the violations on forest protection. In *Preventing and combatting violations against legislations on forest protection* (pp. 7–10). Hanoi: The Department of Preventing and Combating Environmental Crimes (in Vietnamese).

Forest Trends. (2010). *Timber markets and trade between Laos and Vietnam: A commodity chain analysis of Vietnamese-driven timber flows*. Forest Trends & DFID. ISBN: 1-932928-41-3.

FPD. (2012). *Functions and duties of the Forest Protection Department [Chức năng, nhiệm vụ của Cục Kiểm lâm]*. The Forest Protection Department. Retrieved April 11, 2013, from http://www.kiemlam.org.vn/Desktop.aspx/List/Gioithieu/Chuc_nang_nhiem_vu_cua_Cuc_Kiem_lam/

FPD. (2013). *Forest protection force of Vietnam: Forty years of construction and development (1973–2013)*. Hanoi: The Forest Protection Department (in Vietnamese).

FPD. (2014). *Report on the operational performance in 2014 and the working programme in 2015*. Hanoi: The Forest Protection Department (in Vietnamese).

FSIV & FAO. (2009). *Vietnam forestry outlook study*. Bangkok: Forest Science Institute of Vietnam and FAO Regional Office for Asia and the Pacific.

Goncalves, M. P., Panjer, M., Greenberg, T. S., & Magrath, W. B. (2012). *Justice for forest: Improving criminal justice efforts to combat illegal logging*. Washington, DC: World Bank.

Ha An. (2012). Hoàn thiện các quy định pháp luật hình sự về tội phạm môi trường [Improving criminal legislations on environmental crimes] *Dai Bieu Nhan Dan*, 19/9/2012. Retrieved May 27, 2015, from http://daibieunhandan.vn/default.aspx?tabid=81&NewsId=258465

Ha, D. B., & Nguyen, C. T. (2014) Conference paper: Analysis about the influence of Vietnamese cultures and Vietnam's general educational system on Cdio implementation. In *The 10th international CDIO conference*. Universitat Politècnica de Catalunya, Barcelona.

Hanh Nguyen. (2014). Đại gia dưỡng thân: Giường gỗ trăm năm, đá ngàn tuổi [The super rich take care themselves: Hundred-year-old timber or thousand-year-old stone beds]. *VietnamNet*, 4/10/2014. Retrieved December 8, 2014, from http://vietnamnet.vn/vn/kinh-te/200435/dai-gia-duong-than--giuong-go-tram-nam--da-ngan-tuoi.html

Hawksworth, J., & Tiwari, A. (2011). *The World in 2050: The accelerating shift of global economic power—Challenges and opportunities*. London: PricewaterhouseCoopers. Retrieved from http://www.pwc.com/en_GX/gx/psrc/pdf/world_in_2050_jan2011.pdf

Hayton, B. (2010). *Vietnam: Rising dragon*. Yale University Press.

Hoai Nhon. (2012). "Xẻ thịt" rừng làm rẫy ["Butchering" forests for cultivation]. *Nguoi Lao Dong*, 22/08/2012. Retrieved June 2, 2015, from http://nld.com.vn/ban-doc/xe-thit-rung-lam-ray-20120822112765 91.htm

Huong Nguyen. (2010a). Xử lý tội phạm môi trường: "Bó" bởi nhiều cái khó [Handling environmental crime: "Obstructed" by many difficulties]. *Nhan Dan*, 30/11/2010. Retrieved June 2, 2015, from http://www.nhandan.com.vn/phapluat/item/12390302-.html

Hydropower No.7 Management Board. (2006). *The Project of An Khe—Ka Nak Hydropower: Environmental impact assessment report*. Phu Yen: Vietnam Electricity Corporation (in Vietnamese).

ICEM. (2003). *Vietnam national report on protected areas and development—Review of protected areas and development in the lower Mekong River region*. Queensland, Australia: International Center for Environmental Management. Retrieved from http://cmsdata.iucn.org/downloads/vietnam_pad.pdf

International Monetary Fund. (2013). *World economic outlook database—Vietnam*. International Monetary Fund.

Joint Circular. (2009). *Joint Circular No.08/2009/TTLT-BXD-BTC-BKHĐT-BNNPTNT-NHNN on guiding the implementation of the Decision No.167/2008/QĐ-TTg*. Ministry of Construction, Ministry of Finance, Ministry of Planning and Investment, State Bank, Ha Noi.

Kaimowits, D. (2007). Forest law enforcement and rural livelihoods. In L. Tacconi (Ed.), *Illegal logging: Law enforcement, livelihoods and the timber trade* (pp. 110–138). London: Earthscan.

Kieu Thang. (2013). Thực hiện tốt các chương trình nhà ở cho hộ nghèo. *Nhan Dan*, 11/04/2013. Retrieved June 2, 2015, from http://www.nhandan.com.vn/chinhtri/cung-suy-ngam/item/20084202-.html

Kokko, A. (2004). Growth and reform since the Eighth Party Congress. In D. McCargo (Ed.), *Rethinking Vietnam*. New York and London: RoutledgeCurzon.

Lawson, S., & MacFaul, L. (2010). *Illegal logging and related trade indicators of the global response*. The Royal Institute of International Affairs.

Le Chi. (2013). Timber export exceeds $4 million for the first time [Xuất khẩu gỗ lần đầu tiên vượt mốc 4 tỷ USD]. *VN-Express*. Retrieved June 2, 2015, from http://vnexpress.net/gl/kinh-doanh/2013/01/xuat-khau-go-lan-dau-tien-vuot-moc-4-ty-usd/

Le Trang. (2011). Legends of timber as precious as ... pearl [Huyền thoại về thứ gỗ quý… như ngọc]. *Zing News*, 26/08/2011. Retrieved June 2, 2015, from http://news.zing.vn/xa-hoi/huyen-thoai-ve-thu-go-quy-nhu-ngoc/a125063.html

Le, T. T. 2009. Tăng cường hợp tác quốc tế đấu tranh phòng, chống tội phạm xuyên quốc gia ở Việt Nam trong thời kỳ hội nhập quốc tế [Strengthening international cooperation in the fight against transnational crimes in Vietnam during the international integration]. *Tạp chí Cộng sản*, 26/8/2009. Retrieved June 2,

2015, from http://www.tapchicongsan.org.vn/Home/Tieu-diem/2009/2498/Tang-cuong-hop-tac-quoc-te-dau-tranh-phong-chong-toi-pham.aspx

Lin, J. (2005). Tackling Southeast Asia's illegal wildlife trade. *Sybil, 9*, 191.

MARD. (2001). *Five million hectare reforestation program partnership—Synthetic report*. Hanoi: Ministry of Agriculture and Rural Development (in Vietnamese).

MARD. (2009). *Circular No.34/2009/TT-BNNPTNT dated 10 June 2009: Defining the criteria for identifying and classifying forests*. Hanoi: Minister of Agriculture and Rural Development (in Vietnamese).

MARD. (2012). *Overall review on policy arrangement regarding the implemention of plan in forest protection and devellopment in the period 2012–2020*. State Steering Comittee Office of Planning in Forest Protection and Development in the Period 2011–2020, MARD (in Vietnamese).

MARD. (2014). *Decision 3322/QD-BNN-TCLN on the national forest status declaration by 2013*. Hanoi: Ministry of Agriculture and Rural Development (in Vietnamese).

MARD & FSSP. (2014). *Report on the development of the forestry sector in 2013*. Hanoi: Ministry of Agriculture and Rural Development & Forest Sector Support Program and Partnership. Retrieved from http://www.vietnamforestry.org.vn/mediastore/fsspco/2014/04/01/Bao_cao_PT_nganh_2013-ENGLISH.pdf.

McElwee, P. (2004). You say illegal, I say legal: The relationship between 'illegal' logging and land tenure, poverty, and forest use rights in Vietnam. *Journal of Sustainable Forestry, 19*(1-3), 97–135.

Meyfroidt, P., & Lambin, E. F. (2009). Forest transition in Vietnam and displacement of deforestation abroad. *Proceedings of the National Academy of Sciences, 106*(38), 16139–16144.

Minh Thi—Minh Trieu. (2014). Chuyện những ngứời giữ Giáng húóng [The stories on protectors of Giang Huong forest]. *Gia Lai Online*, 2/10/2014. Retrieved June 2, 2015, from http://baogialai.com.vn/channel/1622/201410/chuyen-nhung-nguoi-giu-giang-huong-2342405/

Ministry of Planning and Investment. (2008). *Vietnam continues to achieve the millennium development goals*. Hanoi: Ministry of Planning and Investigation

MNRE. (2010). *National Environmental Report—2010—Overview of the environmental in Vietnam*. Hanoi: Ministry of Natural Resources and Environment (in Vietnamese).

Morris, J., Phi, L. T., Ingles, A., Raintree, A., & Doung, N. V. (2004). *Linking poverty reduction with forest conservation: Case studies from Vietnam*. Bangkok, Thailand: International Union for Conservation of Nature and Natural Resources.

N.Hung. (2014). Về "thủ phủ vàng" nghe chuyện xử lâm tặc [Visit "the capital of gold" to listen the story of treating illegal loggers] *Gia Dinh & Xa Hoi* [*Family and Society*], 3/9/2014. Retrieved February 5, 2015, from http://giadinh.net.vn/xa-hoi/ve-thu-phu-vang-nghe-chuyen-xu-lam-tac-2014090308289738.htm

NAV. (1999). *The penal code*. Hanoi: The National Assembly of Vietnam.

NAV. (2005). *People's Public Security Act*. Hanoi: The National Assembly of Vietnam.

Nguyen, C. C. (2010b). *The theoretical and practical issues on environmental crimes in the Vietnamese criminal law*. Master degree (in Vietnamese).

Nguyen, D. N. V., & Nguyen, T. (2008). *An overview of the use of plants and animals in traditional medicine systems in Vietnam*. A TRAFFIC Southeast Asia Report, Hanoi. http://www.trafficj.org/publication/08_medical_plants_Viet_Num.pdf

Nguyen, D. H., & Nguyen, N. S. (2010). *Đảm bảo an ninh môi trường cho phát triển bền vững [Ensuring environmental security for sustainable development]*. Hanoi: Nha Xuat Ban Khoa Hoc Va Ky Thuat.

Nguyen Le. (2014). Tình hình tội phạm tiếp tục phức tạp [Criminal situation continues to be complex]. *VnEconomy*, 24/10/2014. Retrieved June 2, 2015, from http://vneconomy.vn/thoi-su/tinh-hinh-toi-pham-tiep-tuc-phuc-tap-20141024100925546.htm

Nguyen, M. H., Vu, V. D., Nguyen, V. S., Hoang, V. T., Nguyen, H. D., Pham, N. T., et al. (2007). *Report on the review of Vietnam's wildlife trade policy*. Hanoi: CRES/FPD/UNEP/CITES/IUED.

Nguyen, T. Q., & Tran, H. N. (2011). *How Viet Nam is prepared to meet legal requirements of timber export markets*. Hue City, Vietnam: Tropenbos International Viet Nam.

Nguyen, T. T. T., & Phan, M. T. (2014). *Wood processing industry—Activity code: No-5—Support business associations to prepare studies on development strategy for two selected industries*. Hanoi: EU-MUTRAP & Vietnam Chamber of Commerce and Industry. Retrieved from http://www.trungtamwto.vn/sites/default/files/publications/nghien_cuu_nganh_go.pdf

Nguyen, V. D. (2013). *Muong livelihoods and the role of education in their development: A case study of a Muong community in Cam Thuy district, Thanh Hoa Province, Vietnam*. A thesis submitted to Victoria University of Wellington in fulfilment of the requirements for the degree of Doctor of Philosophy in Development Studies, Victoria University of Wellington.

Pham, D. X. (2008). *Investigation delivered by the investigation police on crimes of economic management and official position over the violations against regulations on forest exploitation and protection*. The thesis of PhD in Law, The People's Police Academy (in Vietnamese).

Pham, Q. N. (2011). Đấu tranh phòng, chống tội phạm và vi phạm pháp luật về bảo vệ môi trường [The work preventing and combatting environmental offences]. *Tap chi Cong san Communist Review*. Retrieved April 30, 2015, from http://www.tapchicongsan.org.vn/Home/Viet-nam-tren-duong-doi-moi/Phong-chong-tham-nhung/2011/11801/Dau-tranh-phong-chong-toi-pham-va-vi-pham-phap-luat-ve.aspx

Pham, T. T., Moeliono, M., Nguyen, T. H., Nguyen, H. T., & Vu, T. H. (2012). *The context of REDD+ in Vietnam—Drivers, agents and institutions*. Occasional Paper 75, CIFOR, Bogor, Indonesia. Retrieved from http://www.cifor.org/publications/pdf_files/OccPapers/OP-75.pdf

Phap Luat Viet Nam. (2014). Gia Lai: Bắt gỗ lậu, kiểm lâm bị côn đồ chém đứt gân tay [Gia Lai: Arresting illegal timber, Kiem Lam is slashed off their hand tendon by gangsters]. *Doi Song & Phap Luat*, 18/09/2014. Retrieved June 2, 2015, from http://www.doisongphapluat.com/phap-luat/an-ninh-hinh-su/gia-lai-bat-go-lau-kiem-lam-bi-con-do-chem-dut-gan-tay-a51269.html

PMV. (2004). *Decision 134/2004/QĐ-TTg—A number of policies to support the households of poor ethnic minorities with production lands, residential lands, residential houses and clean water*. Hanoi: The Prime Minister of Vietnam.

Quang Vu. (2013). Tình hình tội phạm đang diễn biến nghiêm trọng hơn [Criminal situation remains increasingly serious]. *Vietnam+*, 28/10/2013. Retrieved June 2, 2015, from http://www.vietnamplus.vn/tinh-hinh-toi-pham-dang-dien-bien-nghiem-trong-hon/227343.vnp

RECOFTC. (2014). *Vietnam's forestry reforms: A brief overview of forest reform in the country*. The Center for People and Forest. Retrieved December 1, 2014, from http://www.recoftc.org/country/vietnam/basic-page/vietnam%E2%80%99s-forestry-reforms

Saunders, J. (2014). *Trade in illegal timber: The response in Vietnam—A Chatham House Assessment*. Retrieved from http://www.illegal-logging.info/sites/files/chlogging/CHHJ2362_Vietnam_Logging_Research_Paper_FINAL.pdf

SGTT. (2011). Đại gia "chơi" gỗ quý tuyệt chủng tiền tỷ [The well-off enjoys specious timbers worth of about a billion VND]. *Báo Mới*, 26/02/2011. Retrieved June 2, 2015, from http://www.baomoi.com/Dai-gia-choi-go-quy-tuyet-chung-tien-ty/79/5772035.epi

Sikor, T., & To, P. X. (2011). Illegal logging in Vietnam: Lam Tac (forest hijackers) in practice and talk. *Society & Natural Resources, 24*(7), 688–701.

SRV. (2012). *Implementation of sustainable development in Vietnam—National report at the United Nations Conference on Sustainable Development (RIO+20)*. Socialist Republic of Vietnam, Hanoi.

Stretesky, P. B., Long, M. A., & Lynch, M. J. (2014). *The treadmill of crime: Political economy and green criminology*. Oxon and New York: Taylor and Francis.

Sunderlin, W. D., & Huynh, T. B. (2005). *Poverty alleviation and forests in Vietnam*. Subur Printing, Indonesia: Center for International Forestry Research—CIFOR.

Takemoto, H., Ito, M., Shiraki, T., Yagura, T., & Honda, G. (2008). Sedative effects of vapor inhalation of agarwood oil and spikenard extract and identification of their active components. *Journal of Natural Medicines, 62*(1), 41–46.

Thanh Huyen. (2014). Ngủời Việt chủa giàu nhủng "sang": Chỉ vì thích phô trủóng! [The Vietnamese are not yet rich but "posh": Because of exhibitionism! *Dan Tri*, 15/2/2014. Retrieved March 9, 2015, from http://dantri.com.vn/kinh-doanh/nguoi-viet-chua-giau-nhung-sang-chi-vi-thich-pho-truong-1008200.htm

Tien Thanh. (2014). Bị phát hiện, lâm tặc đánh hạt phó kiểm lâm nhập viện [Being detected, Lam Tac beat a vice-captain of Kiem Lam to be hospitalised]. *Doi Song & Phap Luat*, 06/08/2014. Retrieved June 3, 2015, from http://www.doisongphapluat.com/dia-phuong/mien-trung/bi-phat-hien-lam-tac-danh-hat-pho-kiem-lam-nhap-vien-a44685.html

To, P. X., Tran, L. H., & Nguyen, T. Q. (2014). *The overview of the Vietnamese timber sector: Current status and trends*. Forest Trends/VIFORES/FPA Binh Dinh.

To, X. P. (2012). Natural forests and its suffering from timber need [Rừng tự nhiên chịu sức ép từ nhu cầu sử dụng gỗ]. *Con Nguoi & Thien Nhien*, 6/11/2012. Retrieved June 3, 2015, from http://www.thiennhien.net/2012/11/06/rung-tu-nhien-chiu-suc-ep-tu-nhu-cau-su-dung-go/

To, X. P., & Canby, K. (2011). *Vietnam: Overview of forest governance and trade*. Kuala Lumpur: European Forest Institute—FLEGT Asia Regional Office.

To, X. P., & Tran, H. N. (2014a). *Forest land allocation in the context of forestry sector restructuring: Opportunities for forestry development and upland livelihood improvement*. Tropenbos International Viet Nam, Hue, Vietnam.

To, X. P., & Tran, H. N. (2014b). *Rubber expansion and forest protection in Vietnam*. Hue, Viet Nam. Retrieved from http://www.forest-trends.org/documents/files/doc_4671.pdf

Ton, T. S. (2006). Kỳ nam và Trầm húóng. *Tuoitre*, 04/09/2006. Retrieved June 6, 2015, from http://tuoitre.vn/Chinh-tri-xa-hoi/Moi-truong/159942/ky-nam-va-tram%c2%a0huong%c2%a0.html

Tran, C. D., Duong, T. N., Bui, V. Q., & Ly, V. T. (2010). Examining impacts of eco-social factors on forest resources of the Ba Be National Park—Ba Be distric, Bac Kan province. *The Ba Be National Park*, 199–210. Bac Kan: The Management Board of Ba Be National Park (in Vietnamese).

Tran, L. H. (2001). Overview of environmental offences and related issues. *Legal Review—Law University of Ho Chi Minh City*, 4/2001 (in Vietnamese).

Tran, T. L. (2013). *The operation of preventing the offence of illegal trade and transport in forest products delivered by the Investigation Police on the Offences of Position and Economic Management Order of the Quang Binh Province Department of Public Security*. Dissertation of Master in Law, People's Police Academy (in Vietnamese).

UK Customs. (2012). *The Forest Law Enforcement, Governance and Trade regulations No. 178/2012*. HM Revenue and Customs. Retrieved from http://www.legislation.gov.uk/uksi/2012/178/pdfs/uksi_20120178_en.pdf

UNDP. (2013). *Justice index—Assessment of distributive justice and equality from a citizen-based survey in 2012*. Hanoi: United Nation Development Programme.

United Nations. (2007). *Report on the drug and crime in Vietnam*. Hanoi, Vietnam: United Nations.

UNODC. (2004). *United Nations convention against transnational organized crime and the protocols thereto*. New York: United Nations.

UNODC. (2012). *Vietnam country programme 2012–2017*. Vietnam: United Nations Office on Drugs and Crime. Retrieved from http://www.unodc.org/documents/southeastasiaandpacific//vietnam/publication/2014/CP_final_E.pdf

UNODC. (2013a). *Criminal justice responses to the illegal trade in timber in Vietnam*. Bangkok, Thailand: United Nations Office for Drugs and Crime, Regional Office for Southeast Asia and the Pacific. Retrieved from http://www.unodc.org/documents/southeastasiaandpacific//Publications/wildlife/CJS_Response_-_VIETNAM_01_13_Dec_201.pdf

UNODC. (2013b). *Transnational organized crime in East Asia and the Pacific*. Bangkok, Thailand: United Nations Office on Drugs and Crime—Regional Office for Southeast Asia and the Pacific.

UPM. (2013). *Confab provides inputs and collaboration to spur UPM's gaharu researchers*. Faculty of Forestry—Universiti Putra Malaysia. Retrieved March 4, 2014, from http://www.forr.upm.edu.my/ISSA_BI

USAID. (2013). *Vietnam tropical forest and biodiversity assessment.* Retrieved from http://www.usaid.gov/sites/default/files/documents/1861/Vietnam118_119 FinalDraft%2015Oct2013%20(Public%20Version).pdf

Viet Ba. (2011). Thiếu tướng Nguyễn Xuân Lý: "Ý thức bảo vệ môi trường phải được nâng lên" [Major General Nguyen Xuan Ly: "The awareness of environmental protection have to be raised"]. *An Ninh The Gioi Online*, 25/08/2011. Retrieved June 3, 2015, from http://antg.cand.com.vn/Su-kien-Binh-luan-antg/Thieu-tuong-Nguyen-Xuan-Ly-Y-thuc-bao-ve-moi-truong-phai-duoc-nang-len-301312/

Viet Huong. (2014). Ngắm biệt thự bằng gỗ tiền tỷ của đại gia Việt [Seeing the billions-valued villas of the Vietnamese super rich]. *Phap Luat & Doi Song*, 8/9/2014. Retrieved December 8, 2014, from http://www.doisongphapluat.com/kinh-doanh/doanh-nhan/ngam-biet-thu-bang-go-tien-ty-cua-dai-gia-viet-a49822.html

Vo, Q., & Vo, T. S. (2009). *Challenges on the environment and sustainable development in Vietnam.* Research Centre on Resources and Environment—Hanoi National University (in Vietnamese).

VTC News. (2012). Đi tìm lời giải về "thần dược"nấm lim xanh chữa ung thư [Looking for answer on the "magic drug" green Lim mushroom]. *VTC*, 20/4/2012. Retrieved December 8, 2014, from http://vtc.vn/di-tim-loi-giai-ve-than-duocnam-lim-xanh-chua-ung-thu.321.330157.htm

Wells, A., Gatto, F. D., Richards, M., Pommier, D., & Contreras-Hermosilla, A. (2007). Rural livelihoods, forest law and the illegal timber trade in Honduras and Nicaragua. In L. Tacconi (Ed.), *Illegal logging: Law enforcement, livelihoods and the timber trade* (pp. 139–167). London: Earthscan.

World Bank. (2007). Weak forest governance costs US$15 billion a year. *News Release No. 2007/86/SDN.* Retrieved June 3, 2015, from http://web.worldbank.org/WBSITE/EXTERNAL/COUNTRIES/AFRICAEXT/EXTAFRSUMESSD/EXTFORINAFR/0,,contentMDK:21055716~menuPK:2493506~pagePK:64020865~piPK:149114~theSitePK:2493451,00.html

World Bank. (2009). *Country social analysis: Ethnicity and development in Vietnam—summary report.* Washington, DC: World Bank. Retrieved from http://siteresources.worldbank.org/INTEAPREGTOPSOCDEV/Resources/499760ESW0Whit1C10VietnamSummary1LR.pdf

World Bank. (2012). *Well begun, not yet done: Vietnam's remarkable progress on poverty reduction and the emerging challenges.* Hanoi: World Bank. Retrieved from http://www.worldbank.org/en/news/feature/2013/01/24/poverty-reduction-in-vietnam-remarkable-progress-emerging-challenges

Wyatt, T. (2013). *Wildlife trafficking—A deconstruction of the crime, the victims and the offenders*. Basingstoke, UK: Palgrave Macmillan.

Xuan Minh. (2014). Khai thác cây lâm sản ngoài gỗ: Gắn phát triển với bảo tồn [Harvesting non-timber forest products: Linking development with conservation]. *Vietnam+*, 22/11/2014. Retrieved June 3, 2015, from http://www.vietnamplus.vn/khai-thac-cay-lam-san-ngoai-go-gan-phat-trien-voi-bao-ton/292572.vnp

Xuan Trach Comnune. (2013). *Report on the poverty status in the 2005–2012 period and a number of directions in the new context*. Xuan Trach Commune People's Committee (in Vietnamese).

4

Timber Trafficking in Vietnam

This chapter provides a detailed answer to the question: how is timber trafficking currently occurring in Vietnam or what is the current pattern of timber trafficking in Vietnam? It opens with the statistics that reveal the scope and general trend of timber trafficking in Vietnam over the five-year period 2008–2012 (inclusive). It then looks into five different stages or components of the crime: harvesting, smuggling, trading, supporting and processing. Each component is further categorised into distinctive groups of activities that explain in detail where and when the illicit operation takes place, who the offenders are, why they get involved, and what techniques they use to facilitate their crime. The chapter ends by critically discussing the nature and extent of timber trafficking in Vietnam and proposing an alternative approach to the current classification of illegal logging.

Overall Scope and Trend

According to official statistics from the Forest Protection Department (FPD), during the five years from 2008 to 2012 (inclusive), the Vietnamese authorities detected a total of 173,201 (or 34,640 yearly)

N.A. Cao, *Timber Trafficking in Vietnam*, Palgrave Studies in Green Criminology,
DOI 10.1007/978-3-319-64280-2_4

violations against the laws and regulations on forest protection. The main forest offences are: illegal deforestation, violations of forest fire prevention, violations of forestland usage; violations of wildlife management; and trafficking (e.g. illegal harvesting, smuggling, processing and trade) in forest products (FPs) (see Table 4.1 and Fig. 4.1).

Additionally, in this five-year period, among the 173,201 detected cases of forest offences, 150,248 cases were penalised, constituting 86.7% of the detected cases. Among these penalised cases, 148,609 (98.9%) had administrative sanctions imposed,[1] while 1639 (1.1%) were charged with a crime (see Fig. 4.2). Among the 1639 cases handled as crimes, only 222 were brought to the criminal courts, accounting for some 13.5% of the total prosecuted cases and 0.15% of the total penalised cases. That said,

Table 4.1 The number of detected forest offences from 2008 to 2012, Source: synthesised from FPD (2009, 2010, 2011, 2012, 2013)

Forms of violation	2008	2009	2010	2011	2012	Total
Illegal deforestation	6994	4655	3503	3473	3221	21,846
Forest fire violations	439	452	1071	293	483	2738
Forestland usage violations	229	77	104	74	63	547
Wildlife management violations	1406	1285	876	1019	871	5457
Illegal harvesting of FPs	4546	4500	3119	2675	2276	17,116
Smuggling and illegal trade in FPs	20,158	21,221	17,899	14,772	13,022	87,072
Illegal processing of FPs	1942	2036	766	748	874	6366
Other violations	6749	6615	6484	6497	5714	32,059
Total	42,463	40,841	33,822	29,551	26,524	173,201
Changes compared to previous year		−3.82%	−17.19%	−12.63%	−10.24%	

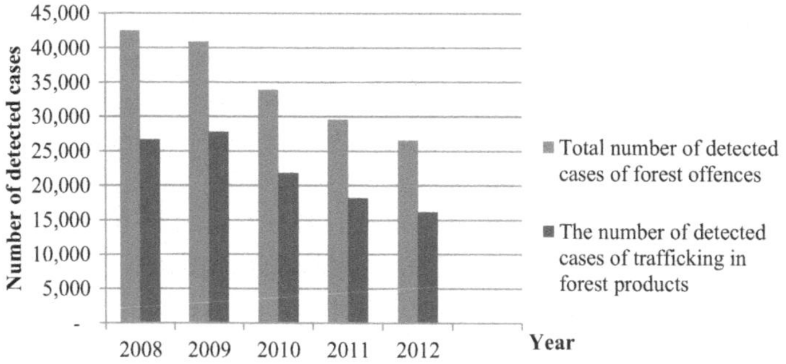

Fig. 4.1 The number of detected cases of trafficking in forest products from 2008 to 2012, compared to the total forest offences, Source: synthesised from FPD (2009, 2010, 2011, 2012, 2013)

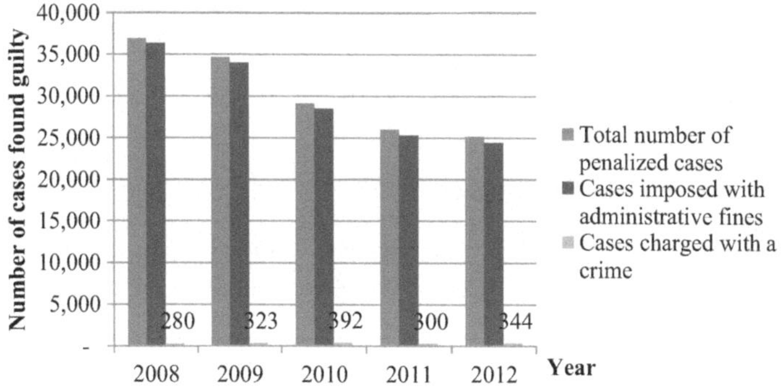

Fig. 4.2 The number of penalised forest offences from 2008 to 2012, Source: synthesised from FPD (2009, 2010, 2011, 2012, 2013)

as with other forms of environmental crime in Vietnam, the proportion of the cases of forest offences charged as crimes is minimal compared to the total cases detected and penalised (see Fig. 4.2).

Furthermore, as the data indicate, among all the detected cases of forest offences were 110,554 cases of trafficking in *forest products*, accounting for about 63.8% of the forest offences (see Fig. 5.1 later). Although there were no separate statistics for timber species, through personal communication with FPD staff, it could be estimated that around 95% of the number of cases of trafficking in forest products in Vietnam were a form of timber trafficking.[2] It could, therefore, be calculated that from 2008 to

2012, the total number of cases of timber trafficking detected by the Vietnamese authorities was about 105,000 cases, which means on average, the anti-trafficking forces detected 21,000 cases of timber trafficking every year. Details about specific illicit acts in timber trafficking are demonstrated in Table 4.2 and Fig. 4.3.

Official statistics also indicate that in this five-year block, the total volume of confiscated timber reached approximately 205,000 m^3, consisting of 24,000 m^3 of endangered timber species and 181,000 m^3 of

Table 4.2 The number of detected cases of timber trafficking from 2008 to 2012, Source: synthesised from FPD (2009, 2010, 2011, 2012, 2013)

Forms of timber trafficking	2008	2009	2010	2011	2012	Five-year
Illegal harvesting of timber	4319	4275	2963	2541	2162	16,260
Illegal smuggling of and trade in timber	19,150	20,160	17,004	14,033	12,371	82,718
Illegal processing of timber	1845	1934	728	711	830	6048
Total	25,314	26,369	20,695	17,285	15,363	105,026
Changes to the previous year		+4.2%	−21.5%	−16.5%	−11.1%	

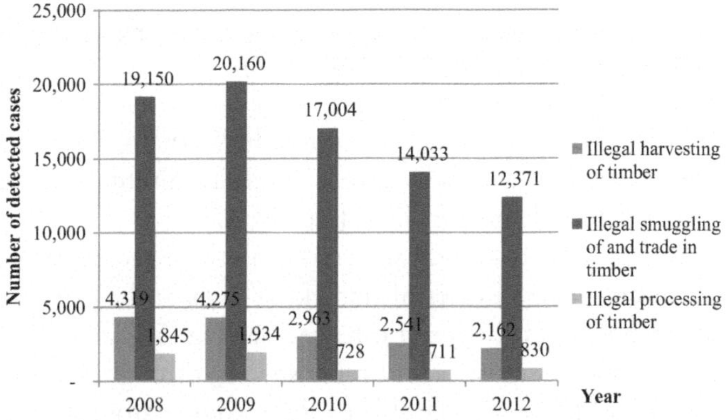

Fig. 4.3 The number of detected cases of timber trafficking from 2008 to 2012, Source: synthesised from FPD (2009, 2010, 2011, 2012, 2013)

plain species.[3] To put it differently, of the 41,000 m^3 of timber confiscated every year in Vietnam, 12% was endangered timber. During this five-year period, the Vietnamese authorities added to the national budget VND954 billion (£31.8 million), equivalent to £6.36 million annually, which was obtained mainly from monetary fines and selling confiscated forest products and vehicles used to commit the crimes such as cars, trucks and motorbikes. These are just official statistics, discovered and declared by the anti-trafficking forces. Presumably, the actual number of cases of timber trafficking would be far higher than the detected figures. Nguyen (2008), for instance, believes that only 3.1% of cases of wildlife trafficking in Vietnam are discovered.

A leading scholar on the subject of timber trafficking in Vietnam (7FO1), having consulted the opinions of other experts in the field, confidently believes that the actual volume of timber illegally harvested in Vietnam is at least 10 times higher than the total timber volume confiscated by official agencies. If this is the case, then the total annual volume of illegal timber is no less than 410,000 m^3 including 48,000 m^3 of endangered timber. It is also observed that the illegal plain timber in Vietnam is mainly from Group IV and Group V with average prices worth £250/m^3, while the average prices of endangered timber species are worth £830/m^3.[4] It is thus estimated that the annual value of timber trafficking in Vietnam is at least £130 million. To conceive how high this value is in the context of Vietnam, a simple calculation shows that Kiem Lam forces in all of Vietnam have nearly 12,000 officers whose average annual salary is £1200; thus, the actual monetary value of timber trafficking in one year is enough to pay for the salaries of all the Kiem Lam officers throughout Vietnam for at least nine years.

It can be generally observed from the foregoing statistics that over the five year period under consideration the number of forest offences generally and timber trafficking particularly appears to be gradually decreasing. Indeed, the annual decrease of detected forest offences is 3200 cases; while the number of detected incidents of timber trafficking is falling by almost 2000 cases every year. Observations by the majority of interviewees in this research and formal evaluations in the reports of law enforcement forces also share a consensus that overall the problem of timber trafficking has been better controlled recently.

However, it must be noted that although the number of detected cases shows a gradual decline, the number of serious cases appears to increase. In the five-year block, from 2008 to 2012, the number of criminal cases increased from 280 to 344, equivalent to a rise of 23%. This points to the likelihood that although the overall quantity of detected cases might be shrinking, the increase in criminal cases means that the consequences of the crime may be increasingly severe. Furthermore, irrespective of the declining pervasiveness, it is firmly believed by the interviewees, addressed by the media, and confirmed by official reports that, in one form or another, timber trafficking still exists in all forested parts of Vietnam, and that in many provinces such as Son La, Bac Kan, Thai Nguyen, Quang Binh, Quang Tri, Binh Phuoc and in five provinces in the Central Highlands, the crime remains alarmingly prevalent. In Bac Kan, for example, from 2009 to June 2013, timber trafficking accounted for 81% of the total of 91 cases that were criminally charged for economic crimes by investigative police (11IP02). Likewise, in the region of the Central Highlands, the Government Office (2013: 3) concludes that "illegal harvesting of endangered timber in special-use and protected forests is still formidably happening in a mounting trend". Ultimately, it is assessed that timber trafficking "remains a serious problem" (Saunders 2014: 2), is "still a pressing concern in Vietnam today" (Forest Trends 2013: 1), and that the results of the mission of forest protection are not comprehensive and stable, while the control of timber trafficking has not fundamentally improved (Department of Legal Affairs—MARD 2012: 24).

General statistics on timber trafficking have been provided, which show the scope and trend of timber trafficking in Vietnam over the five year period 2008–2012. The following sections will examine in detail the current typology of timber trafficking that uncovers who is engaged in the crime, how intense their engagement is, what techniques are used in the criminal acts, where and when the acts often take place, and why it is these particular offenders, techniques, locations and times that are of concern. All of these aspects will now be examined in each of the five steps in timber trafficking: harvesting, smuggling, trading, supporting and processing. As mentioned, existing literature offers little information on the pattern of timber trafficking in Vietnam, thus, the vast majority of the information on the typology formulated below is collated from the research data including criminal case records, official reports, newspaper articles and especially in-depth semi-structured interviews. Where relevant, the information from existing literature will be used to add to, compare and contrast with, this research's evidence.

Harvesting

In the criminal chain of timber trafficking, illegal harvesting of timber or illegal logging is the first stage, and plays a crucial role in the entire chain. This is simply because the crime would probably collapse without the supply of the illegally harvested timber. In the academic domain of timber trafficking, illegal timber harvesting receives far more attention than the subsequent stages such as smuggling and trading. Tacconi (2007) presumes two reasons for the particular focus on the illegal harvesting step. First, harvesting has significant harmful impacts on the forests, perhaps more tangible than other steps. Second, statistics on timber harvesting are more widely available, making access to statistics on illegal logging easier than those for other illicit activities (Tacconi 2007).

Given the pattern of illegal logging, as reviewed in the section on "The Sophistication of Timber Trafficking" in Chap. 1, literature pertaining to this step often focuses on the techniques of illicit cutting (e.g. logging in protected areas, of protected species, without permits or exceeding the permitted volumes) to classify different types of illegal logging. Nonetheless, investigating illegal timber harvesters and their actions in the present context of Vietnam reveals strong evidence substantiating the argument that different types of illegal logging are not simply different in terms of the logging tactics, but also in other important features such as the demographic background of loggers, their motivation, the intensity of the criminal engagement and the organisation of the logging. It is then suggested that these harvesters and the characteristics associated with their logging can and should be classified into three different categories, consisting of small-scale, medium-scale and large-scale illegal timber harvesting.

Small-scale Illegal Timber Harvesting

The terms "small-scale logging", "small-scale timber extraction", "informal logging" and occasionally "artisanal logging" are mentioned in a number of studies on illegal logging that essentially refer to the illegal harvesting of small volumes of timber. At international levels, although small-scale illegal logging receives much less academic discussion than large-scale logging, this unlawful practice has been examined in some countries such as Kalimantan in Indonesia (Casson and Obidzinski 2007), Ghana and Cameroon (Lawson and MacFaul 2010; Lescuyer 2007), South East Albania (Stahl 2010) and Uganda (Jagger et al. 2012). These studies commonly show concern over the legality of the logging, but jointly recognise that it is relatively popular among forest-based communities, and has an important role in supporting the livelihood of the loggers.[5]

In Vietnam, some studies also mention the term "small-scale illegal logging" (Forest Trends 2013; McElwee 2004; Sikor and To 2011; Sunderlin and Huynh 2005). These studies offer general descriptions of the logging, characterised by small volumes of timber illegally cut down and by the loggers who are often local residents. However, the understanding of this unlawful operation remains a "rudimentary picture based on fragmentary knowledge" (Sunderlin and Huynh 2005: 24). It is observed that other defining features associated with this illicit logging such as where, when, how and why it occurs still remain comparatively unknown. More fundamentally, these studies do not examine other parallel forms of illegal logging, thus overlooking the possible connections with, and meaningful comparisons between, different types of illegal logging concurrently occurring in Vietnam.

From the systematic analysis of data, it can be defined that:

> Small-scale illegal timber harvesting (SSITH) in Vietnam is any acts of illegally harvesting trivial volumes of timber that is opportunistically and/or occasionally committed by individuals or small groups of poor forest-based residents using rudimentary methods under subsistence-based motivations.

Analysing evidence pertaining to SSITH indicates several archetypal features of this form of logging as follows. First, regarding the demographic characteristics of the loggers, typically participants in SSITH are indigenous poor forest dwellers for whom forests are an inherent part of their life (1EP01, 3EP02, 13LA01, 14LR01). They are, for instance, the Tay, Thai, Nung, Dao and/or Muong in the North; the Pa Ko and/or Van Kieu in the Centre; and the Giarai and Ede in the Central Highlands who, as indicated in Chap. 3, have exceedingly limited land and production tools to carry out farming, thus harvesting forest resources becomes their everyday job.

These illicit harvesters might carry out the work individually, but small groups are preferred because of the physical risks and the arduousness of logging inside the forests (01EP01, 10IP01, 15FO04, 25LR03, Ba Be Investigation Police Agency 2013aa, b). These groups are either family-based, village-based or peer-based and routinely have three to five members. The family-based groups are formed from young male members of extended families who are sons or cousins, but occasionally they are young husbands and wives (01EP01, Xuan Tho 2014). Indigenous villagers create the village-based groups (25LR03; Ba Be Investigation Police Agency 2013aa, b), while gangs of youths of a similar age can gather together for the purpose of illegal logging; this is known as a peer-based logging group (01EP01, 10IP01, 15FO04).

Second, the main purpose of SSITH is often non-commercial, that is mostly subsistence-led rather than commerce-led, and the timber harvested from this form is important for the harvesters' subsistence. Typically, local residents cut down trees for the purpose of building their house, kitchen, furniture and cages for cattle ranching. In fact, the forest-based people have been enjoying houses and furniture made from timber for generations, believing that maintaining this form of housing is also a way of keeping their ancestor's assets (14LR02). More fundamentally, in many remote forested areas of Vietnam, it is often extremely expensive for local people to build decent houses made from bricks and cement because the cost to transport such building materials from the lowlands via poor, winding and uphill routes is very high (12EP04, 13LA01). While the local people are, as indicated above, inherently poor, a sizable number of households construct and repair their houses, furniture and kitchens with lumber unlawfully logged from the forests close to their communities[6] (12EP04, 13LA01, 39NG05, MARD and FSSP 2014; Xuan Tho 2014). Although this form of illicit logging is acknowledged as "widespread" among local forest communities (USAid 2013: 49) and accounts for most of the detected cases of illegal logging in some forested localities (11IP02, Ha Thanh 2014a), it is currently out of the control of the local governments (27FO06, MARD and FSSP 2014). This is explained by a forest protection official:

> Because the local loggers live inside or close to the forests, it is virtually impossible to know when they come to the forests and when and where they illegally cut the trees. The forests, as you see, are often secluded. With the current policy, in which one Kiem Lam officer is assigned to monitor the forests covering a whole commune, I think it is very difficult to prevent this logging. (27FO06)

In SSITH, loggers may perceive their conduct as a both illegal and environmentally harmful, but because of the difficulties in guaranteeing subsistence, they decide to carry out SSITH anyway. A study by Pham (2008: 108) indicates that for many remote forest-based communities in Vietnam, since forests are the unique source of their livelihood, "in order for survival they ignore the law, do whatever that gives them something to eat, to wear in everyday life". A case study by McElwee (2004) in Ha Tinh province indicates that while, on average, a forest household needs 4–5 m^3 of timber per year for subsistence, 75% of these households have no legal access to forests nearby; consequently, many of them log illegally.[7] A local resident in Bac Kan firmly and repeatedly asserts that in order to build or fix his house, he must log:

> We know logging is not good for the forests. It is forbidden by the state and we want to protect the forests for our descendants. But due to the pressing needs, we now have to log. When my son is going to get married, he needs a new house, right? When our house has been destroyed by landslides, flooding or storms, we need to fix or build a new house, right? Even though our logs may be seized by officials, we still have to log again and again until the houses get built or repaired. (14LR02)

In parallel to the motivation of meeting practical subsistence needs, SSITH is also motivated and justified by the traditional attitude of the indigenous loggers. It is their belief that their harvesting of forest resources like timber, which has existed "for thousands of years", should be recognised as a legitimate right (14LR02). Research by Forest Trends (2013: 4) highlights that the long-standing idea that "forests belong to villagers" is a popular belief that rationalises the villagers' logging regardless of the logging ban by the government. An ethnic minority in Ba Be National Park stresses:

> It is these forests that were owned by our ancestors. The forests are where we have been living for many, many generations. They are also where our ancestors have been laid to rest. So it is clearly our own duty to protect and our right to use these forests. (14LR02)

Third, it is typical that SSITH actors take part in the logging in an occasional and opportunistic manner. This is explained by the primary purpose of SSITH to meet subsistence needs that occur infrequently. A study by To et al. (2014) estimates that, on average, each forest-dwelling household needs to build a new house every 10 years. Moreover, as seen earlier, since timber sources in Vietnam now are no longer abundant, it is frequently not easy for unskilled loggers to find large trees and illicitly log. The occasional nature of SSITH is also explained by farming that remains the main job, and consumes most of the time, of SSITH participants.

> The farming does not span continuously throughout the whole year. It has slack intervals of two months or so in between the busy periods of planting and harvesting. During these intervals, we have nothing to do but come to the nearby forests to gather some wood, birds or orchids. It becomes a common practice. The amount of logs, animals or plants collected is not much. But a little is better than nothing. (05LR01)

Fourth, loggers often use rudimentary techniques to carry out SSITH. Indeed, their initial talking about logging and further activities, such as looking for and cutting trees, and bringing logs out of the forests, are simple and not well-planned. The idea of carrying out an incident of illicit logging, for example, comes up by chance after a drink with dinner (Ba Be Investigation Police Agency 2013ab). Another example is when a group of local residents goes to a forest to collect NTFPs, accidentally finds one or two big or high-value trees, and then decides to illegally cut the trees (2FO01). For the cutting, loggers may still use handsaws, but in most cases they now use chainsaws to cut down one or two trees that can then be left in the forest for several days before being carried back to village (8EP03, Kon Tum Investigation Police Agency 2009).

These elementary methods are also visible in the task of carrying illegal logs from cutting sites to the village. Depending on the forest terrain, porters employ different carrying methods, but they are mainly manual. If the forest is steeply sloped, the best way is to carry the logs on shoulders or backs (20EP07, 29IP04, Kon Tum Investigation Police Agency 2009). Using this method, local loggers are able to carry 10 × 40 × 50 cm lumber that weighs about 70 kg, walking effortlessly in rugged cliffs (20EP07). If there is a small pathway, they can use water buffalos, bulls or special bicycles for transportation. If detected, they disconnect the logs from the buffalos and bulls and then beat them to run fast to escape (15FO04). In short, characteristics typical of SSITH are the involvement of poor forest-based harvesters, subsistence-led purposes, small volumes of illegal timber, opportunistic and/or occasional involvement and rudimentary methods. These attributes are largely different from medium-scale illegal timber harvesting that will be investigated in the next section.

Medium-scale Illegal Timber Harvesting

It seems that the term "medium-scale illegal timber harvesting" or "medium-scale illegal logging" is not used in the existing literature on illegal logging and illegal trade in timber. To some extent, both small-scale and large-scale illegal logging, as addressed by other studies, have some overlap with the idea of medium-scale illegal timber harvesting in this research. However, this research argues that overall the two forms of small-scale and medium-scale illegal logging are fundamentally different.

The overlaps and distinctions between the two will be clarified in this section.

It can be defined that:

> Medium-scale illegal timber harvesting (MSITH) in Vietnam is an illegal form of logging that involves specialist loggers who, motivated by profit, illegally harvest relatively large volumes of timber in a well-planned manner.

It is observed that MSITH and SSITH probably accounts for the highest proportions of the total cases of timber trafficking captured by official statistics in Vietnam. This is because unlawful large-scale timber harvesting, as will be examined later, often falls outside the official statistics as it is apparently labelled as a legal act. There are several distinctive attributes associated with MSITH.

First, like small-scale illicit loggers, medium-scale counterparts are often poor, unemployed people. However, what differs between SSITH and MSITH is that while SSITH participants are mainly indigenous residents who harvest nearby forests, a sizable number of MSITH actors are migrants coming to different forests for illicit logging (25LR03, 32IP05, 33LA03). These migrants move from high-density regions of Vietnam, particularly from Northern provinces, to highland areas such as the Central Highlands where far more cultivable lands and forest-based resources are thought to be available (Pham 2008). National statistics show that from 2005 to 2009, 161,000 migrants moved to the Central Highlands (GSO 2009); many of them, however, found no stable jobs and believe that logging is far more well-paid than other occupations such as rice cultivation and cattle ranching (32IP05, Pham 2008). A police officer in the Central Highlands explains the idea that "logging in one year is enough for eating for ten years" leads the needy migrants to be vulnerable to illegal logging (32IP05). Whether aboriginals or migrants, by and large, these loggers share a similar personal situation of an underprivileged background, demonstrated by their poverty, poor education and unemployment. This makes loggers different from other actors such as traders and processors who are commonly better-off (3EP02, 32IP05, 33LA03).

Second, MSITH involves professional loggers who carry out the illegal harvesting in an intensive and well-planned manner. Indeed, the loggers may stay in forests for long periods of time, often weeks or even months, and the earnings from illegal logging make a major contribution to their income (3EP02, 5LR01, 26EP09, 32IP05). Like small-scale loggers, medium ones also form groups, but these groups, known as cutting groups ("nhóm thợ cưa"), are much more durable, and the number of group members is larger, typically consisting of 7–10 people. It is common that these cutting groups exist for relatively long periods of time, often for years. One of the major traits of the cutting groups is that they know the forest topography very well, which helps them locate the largest trees and escape arrest by the anti-trafficking forces[8] (12EP04).

The loggers are self-employed or employed by a local trader who assigns each member to specific tasks such as preparing logging tools, cutting and carrying logs, and observing the anti-trafficking agencies (25LR03, 26EP09, Kon Tum Investigation Police Agency 2011). In areas with large forests, the loggers may also be tasked with locating rich timber zones and estimating the accessibility to, and timber capacity of, the forests (26EP09). When a cutting group teams up with a trader, if the trader pays too slowly or poorly, the whole group can work for another trader (25LR03, 26EP09). The close connection between logger groups and timber traders indicates an apparent distinction between SSITH and MSITH actors. While the former has almost no links to other actors in the entire chain of timber trafficking, such as professional traders and transporters, the latter has the opposite. This connection can be seen more clearly in the later sections on the other stages of timber trafficking.

Medium-scale loggers employ a wide range of rather devious tricks to conduct their harvesting. To have access to protected forests, chainsaws are taken apart and hidden inside backpacks, while petrol is passed off as drinking water (15FO04). When using chainsaws, to reduce the noise, their exhaust is soaked in water (15FO04) or they are reassembled with long plastic pipes with the end buried underground (Minh Thi—Minh Trieu 2014). In dense forests, strong ropes are used to tie the top parts of a targeted tree to other surrounding trees, so that when the main body of the targeted tree falls down, it makes less noise while the top part of the cut tree remains in a position similar to that before the cut (22EP08). This makes it difficult for Kiem Lam officers, who observe the forests from the outside, to see the trees that have already been cut (22EP08). In some cases, if the trees are big, loggers just cut two-thirds of the trees' bodies, so that the trees remain alive (Minh Thi—Minh Trieu 2014), whereas if trees are positioned in such a way that makes their cutting difficult, the trees can be poisoned or slowly burned to death. The trees will then be felled by winds (12EP04).

Normally, trees are cut into long planks that are more valuable than small ones, but recently, because of tightened law enforcement, in some areas, endangered trees can be cut into smaller pentagonal or hexagonal boards to ease their transportation and avoid detection from authorities (12EP04). Occasionally, highly valuable trees are moved to loggers' forest gardens. After that, they ask the local authorities for permission to harvest the garden trees, which makes it easier to receive valid logging documents (5LR01).

In cases of logging teams being arrested, a very common technique is to divide illicit timber volumes into small fractions for each individual logger; consequently, the divided portions do not qualify for the criminal threshold, but instead incur only administrative fines.[9] As discussed in Chap. 3 on legal loopholes, while habitually, administrative fines substantially exceed the value of the logger's property, making the authorities unable to seize any of their possessions, the illegal loggers, who are not criminally charged, would bear little or no legal liability. This legal loophole, as widely mentioned by police officers in this research, is one of the major reasons for the extremely low rate of criminal cases among the detected cases of timber trafficking.

With regard to location, for SSITH, popular locations are often forests close to the local communities, for MSITH, loggers target any timber species in any types of forests. However, to reduce the likelihood of detection, forest zones near the boundaries between different districts and provinces, where the duties of the anti-trafficking forces overlap, are particularly targeted. In addition, as a result of the degraded quality of forests in most parts of Vietnam, some of the remaining rich forests are still hotspots for MSITH (2FO02, 8EP03, 29IP04; FPD 2014; Quang Thai 2015; Xuan Dung—Trong Loi 2014).

As can be observed, both small- and medium-scale timber illegal harvesters have a common feature in that they are predominantly poor people who log in a clandestine manner. In this sense, there is in Vietnam another type of harvester who is almost entirely different from the first two types. The third type of illegal timber harvester will be scrutinised in the next section.

Large-scale Illegal Timber Harvesting

Albeit under varying definitions as large-scale or commercial or industrial illegal logging, largely characterised by the illegal removal of numerous volumes of logs, has been documented in at least 30 countries (Lawson and Macfaul 2010).[10] However, Tacconi (2007: 4) believes that these "extensive illicit operations have been revealed whenever and wherever authorities have tried to find them". It is observed that this form of logging has attracted the bulk of international discussion on illegal logging and the majority of studies on illegal logging discussed in this book so far also refer to this category. However, in the recent context of Vietnam, due to the substantial decrease in the timber resources, in tandem with the improvements in political will, legal provisions and law enforcement in the forestry sector (see Chap. 3), this form of logging is no longer able to yield very large volumes of illicit timber. To obtain such large amounts, the technique of unlawfully abusing forestry policies has been employed over the past decade in Vietnam, which is the focus of the definition of large-scale illegal timber harvesting in this research. It is defined that:

> Large-scale illegal timber harvesting (LSITH) in Vietnam includes any acts committed by large companies or timber barons who abuse forest-based policies and violate relevant legal requirements to harvest large volumes of timber for commercial purposes.

The case of the Minh Phuc Company in Dak Nong province is a typical example of LSITH. In April 2009, the company was approved by the Dak Nong People's Committee to rent 402 hectares of forestland for the purpose of forest protection and plantation, in which 173 hectares were planned for a rubber plantation. Taking advantage of the permission for logging in the areas designed for the rubber plantation, the firm illegally logged outside the anthorised zone, moved the illegal logs to the permitted area and mixed them with logs legally harvested there. In January 2011, anti-trafficking forces inspected six log-gathering locations and detected 149 illegal logs totalling 177 m^3. The case was then criminally charged by the provincial investigative police agency (Dak Nong Environmental Police 2012b).

Another notable example took place during 2005–2006 in the Khe Dien hydropower plant in Quang Nam province. From November 2005 to September 2006, taking advantage of the logging permission in the hydropower reservoir, Ngoc Son Construction Firm illegally collected, logged and transported almost 1000 m^3 of timber from the 444 Zone, 446 Zone and Ca Tang area (Tuong Vy 2009). The illicit practice by Ngoc Son was heavily supported by other companies and a series of provincial and district officials. The complicit companies helped Ngoc Son to falsify the logging plan, whereas the involved officers abetted Ngoc Son in certifying the fabricated documentation. It took more than two years to complete the criminal investigation, in which nine governmental staff including the chief of the Department of Agriculture and Rural Development of Quang Nam province and directors of the companies were prosecuted and adjudicated; 20 other people were also summoned to the courts (Tuong Vy 2009).

While the Minh Phuc and Ngoc Son cases ably demonstrate LSITH, it is believed that they are not isolated cases. However, only a handful of similar cases are brought to the criminal courts. Examining LSITH in Vietnam at present reveals a number of its distinctive attributes as follows.

First, LSITH is often organised by those who are commercial company owners or powerful timber traffickers. These corporations can influence policy-making processes in the forestry sector, which results in forestry policies that create opportunities for the occurrence of both large-scale legal logging and LSITH (33LA03). In the harvesting operation, these LSITH harvesters do not directly take part in cutting trees, but organise and manage the logging by making logging proposals to law enforcement authorities, bribing these authorities if necessary, falsifying the required documents, employing loggers, and integrating the logging with further steps such as smuggling and trading (3EP02, 33LA03). A police officer stresses:

> I think the traditional type of illegal logging, which is secretly carried out, still happens now, but it cannot not get large amounts of logs, especially big ones. At the moment, only timber firms or influential businessman can get these logs. They are non-traditional harvesters. They have good relationships with forestry authorities, so that that they know when there are logging opportunities and how to get most out of them. (3EP02)

Second, LSITH is carried out by unlawfully taking advantage of forest-based policies. This is a crucial point in LSITH, making it fundamentally different to the two other forms of illegal logging. There is a variety of sources providing strong evidence for the claim that some of the forest-based policies are being abused to unlawfully harvest large amounts of timber in Vietnam. Indeed, many interviewees show their deep concern over the abuse; meanwhile it is not difficult to search on a large number of online newspapers to see articles addressing the problem.[11] Additionally, in the recent Conference Proceedings on Preventing and Combating Violations against Legislation on Forest Protection organised by the central Department of Environmental Police, the acts of mistreating forest-based policies, like unlawful logging, are officially addressed by a number of provincial environmental police agencies such as Yen Bai, Quang Binh, Dien Bien, Thua Thien Hue, Gia Lai, Quang Nam and Lam Dong (Environmental Police Department 2012). A paper by Lam Dong Environmental Police (2012c: 138), for example, describes the abuse of forestland policy for the purpose of forest exploitation in the province as an "alarming" challenge that "has been occurring rampantly anywhere, anytime at various forms".[12]

In a notable investigation on the forestland policy in five provinces in the Central Highlands, Uyen Thu (2013a: 3) evaluates that "the majority of the rubber plantation projects occupy forestland, exploit forests, and then abandon them"; among the forest areas devoted to rubber plantation projects, over 7400 hectares have been illegally logged (Uyen Thu 2013a). A report by the Gia Lai Forest Protection Sub-department indicates that in the period from 1997 to 2005, 90% of inspected logging companies were guilty of technical misconducts (Pham 2008). Still in the Central Highlands, some of the latest violations can be seen in two Inspection Conclusions announced by the Government Inspectorate (2014a, b) that indicate problems associated with the mismanagement of projects of forestland conversion in Dak Nong and Gia Lai provinces.[13] Consequently, a number of the owners of such projects have been criminally charged (Dang 2014).

It is argued in this research that there are now three specific policies in Vietnam that are notoriously noted as an effective means of illegally acquiring sizable quantities of timber, comprising forestland conversion, planned logging and housing policy for poor forest-dwelling households. The first and perhaps most notable policy is of forestland conversion. As mentioned in Chap. 3, in the past decade, some 363,500 hectares of forest throughout Vietnam have been used for almost 2400 projects of forestland conversion via various forms such as plantations, irrigation construction, road building, hydropower plants, exploitation of natural resources, resettlement and building of spiritual sites (FPD 2014). In parallel with economic and social benefits (Chap. 3), each of these forms brings about certain opportunities for the illegal acquisition of large amounts of timber. Among these forms, as suggested by the majority of the interviewees, two forms of forestland conversion into hydropower construction and rubber plantation are the most abused. These two practices also receive the largest proportions (around 73%) of the total area of converted forests in Vietnam (To et al. 2014).

The most sophisticated tactic employed in LSITH occurs in the step of illegally obtaining logging permits via the misrepresentation of the status of the forests, particularly the timber volume and the diversity of timber species (20EP07, 32IP04, Lam Dong Province Inspection 2013). By law, specifically stipulated in Circular 58/2009/TT-BNNPTNT, issued by the MARD to guide rubber plantations on forestland, one of the prerequisite conditions to conduct the forestland conversion is that the targeted forests must actually have poor timber capacity, normally less than 10–100 m^3 of timber per hectare of forest, depending on type of forest and species of timber.

However, the research evidence suggests that many of the logged forests are not poor; this is to say, at times the evaluation of forest timber productivity is problematic (20EP07, 32IP04, 35FO07). A member of staff in the Forest Inventory and Planning Institute observes that the majority of the shifted forests are rich (Bich Ngoc 2013). An intensive study on forest conversion into rubber plantations in Vietnam by To and Tran (2014: vii) observes that 79% of the total area of 116,000 hectares of converted forests in the Central Highlands are natural forests; and not all of these forests are poor, whereas almost 400,000 m^3 of timber from the salvage harvest obtained during this implementation "only partially reflects the true volume of harvested timber". At the same time, over 3000 hectares of special-use and watershed protected forests in the Central Highlands, that are usually rich forests, have been converted to build hydropower plants (Nhien Di 2013).

As presumed by a number of interviewees, there may have been several ways to misrepresent rich forests as poor ones, mainly including deliberately selecting actual poor zones in the forests for the sampling of forest quality, skipping some of the required procedures, influencing the forest evaluators and bribing officials who serve as evaluation supervisors (7NG01, 20EP07, 32IP04, 35FO07). A forest officer indicates that one problem with the current regime of implementing forest conversion policy in Vietnam is that there are no organisations undertaking the task of evaluating the forest status which are independent from timber harvesters (35FO07). It is now the evaluative organisations that are employed by timber harvesters, which leads to the evaluation of forest status being susceptible to the influence of the harvesters (35FO07). Another problem is that, in reality, the forest evaluators only take excessively small zones of forest for sampling, which often account for only 2% of the entire logging areas, meaning that it is difficult to reach accurate conclusions (Lam Dong Province Inspection 2013).

At the same time, by law, local forestry officials are required to take part in the supervision of the evaluation of forest statuses. However, forests are often located in remote places isolated by poor roads, which means these evaluators are disinclined to follow proper procedures. Thus, the easier method is to, at best, take a glance at the forest outskirts, or, at worst, look at the forest map, listen to the harvesting companies, and then these officials just issue the final assessment (20EP07, Lam Dong Environmental Police 2012c). It is also plausible that these officials are bribed by the harvesting companies, and then the status of isolated forests becomes a justification for not conducting proper supervision. Furthermore, many of the companies delivering the projects are not specialists in rubber plantations, but rather timber companies whose principal aim is to harvest timber (Bich Ngoc 2013; Government Inspectorate 2014a). The above-mentioned techniques used to falsify the timber capacity of forests are thought to be similar to those employed to fabricate forest diversity in timber species, when the species listed in the logging permits are usually less valuable than the ones harvested in reality.[14]

Another common tactic found in the logging from forestland conversion projects is to illegally log or buy illegal timber outside the permitted areas, then move and mix it with lawfully harvested timber. Through this technique, the permitted areas can be used many times to launder illegal timber. The cases of Minh Phuc Company in Dak Nong province and Khe Dien Hydropower in Quang Nam province, as mentioned above, are prominent examples demonstrating this technique in which hundreds of m^3 of illegal timber were mixed with legal logs. In some projects that are carried out near the border areas in the Central Highlands, timber traffickers smuggle illegal timber from Cambodia and mix it with legal logs from the projects (Duong 2012).

The second policy that can be abused to log large volumes of timber is the policy of planned logging.[15] As indicated in Chap. 3, by 2012 the practice of planned logging yielded as much as 5 million m^3 of logs in Vietnam (Nguyen and Phan 2014). However, there is some evidence to suggest that many harvesters fully comply with the statutory requirements. In Bac Kan province, for instance, over the five year period 20??–20??, the provincial authorities approved more than a thousand logging permits every year; but only several violations of abusing the permits were detected (FPD Bac Kan 2009, 2011).

However, despite this very small number of cases officially detected by the law enforcement agencies, some interviewees suspect that many forestry companies abuse the policy to illicitly gain large amounts of timber in the forests which they are allocated to manage (3FO02, 18EP06,

24LA02). In Directive No.12 of 2003 regarding "Implementing urgent solutions in forest protection and development", the Prime Minister of Vietnam confirms:

> SFEs, which undertake the role of forest owners, have not delivered well their functions. Some even collude with illegitimate entities to illegally harvest rare and precious timbers and other forest products. For their own benefits, some SFEs do not adhere to approved logging schedule. (PMV 2003: 1)

It is not possible to count precisely in the whole of Vietnam how much timber has been illegally cut down from planned logging, but there is a consensus among some interviewees that the actual volume of logged timber may well exceed the amounts and zones allowed in the logging proposals (3FO02, 18EP06, 24LA02). A police officer observes:

> It is the current situation that the forest zones allowed for harvesting are small, but the areas logged in practice are very large. Besides, the logging permits require that the harvesters must cut the trees that are broken, fallen or unhealthy. But no, they only cut the big and healthy trees. By law, Kiem Lam forces are required to check the logging sites and mark the log with seals. But to be honest with you, in many cases, basically the Kiem Lam has contributed to the illegal logging. (3FO02)[16]

In addition to the technique of exceeding the approved quantities, another method used to illicitly cut down a large number of trees is by justifying the constructing roads to transport timber from the harvesting locations out of the forests. This technique is found in some provinces such as Bac Giang (Anh The 2013), Binh Dinh (Vu 2014b) and Quang Nam (Tan Vu 2014a). In the case of Son Dong Forestry Company in Bac Giang province, for example, after illegally harvesting timber in 23 hectares of natural forests that were not permitted by the logging plan, the company illegally opened a 4-km road through the forests (Anh The 2013). Although the company claimed that the road was built for the purpose of planting in a barren zone within the forests, it was in fact allegedly for harvesting timber and transporting the illegal timber out of the forests (Anh The 2013). The problem is that, in some cases, the loggers deliberately locate the roads in rich timber areas and make the roads curved, so that the timber amounts obtained are maximised (3EP02), and that once the roads are opened, further illicit logging is likely to occur alongside these new roads (35FO07).

The third policy that timber traffickers unscrupulously take advantage of to procure a relatively large amount of timber is the housing policy for the disadvantaged inhabitants living in forested areas. As shown in Chap. 3, to support poor forest-based households in having stable houses, the Vietnamese government, via Project 134 and Project 167, allows these households to harvest certain amounts, usually less than 10 m^3 of non-endangered timber in nearby forests, to build their residential houses. In parallel with the beneficial results, it is highly likely that this policy is being abused by powerful timber traffickers to harvest and/or to launder large amounts of illegal timber[17] (3EP02, 5LR05, 12EP04, 30LR04, Ha Vy 2011; Hoang Nam 2011; Ngo 2008).

There seem to be three main scenarios when abusing this policy. First, these timber traffickers urge, and then guide, several ethnic minorities to make logging proposals for the purpose of house-building. The reasons given are seemingly convincing, such as having a severely damaged house, an overcrowded house or having a mature son planning to get married; all reason for which it is imperative to get a new house. Once the proposals are approved, these traffickers invest the money required for the ethnic minorities to log amounts of timber that often substantially surpass the permitted volumes and then build temporary houses (3EP02, 5LR01, Phung 2011).

The second plausible scenario as revealed by a local resident is that:

> The timber barons often collude with the owners of rich forests, especially SFEs. They can then freely bring cutters and equipment to log illegally. They then carry the logs to nearby villages, and then build large houses there. After the houses have been roughly completed and made to appear fairly old, the timber barons ask several ethnic minorities to be the temporary owners of the houses and pay them some money in return. (5LR01)

In both scenarios, as soon as the temporary houses are built, some convincing reasons are given for the purpose of selling the houses and transporting them to destinations in lowland areas. The owners claim, for example, to be in desperate need of money to overcome starvation, or that they are planning to move to reside in their original hometown in the lowlands (03EP02, 05LR01). The transported materials are no longer classed as timber, but rather house materials, with documents confirmed by local authorities, which makes it easier for the transportation to get through checkpoints. It appears that in order for these plans to be successfully achieved, seeking support from anti-trafficking officers is necessary (03 EP02, 05 LR 01, 06FO02, Hoang Quan 2012).

The third scenario is when companies employed by local authorities to harvest timber and build houses for the local families abuse logging permits to illicitly harvest large volumes of timber. The case in Quang Nam province is an example. To implement the housing policy under the 167 Programme, a timber company was employed by district authorities to log and construct 136 timber houses for the poor indigenous households. Taking advantage of the logging permit, this company modified the logging design, falsified documentation and illegally logged over 220 m^3 of high-quality timber. The company director was arrested and charged with timber trafficking (Ha Vy 2011).

Thanks to the method of abusing house-building policy, the total volume of timber in each house may reach many dozens of m^3 of timber, worth several tens of thousands of pounds (5LR01, Phung 2011). A local resident who used to witness this technique describes how to maximise the house's timber amount:

> Often the big stilt houses (nhà sàn) are built from good timber types. The main pillars can be made from round logs 9 m in length and 60 cm in diameter. Its 80 m^2 floor is paved by hordes of timber planks with 10 cm in thickness, 80 cm in width and limitless length. Walls, ceiling and furniture are also made from the large planks. (5LR01)

A senior police officer provides another insight into this method:

> This is not a new technique because it has been used in Vietnam since 1980s. But at this moment, it has become more commonly used and particularly preferred in many areas of Vietnam. This is because it appears

> legal, so it is easier to team up with the authorities. Any obstacles can be handled by money. Moreover, this technique helps get quite large amounts of long, large planks of timber. In Vietnam now, these planks are far higher valuable than those with the same quantities and timber type, but with smaller and shorter sizes. (03EP02)

This section has investigated in detail the sophisticated techniques employed by large-scale timber traffickers to illegally harvest large amounts of timber. The research findings on the typology of LSITH indicate that illegal logging in Vietnam occurs in a similar manner to the changing pattern of illegal logging currently found in many other parts of the world. That said, it is likely that major timber traffickers "are increasingly turning to clearance licences as a means to gain access to valuable timber resources" (Lawson and MacFaul 2010: 83), and moving from "direct illegal logging to more advanced methods of concealment and timber laundering" (Nellemann 2012: 6).

Outside the literature on timber trafficking, the empirical evidence on LSITH in Vietnam points out that LSITH comprises at least four defining attributes of corporate crime: (1) the involvement of powerful businessmen; (2) the close state–corporate relationship; (3) the blurred line between legality and illegality; and (4) the severe consequences resulting from the corporate crime. First, as clearly seen in this chapter, LSITH is committed by large commercial companies who abuse forest-based policies such as forestland conversion and hydropower construction to harvest large volumes of timber. The involvement of the business entities is the starting-point of corporate crime.

Second, the powerful LSITH actors are frequently deep in collusion with corrupt state officials. As demonstrated in this chapter, and further documented in Chap. 6, it is highly likely that corruption permeates through all the steps of LSITH ranging from the planning of forest exploitation, the evaluation of forest status and the issuing of logging permits to the harvesting activities, which combine to create a formidable obstacle to efforts at combatting the crime. Large timber companies, for example, invest heavily to influence forestry policy makers who then formulate forest-based policies which create opportunities for large-scale timber harvesting. The LSITH actors and state officials establish mutual connections that are conducive to harvesting operations, and concurrently beneficial to the state actors' own interests.

What occurs with LSITH reflects two ways corporations are connected to states as revealed by Kramer et al. (2002) and Michalowski and Kramer (2006), in which the former is *initiated* and/or *facilitated* by the later. This connection is termed by Tombs and Whyte (2009, 2015) as the *symbolic state–corporate relationship*, in which "the power of the corporation rests upon the power of states, and vice-versa" (Tombs and Whyte 2015: 162). This means that corporate crime like LSITH "is not a result of the success or lack of success of the state acting as 'a policeman', but is produced as a result of the symbiotic relationship between states and markets" (Tombs and Whyte 2009: 114).

Third, recent corporate crime literature such as Glasbeek (2002), Hartley (2008), Tombs and Whyte (2015) and Yokoyama (2007), which is also applicable to LSITH in this research, suggests that there is a blurred line between legality and illegality; and the legality is sophisticatedly abused by corporations. It is observed that "corporate crimes tend to have an ambiguous relationship to (il)legality" (Tombs and Whyte 2015: 136), and that that "corporations that hide behind the veil of legality to callously put profits before lives do so on behalf of people who, in turn, hide behind the law" (Glasbeek 2002: 222).

In this research, LSITH is apparently legal since it is conducted through the logging approvals of competent authorities; and the entire harvesting activities are monitored by these authorities. However, in most cases, it is, in essence, illegal because it employs many sophisticated techniques which essentially contravene laws and regulations pertinent to the logging, in addition to other statutory operations required before and after the logging. It is, nonetheless, extremely tricky to identify these techniques. The abuse of the intersection between legality and illegality by LSITH in Vietnam, together with the discussions of "quasi-legitimate" by Liddick (2011: 99) and "legal-illegal interfaces" by Bisschop (2012: 191) on the thin boundary between legal and illegal logging provide further examples of the technique used by corporate criminals when manipulating existing laws.

As a result of their close relationship with state officials and their methods of abusing legality, corporate actors are relatively immune from criminal punishment. In this research, it is argued that the Vietnamese criminal justice agencies largely fail to hold LSITH actors accountable. This reinforces the conclusion highlighted by Tombs and Whyte (2015: 50) that corporate operations like occurrences of LSITH "are unequivocally crimes, but they have been hardly treated as such".

The fourth feature of corporate crime is the severest damage it causes compared to other criminal forms of its kind. Indeed, the official statistics indicate that SSITH and MSITH together cause only 5400 hectares of forest loss annually, whereas LSITH alone arguably accounts for the majority of the annual loss of 45,000 hectares of forest in Vietnam (Environmental Police Department 2012; FPD 2014). The other forest damage that the official reports fail to encompass is believed to be significant. It is, therefore, reasonable to suggest that as opposed to SSITH and MSITH, LSITH generates the majority of the environmental impacts, which undoubtedly leads to further social, economic and cultural implications. This corresponds to the remark made by corporate crime scholars on the especially destructive attribute of corporate crime. When addressing the issue of air pollution, Tombs and Whyte (2015: 50), for example, emphasise that "corporations certainly produce most of the air pollution that threatens our health, economy and the environment".

When proposing that corporation are by their nature inherently harmful, Tombs and Whyte (2015: 158–159) conclude that "corporations, as currently and historically constructed, cannot be effectively reformed…It is an essentially destructive and irresponsible phenomenon. In short, the goal of corporate opposition must be the abolition of the corporation". This conclusion suggests that in order for the fight against timber trafficking in Vietnam to succeed fully, the control of LSITH featuring typical characteristics of corporate crime must be a priority. Potential solutions for limiting LSITH will be recommended in Chap. 6.

This research has thus far examined three forms of illegal timber harvesting in Vietnam: SSITH, MSITH and LSITH. In addition to the qualitative distinctions among the three forms as analysed above, it is important to clarify the volume of illegal timber considered under each category. This task will be undertaken in the next section.

An Overall Framework of Illegal Timber Harvesting

It is important to note that illicit loggings that are conducted in different forest types (e.g. special-use, protected or production forests), with different timber types (e.g. endangered or plain timber) require different amounts of timber to fit each category of the logging (SSITH, MSITH or LSITH), and therefore should bear different levels of sanction (e.g. criminal or administrative). It is, therefore, necessary to use legislative, theoretical and practical grounds to determine the specific starting point of each category of logging in each forest type and timber type.

Legislative provisions help identify three types of forest (based on the Forest Protection and Development Act in 2004), two types of timber species (based on Decree 32/2006/ND-CP in 2006), and two levels of sanctions for specific illegal timber volumes (based on Penal Code in 2009, Decree 157/2013/ND-CP and the Joint Circular 19/2007/TTLT/BNN&PTNT-BTP-BCA-VKSNDTC-TANDTC). Theoretical grounds, in combination with legislative ones, can help determine the maximum threshold of SSITH for plain timber in production forests; this then becomes the minimum level of MSITH for the same timber and forest types. Indeed, as examined earlier, SSITH loggers are often poor forest-dwelling people who log principally for subsistence purposes, especially for residential house-building. At the same time, Article 7 of the Joint Circular No. 08/2009,[18] that guides the implementation of housing policy, specifies that the maximum amount of plain timber which each household can harvest to build its house, is 10 m^3 in production forests. It can thus be suggested that logging less than 10 m^3 of plain timber in production forests can be considered as SSITH.

The next step is to determine the maximum volume of endangered timber in different forest types for SSITH, which again becomes the minimum level of MSITH for the same timber and forest types. It is believed that SSITH should not be judged as a criminal offence as defined in the Penal Code of Vietnam. Article 175 of the Penal Code prescribes that an act of timber trafficking qualifies as a criminal offence when it causes *"serious consequences"*. Meanwhile, the Joint Circular 19/2007 stipulates the measurement of *"serious consequences"* of timber trafficking, namely that the logging of more than 2 m^3, 1.5 m^3 or 1 m^3 of the endangered timber listed in Group IA in production, protected and special-use forests, respectively, will be considered as causing *"serious consequences"*. This suggests that an act of harvesting less than 2 m^3, 1.5 m^3 or 1 m^3 of endangered timber in production, protected and special-use forests, respectively, can be judged as SSITH.

As shown by the results, it can now be seen that with the same type of production forest, to fit into SSITH, the amount of plain timber (e.g. less than 10 m^3) is five times higher than the amount of endangered timber (e.g. less than 2 m^3). This rate can be applied to discover the amounts of plain timber in protected forest and special-use forest under the category of SSITH. This method, in the end, results in the quantities of timber in SSITH: the logging of less than 10 m^3, 7.5 m^3 or 5 m^3 of plain timber, or less than 2 m^3, 1.5 m^3 or 1 m^3 of endangered timber, in production, protected and special-use forests, respectively, can be considered as SSITH.

Finally, practical considerations, taken from the research data, show that in reality there is large variety in the quantities of illegal timber in SSITH, MSITH and LSITH. As estimated above, in SSITH, the amounts of plain timber logged in production forests are less than 10 m^3. Looking at the amounts of illegal timber in timber trafficking cases detected so far in Vietnam, with reference to the classification of MSITH and LSITH in this research, quite often it is observed that the amounts of plain timber logged in production forests range from 10 to 100 m^3 for MSITH and more than 100 m^3 for LSITH. Thus, to cover all the quantities in the three levels, it is suggested that with the same type of timber harvested in the same type of forests, illegal timber amounts harvested in larger scale forms should be 10 times higher than the smaller scale.

From all the above-analysed considerations concerning both qualitative and quantitative aspects, Table 4.3 is proposed as a framework that can be used to identify each form of illegal timber harvesting in the current context of Vietnam.

PdF Production forests, *PtF* Protected forests, *SuF* Special-use forests

It is emphasised that while the framework reflects typical and general features of three concurrent forms of illegal timber harvesting in Vietnam, there are necessarily connections and overlaps among the three. For example, the volumes of illegal timber for each type ought not be abso-

Table 4.3 A proposed framework of SSITH, MSITH and LSITH in Vietnam

Set of criteria		SSITH			MSITH			LSITH		
		PdF	PtF	SuF	PdF	PtF	SuF	PdF	PtF	SuF
Quantitative aspects (m^3)	Plain timber	<10	<7.5	<5	10–100	7.5–75	5–50	>100	>75	>50
	Endangered timber	<2	<1.5	<1	2–20	1.5–15	1–10	>20	>15	>10
Qualitative aspects	Key actor	Indigenous poor people			A blend of various actors			Commercial organisations and timber barons		
	Main motivation	Subsistence-oriented			Profit-oriented			Commercial-oriented		
	Techniques	– Rudimentary – Ad hoc			– Professional – Well-planned			– White-collar – Outwardly legal		

lute and rigid since there are a few cases where illegal timber volumes harvested from SSITH and MSITH exceed 10 $m^3$and 100 m^3 of timber, respectively.[19] Additionally, while loggers in SSITH can take part in MSITH and LSITH, LSITH offenders, once logging is approved, can collude with medium- and small-scale loggers to take the most advantage of the logging permits. In the case of Mang Den Enterprise, for instance, as shown above, once the logging plan was accepted, Mang Den Enterprise brought six different logging firms with hundreds of local loggers into the forests (My Ngoc and Nguyen Tran 2004).

Furthermore, LSITH may both tangibly and intangibly create conditions for MSITH and SSITH to take place. Indeed, LSITH emerging in forest-based projects may enable easy access to remote but rich forests via the building of roads and other ancillary infrastructures, which worsens the problem of timber harvesting as a whole. In Lam Dong province, for example, following the approval of 451 projects of forestland conversion spreading over 77,438 hectares of forest, some of the project owners, after receiving forestlands and harvesting timber, do not deploy sufficient measures to protect forests as required in their approved project proposal, which results in rampant illegal logging being carried out by local loggers (Lam Dong Environmental Police 2012c). In Quang Nam province, since the Bung 4 Hydropower Plant, which was allowed to harvest over

1000 m^3 of timber, started to operate, the problem of illegal logging has emerged suddenly and prolifically with 35 motorboats available for smuggling illegal timber (ANTV 2014; Tan Vu 2014a). Similar situations are addressed in other provinces such as Son La (Xuan Tho 2014), Kon Tum (Quang Thai 2015) and Dak Nong (Government Inspectorate 2014a).

Smuggling

As observed by Crow et al. (2014: 190), most previous research on wildlife crime "speak[s] broadly of poaching, without providing a full understanding of the range of behaviour that falls under that general term". Similar observations are found in the literature on timber trafficking that tends to pay attention to the harvesting stage, while the subsequent stages, particularly smuggling or transporting activities, seem to attract far less intense examination (Tacconi 2007). Accordingly, understanding of the stage of illegal timber smuggling remains very limited. It would be the case that this limited understanding is one of the restrictions to effective law enforcement in timber trafficking. Indeed, even if timber has already been harvested in logging locations, if law enforcement is to be effectively delivered in the smuggling stage, there would be little opportunity for bringing the illegal timber to the markets, thereby significantly curtailing the crime as a whole. Adequate understanding of the smuggling step would greatly assist this enforcement, and is the focus of the following section.

After the trees are cut down, illegal timber is handled in different ways depending on the form of logging. In LSITH, because the harvested timber, as mentioned earlier, is typically labelled as or laundered as legally sourced, in many cases it has full valid documentation for the transportation. In this instance, it can be inferred that illegal timber is transported in a way similar to legal timber, which means no criminal tactics are particularly needed to pass checkpoints. With regard to timber cut down by SSITH, unlawful timber is carried to the loggers' villages close to the forests using simple methods that are already examined in the section on SSITH earlier. In sharp contrast, for timber smuggling in some cases of

LSITH that do not have valid documents, and in almost all cases of MSITH, various methods are often used to arrive at the final destination. These methods can be categorised into two types: advanced and traditional.

Advanced Methods

Unlike logging activities that take place clandestinely inside forests, timber transportation on main routes appears to prove immensely difficult to hide from the authorities. Indeed, in Vietnam it is now uncommon and virtually impossible to transport large vans of timber without any valid documents or at least apparently valid ones. This is because there are thousands of checkpoints, both fixed and mobile, along all routes potentially used for illicit timber transportation (Sikor and To 2011). Furthermore, the law enforcement agencies now have increasing informant networks to detect incidents of timber smuggling (5LR01). Thus, to transport large amounts of timber to long-distance destinations, the most common technique now is to use forged, partly forged or re-cycled documents, named by a police officer as “talisman” (26EP09), in conjunction with strong support from corrupt officers (5LR01, 20EP07, 26EP09). In other words, similar to an observation by Wyatt (2013b) on illicit wildlife transportation, actual illegal timber is not physically hidden, but it is labelled, or partly labelled, to look legal. In light of this, an interviewee observes:

> More than ten years ago, the incidents of Kiem Lam officers noisily chasing Lam Tac on streets were widespread. But such situations are now no longer common because almost all big vans of timber are “lubricated” already. There are now still volumes of illegal timber that are seized by the authorities. But they are little. The volumes of illegal timber that “legally” pass the checkpoints are much larger. (5LR01)

Any method of using either forged, partly forged or re-cycled documents begins with the buying of a volume of legal timber from legal sources such as legitimate wood firms or official timber auctions. So a timber smuggler has, for example, 100 m^3 of legal timber with fully valid certification files. From this beginning, different specific tactics are deployed. When using partly forged documentation, during each transport, only a part of the certified timber, say 10 m^3, is loaded in the vehicle such as a truck or a train wagon that may contain, for example, a total volume of 40 m^3 of timber; thus 30 m^3 of timber is actually illegal (17EP05, 20EP07, 29EP04). Alternatively, if there is more than one truck or wagon, several of these carry the illegal timber and lead the timber convoy (Pham 2008).

Thanks to prepaid bribes, illegal timber, which is transported with, and hidden inside, legal timber, will be ignored by the corrupt officials. If there is a number of trucks or wagons in the convoy, the leading trucks and wagons carrying illegal timber can be overlooked; only the final trucks containing legal timber with valid documents are checked; and finally the entire convoy will pass the checkpoints (17EP05, 20EP07, 29EP04, Pham 2008). An example of this technique can be seen in the biggest illegal timber-carriage haul discovered in Vietnam so far. It was delivered by train in 2011 when 400 m^3 of endangered timber was found in 15 goods wagons. The illicit timber was collected from the Central Highlands, some of it destined for Bac Ninh markets, and some for China (FPD 2012; Tienphong 2011). The main technique, as admitted by the transporters, was to forge documents, in which 70–80% of the total seized timber had no valid documentation (Tienphong 2011). After this case was detected, the use of trains to carry high-value timber over long distances has increased to avoid the highway roads that have recently been more tightly policed by law enforcement agencies (Hoang Nam 2014b).

When prepaid corridors have not yet been formed, haulers use fake timber bills collected from fictitious firms in far off provinces. Thanks to this trick, if the timber is detected, it takes the anti-trafficking authorities a long time using complex procedures to clarify its validity. In the meantime, the transporters can find ways to counteract their arrest. A common method is to bribe the officers who will then either make decisions beneficial to the timber transporters or help them falsify documents, including by directly providing their signature to certify the documents' validity (17EP05, 18EP06, 29IP04, Dang 2013b). Additionally, to facilitate the collaboration, transporters and officers take advantage of the discretion in identifying between timber and firewood or between raw timber and processed wood products (Hoang and Nguyen 2012). It is admitted that although the use of fabricated documentation is currently widespread, it is not yet possible for anti-smuggling agencies to thoroughly detect and prevent such falsifications (Dang 2013a).

In the approach of re-utilising documents, it is again essential to have some legal documents and officer–trafficker collaboration. To mask the partnership from honest officers and the public, these abettors have to reach a consensus regarding the methods of manipulating the tracking information in the documents, passing the checkpoints, and coping with the random mobile inspections during the smuggling process (20EP07).[20]

Traditional Methods

In parallel with the advanced tactics of timber transportation that are inseparable from corruption elements, there are in Vietnam still some "traditional" ways to smuggle smaller quantities of timber especially endangered and valuable species. A recurrent consensus among the interviewees, supported by a number of FPD reports (FPD 2010, 2012) and newspaper articles (Hoai Nam 2014a; Thanh Nguyen 2014; Thien Nhan 2013; Van Thanh 2014b; Xuan Huy and An Bang 2011), is that carrying illegal timber by modified motorbikes is a practical smuggling choice in the following four circumstances: (1) no valid documentation is available, (2) timber volume is not very large typically around 0.3–1 m^3, (3) the roads are in good condition and (4) the transporting distance is not too far, preferably inside a territory of a district or province.

In this situation, transportation is often carried out by motorbikes with a skilful rider (*người lái*) and another observer (*người lơ*). The observer often sits on top of the timber boards, facing backwards and holding a knife, always ready to cut the ropes that are used to secure the timber in cases where they are being chased by the authorities and cannot escape without unloading the timber. The motorbikes used to carry timber are modified, old, unregistered and low-valued, perhaps costing £60–80, so that if arrested, the drivers are willing to abandon them (2FO01, 12EP04, Hoai Nam 2014a; Thanh Nguyen 2014; Xuan Huy and An Bang 2011) (see Figs. 4.4 and 4.5).

Fig. 4.4 A motorbike specially used to transport timber, Source: Own photo taken in the Central Highland in September 2013

Fig. 4.5 A case of using modified motorbike to transport illegal timber, Source: Thien Nhan (2013)

To prevent Kiem Lam cars from chasing timber vans, the drivers may scatter big three-edged nails or small boards containing sharp metal nails on the streets (6FO02). If there is a stream, river or lake near the logging sites, small rafts with logs fixed underneath by ropes are preferred because they are often cheaper and safer. If timber is heavy, the rafts have car inner tubes attached and are pulled by motorboats; and if they are noticed by authorised forces, transporters can cut the ropes and let the air out of the inner tubes, allowing the logs to submerge in the river or lake bottom where the transporters can come back later to collect these logs (1EP01, FPD, 2010, 2012; Van Thanh 2014b). What is more, there are some cases where, in order to hide illegal timber, especially endangered species, a small number of the timber planks are wrapped in blankets and placed in large confectionary boxes or even in steel cribs as used by itinerant dog buyers or scrap dealers (12EP04). A forest protection officer describes another way of taking advantage of the support from the carrier's community:

> If the carriers are riding near their village, once chased by anti-trafficking forces, they try to run close to their house. Then they call for the support of their family members and villagers. In some situations women and children lie on and hug the timber planks and motorbikes. At the same time, the villagers are asked to strongly criticise the enforcement operation, saying "please stop, don't be excessively immoral, let's give them a way to survive". (2FO01)

As it has been observed that official, army, wedding or emergency cars are used to transport valuable timber, two divergent discourses emerge. Some officers insist that such means of transport are still relatively common in use (02FO01, 06FO02, 20EP07, 21IP03), whereas others do not concur with this (11IP02, 17EP05, 18EP06). The former indicates that to distract the anti-trafficking agencies, the transporters employ every tactic possible to achieve their task including the use of special cars as listed above, using many different number plates for one car and using a number of similar-looking cars. However, the latter believes that, in recent years, the technique has become outdated and cannot be employed anymore because it is complex, costly and sometimes counter-productive.

There are though tactics commonly agreed on by all the officers: using vans with two-layered trailers or modified chassis, interlacing timber with other legal commodities such as building materials, agriculture products and fertilisers, and even placing timbrt inside packages sealed with lead (see Fig. 4.6).

Another tactic that can be applied in a situation where either avoiding arrest or bribing officers is unachievable is the strategy of accepting the possibility of being arrested several times in succession. These consecutive arrests put pressure on police officers to be more lenient on future occasions and create an opportunity for the smugglers to achieve greater success in their next transport[21] (05LR01).

In any case of being arrested, one of the common ways to reduce legal liability, as largely shared among the anti-trafficking officers, is to claim that the illegal timber volume is the property of many people, so that each person is responsible for only a minimal amount of illegal timber. This method is the same as that used in illegal logging. By using this method, in many cases the illegal transporters can avoid criminal penalties, instead only receiving administrative fines, which means no further long and complex criminal investigative procedures will be launched. With the administrative sanctions, the transporters repeatedly ask the

Fig. 4.6 The arrest of a van covered by sweetcorn sacks but contained 1.4 tonnes of Trac timber in the core, Source: Own photos taken in Quang Binh province in September 2013

Kiem Lam office to buy the seized timber. Furthermore, by law, the people whose property is confiscated are given priority in buying the confiscated property through official channels such as at auction (17EP05). This helps the illegal transporters buy the captured timber and acquire the valid documents that subsequently may be utilised for advanced methods of transportation (17EP05). This shows a connection between traditional and advanced methods of transporting illegal timber.

The connection between traditional and advanced methods is also seen in cases where the traditional methods are used to smuggle smaller amounts of illegal timber within communes or districts. After accumulating larger timber amounts, advanced methods are employed to deliver the timber to further destinations in other provinces. Additionally, there have been recently techniques of transporting illicit timber that fall between the two types of smuggling methods. An example is the use of high-quality cars and airplanes to carry fairly small amounts of highly valuable timber such as Sua, Ky Nam and Trac. This is seen as an advanced method in the sense that it makes use of modern means of transport for long distances, but traditional because only small volumes of timber are transported without either valid documents or the collusion of law enforcers.

It seems that these "hybrid" methods are currently increasingly popular because it is now not only difficult to accumulate, but also highly risky to carry, large amounts of these "golden" timber species. Another trick to transport these timber species is to use small bars of the endangered timber that can be fixed as fragments of whole large wood products such as tables or beds. These timber portions will be carefully removed from the commodities as soon as they get to their final destinations (3EP02).

With regard to the temporal dimension of transporting, all law enforcement officers agree that there are no rigid temporal principles for conducting timber trafficking in general and illicit transport in particular. This means that the perpetrators can commit their crimes at any time, day or night, weekday or weekend, rainy or dry season, as soon as they believe it might be more advantageous for their operation. However, a broad theme emerged from the analysis of the research data. In order to minimise the chance of being detected, smugglers choose particular times for transportation and harvesting. These are at night, outside office working hours, during extreme weather days, national holidays and the Lunar New Year. They may also include the days that the anti-trafficking agencies have plenary meetings, celebrations of founding days and sporting events.

In addition, the dry season is also more preferable than the rainy season for harvesting and carrying timber. This is because in rainy weather, it is too risky to enter the forests due to the harsh geographical terrain, along with poorly built roads, which results in danger from landslides, flash floods and accidents. There may be an exception when heavy rain facilitates the movment of timber on the rivers or streams located inside or alongside the forests. Timber traffickers also pay heed to and avoid periods of intensive raids deployed occasionally by anti-trafficking forces.

With regard to geographical locations, illegal timber can move freely anywhere, but it is understandable that within a province, illegal timber flows tend to move from highland and forested districts to lowland localities. Meanwhile, at the national level, large illegal timber flows tend to move towards the North. This is because the demand for timber in this region is higher, especially in the nearby Chinese markets; whereas timber resources in northern natural forests are rapidly depleting.

As can be seen thus far, there are in Vietnam a myriad of techniques that can be used to harvest and smuggle unlawful timber. Timber trafficking does not stop at the harvesting and smuggling stages, but leads on to the activities of buying and selling illicit timber. These trading activities will be examined in the next section.

Trading

A major theme emerging during the course of data analysis is that illegal timber traders (*các đầu nậu*), who basically buy and sell illegal timber to gain profit, are frequently the central actors, playing a vital role in the entire timber trafficking chain. At the same time, these traders may also work as orchestrators of different activities in the trafficking chain, ranging across harvesting, transporting, laundering, processing and seeking support from the relevant authorities. The traders are the actors who receive the greatest blame, not only from other actors engaging in the crime but also from the frontline officers mandated to combat timber trafficking. This is because the traders frequently receive the highest share of profit in the entire illicit chain, yet encounter the lowest probability of detection.

Indeed, local loggers and porters undertake the most physically arduous and dangerous tasks, but gain the least profit. For example, normally the price of a commonly used timber at the cutting location in the Central Highlands is only £30–40/m^3 (26EP09). Meanwhile, in the final markets in lowland cities, the same timber species can be sold by traders for £1000–1200/m^3 (26EP09). Likewise, for indigenous porters who are employed to carry illegal timber from the cutting points to outside the forests in the northern province of Bac Kan, when successfully carrying a 70 kg hexagonal Nghien board, they only gain £1.80 and sometimes £3 for longer distances (12EP04). In the central province of Quang Nam, the highest price in cutting locations is only £65/m^3, but at the markets in lowland districts of the same province, over £1600/m^3 is not unusual (Xuan Hoai 2014).

Despite their high earnings, the irony is that many timber traders remain untouched. Five possible explanations have emerged in this research. First, they do not visibly and directly engage in the illicit activities of cutting, storing or transporting; instead, "to commit the crime, all they have is a mobile phone" (29IP04). In other words, they simply stand behind, induce, guide and equip the loggers and transporters (Le Minh 2014; Minh Dat 2014). Second, the traders are adept at persuading their associates and employees not to testify against them if arrested. To achieve this, the traders often demand employees adhere to firm principles such as a "confidential principle" and/or a "voluntary principle" that requires employees to voluntarily accept their involvement and if caught, they should never testify against traders and other co-offenders (29IP04, 32IP05, Hoang 2012). Furthermore, the traders may provide their loggers and transporters with payment for their work in advance, and even provide them with financial aid during harsh periods in the year such as in the heavy rainy season or after natural disasters, which induces local loggers be more loyal to the traders (25LR03, 26EP09).

Third, by having much experience in the illicit business, traders often have a deep understanding of the pertinent legal provisions, and where possible, they are competent at circumventing these provisions. A police officer observes:

> In this province, the majority of timber traders are originally professional loggers. They gradually become richer, more experienced, establish more helpful relationships, and then become traders. They know very well the potential locations for timber resources, promising markets and easy routes of transport. They know the pieces of legislation that can be possibly exploited and the specific law enforcement units that can team up more easily. (29IP04)

Fourth, they have lasting relationships with influential leaders, who have both formal and informal influence on frontline officers. The highly lucrative profits in the timber business embolden traders to invest vast amounts in establishing relationships with the leaders of local authorities who are necessary for their operation. A police officer states:

> In some cases, if there are new officers, the traders might even intentionally surrender certain volumes of illegal timber for confiscation by the officers. The traders then have a pretext to contact and build up a rapport with these authorities. (20EP07)

Finally, to facilitate the illicit business, as commonly found in timber trafficking worldwide such as in Madagascar (Global Witness and EIA 2009), Indonesia (Luttrell et al. 2011) and Russia (Wyatt 2013a), the large-scale traders often run legal timber businesses in combination with illegal operations. As found in this study, there are clear advantages in simultaneously running legal operations. To begin with, a legal operation is an ideal location for timber laundering. A prominent example is a serious criminal case that took place in the three provinces of Dak Nong, Binh Phuoc and Tien Giang in 2012 where six out of nine timber traders who were charged with illegally trading approximately 2500 m^3 of timber were directors and deputy directors of commercial companies (Investigation Security Agency 2013b). These defendants employed similar tactics: they bought illegal timber from local people and laundered the timber in their processing factories before bringing it to various timber markets (Investigation Security Agency 2013b).

Alternatively, once contracts with valid documents have been obtained from the legitimate business, the details about quantities, plank sizes and timber types in these documents will be given to the loggers. The loggers then look for, cut and modify the illegally harvested trees following exactly the details shown in the legal contracts, which will help the illegal transporters to achieve their task (3EP02, 17EP05). The well-known scandal of the trader Chien in Quang Ngai province is an example of this technique. Chien, previously an infamous forest hijacker, became the richest man in Son Ha district. He regularly employed about 10 key veteran loggers, each of them being asked to recruit another 10–15 specialist loggers, thus forming a massive logging ring with hundreds of members. The whole ring entered the Nuoc Nia Protected Forest, built 10 shelters with cooking equipment and harvested timber for years. The logging scene was described as a "big construction site". To launder the illicit timber, Chien opened a large sawmill in Di Lang town near the forest where timber contracts were collected. Details of the quantity and size of planks from the contracts were given to the loggers working inside forests in order to produce illegal timber planks similar to the legal ones. The strong determination of the law enforcement agencies in Quang Ngai province has finally brought Chien and his co-offenders to court (Nhan Dan 2005; Viet Bao 2005).

In brief, professional traders in illicit timber are typically the biggest profit recipients, but the least likely to enter the criminal justice system because of their "invisible" involvement, ability to persuade other actors to take the blame and to circumvent legal provisions, their close relationships with police officers and especially their establishment of parallel legal businesses. Three key actors in timber trafficking—harvester, smuggler and trader—have thus far been examined in this chapter. In the current context of Vietnam, as seen in Chap. 3, these actors are finding it more difficult to commit their crime particularly due to the loss of massive cheap timber supplies and an increase in political will to address the crime. They, therefore, must find as much support as possible from different sources to maintain their illicit businesses. The next section will investigate these supporting activities.

Supporting Activities

It is proposed that there are three types of supporters currently engaged with the crime of timber trafficking in Vietnam: scouts, protectors and officers. The third type of supporter will be discussed in the corruption section of Chap. 6; the following pages examine the first two types.

Scouting

This type of supporter might be called a scout ("nhóm chim lợn" [slang]) or navigator ("hoa tiêu") who undertakes the tasks of finding out, observing, following the authorities' operations in both workplaces and officers' houses, and then informing their abettors about these observations. From this information, their accomplices will find it easier to counter the law enforcement agencies. With regards to the desired information, working timetables and raiding plans, faces and vehicles of anti-trafficking officers, and even undercover informants of the anti-trafficking agencies are carefully noted (1EP01, 2FO01, 6FO02, 21IP03, 29IP04, 32IP05, Kon Tum Investigation Police Agency 2011). A Kiem Lam officer states:

> The scouting of law enforcement agencies has recently become popular. The scouts are very active in observing our staff numbers and timetables. They know, for example, that a Kiem Lam station in the forest has three officers. So that, if one officer is going to have a couple days off, then they also know that the two others will stay in the station rather than conduct forest patrols. That is a good time for logging. (2FO01)

To obtain the desired information, scouts may be stationed close to an agency's premises, the officers' houses, or in a scattered fashion surrounding logging sites or along transportation routes to detect itinerant officers (2FO01, 21IP03, 29IP04, Xuan Huy and An Bang 2011). Due to mobile phone networks that recently cover most forest zones, associates find it easy to inform each other if they notice any signs of anti-trafficking officers. This information helps the offenders escape from the crime scene before the officers arrive. This explains why in many cases, although the law enforcement officers detect trees that have been illegally cut down inside forests, or illegal logs that have been loaded on roads, they cannot detect the illegal loggers or transporters (2FO01, 20EP07, 21IP03, Xuan Huy and An Bang 2011). A police officer also mentions:

> The navigators are particularly keen on tracking our official cars that are frequently used for forest-related operations such as patrol and arrest. We often then have to rent different private vans with the staff concealed inside the cargo area to approach the logging sites. Otherwise, we would arrest no one when arriving. (21IP03)

Sometimes, children and women are involved in scouting by standing near forest gates pretending to herd or fish (6FO2, Duy Hau and Quang Tao 2011). In some cases, scouts follow the officers' movements, usually making attempts to hide, but if they are discovered they are willing to overtly continue their task because they believe the officers have no right to arrest them based on their actions (29IP04).

Protecting

The second type of supporter that sometimes participates in the crime is the violent actor ("đối túợng hình sự" or "xã hội đen") who is in charge of physically protecting the illegal logs and sometimes seizing harvesting areas from other groups. A police officer explains:

> Because many timber trafficking groups scramble to get areas for illicit operation, they may need some physical protectors. These protectors should be reckless enough to safeguard their illicit activities, and to compete with other groups. It becomes more important in areas where timber sources are increasingly scarce. (26EP09)

These violent supporters often have criminal histories involving serious violent crimes or may be heavy drug addicts. This is explained by an interviewee:

> The traffickers know that heavy junkies, especially those with HIV, are surely desperate for money to buy drugs. If the junkies are given for example

> [VND] 10 million [£330] and asked to hack off the arms of other traffickers, they would do so with little hesitation. They just lurk somewhere, wait for the target, suddenly get close, axe [the victim] and then run away. After the sudden attack, it is then difficult to find them out. (5EP01)

The violent actors may be long-time members of timber trafficking squads, but more commonly they are separate from the squads. In the latter scenario, these protectors are employed from time to time to carry out necessary actions such as seizing logging zones and threatening other offenders with physical violence; consequently once these tasks are completed, they withdraw from the squad (5LR01, 26EP09, 32IP04, Minh Dat 2014; Quang Thai 2015). In addition to these tasks, sometimes these violent criminals directly undertake the task of smuggling relatively small volumes of timber on "impossible routes", as described by a police officer:

> They are named as "impossible routes" because the routes are strictly controlled by strong and proper authorities. On these routes, "benign" transporters often find it impossible to carry any illegal timber. In this case, the task would be assigned to the very heavy junkies with HIV. Taking advantages of their tremendous audacity, they run incredibly fast. This may help them successfully pass the checkpoints. (20EP07)

The violent actors are also, at times, assigned the task of threatening anti-trafficking officers and whistle-blowers, which will be examined in the section on personal security in Chap. 6. Four types of illicit activities involved in timber trafficking in Vietnam have thus far been examined: harvesting, smuggling, trading, and supporting. In order for illicit timber to reach final consumers, it needs to enter the stage of timber processing, which will be looked at in the next section.

Processing

One of the features that distinguishes timber from other wildlife species such as wild animals is that while many wild animals, after being illegally poached, can be sold directly to buyers such as collectors or bush meat consumers, virtually all trees illegally cut down need to be processed to a certain degree into wood products before reaching the final timber purchasers. Hence, illegal timber processors have an important role in the whole chain of timber trafficking.

However, identifying and evaluating the illicit operation of timber processing tends to be a tricky task. As Schloenhardt (2008: 92) argues, illegal timber processing is "one of the most complex steps" used to disguise the origin of the illegal timber, making the final products indistinguishable from products involving lawfully gained materials. The official statistics in Vietnam, to some extent, confirm the difficulty of detecting illegal incidents of timber processing. During the five-year block from 2008 to 2012, the Vietnamese anti-trafficking forces detected a total of over 6000 cases of illegal timber processing, which means that every year, some 1200 cases of illegal timber processing were disclosed, accounting for only 6% of the total number of cases of timber trafficking.[22]

With regard to the pattern of illegal timber processing in Vietnam, it is observed that two main types of activities are involved: processing without a licence and, more commonly, processing of illicitly obtained timber. Processing without a licence often occurs in the small factories located inside or close to forests. As indicated by the MARD (2014), this processing occurs in a number of provinces in Vietnam. By 2011, Ha Tinh province, for example, had a total of 289 sawmills in operation, but 171 (59%) were started without a licence (Ha Tinh People's Council 2012).

It is also highly likely that the close-to-forests sawmills, either with or without a licence, serve as ideal points to launder illegal timber. As examined in the section on the timber trade, illegal timber may be laundered in the close-to-forests processing factories that have a legal licence. Indeed, the sawmills may well be the points where timber unlawfully cut from nearby forests is quickly and initially processed into rough wood products (Ha Thanh 2014a; Kon Tum Investigation Police Agency 2011; MARD 2014; VTV 2014). The FPD confirms:

> Some timber processing factories near forests do not follow the planning of local authorities, have no stable supplies, and are not being regularly inspected. They then use illegal timber and become locations for illicit timber processing. (FPD 2012: 21)

In the Central Highlands, for example, there are more than 1500 timber processing factories, many of them close to or even inside forests, using the suspicious timber mainly from natural forests (Uyen Thu 2013b). The same situation is found in the provinces of Bac Kan (Ha Thanh 2014a), Ha Tinh (Ha Tinh People's Council 2012) and Quang Nam (VTV 2014).

In addition to the forest-located workshops, some processing factories in lowland locations are also suspected of being processing points for illegal timber. The criminal case investigated by Quang Binh Investigation Police Agency (2012) provides a piece of firm evidence for this suggestion. The Investigation Conclusion reveals that the director of a carpentry company in Nam Dinh, a lowland Northern province, came to Quang Binh, a densely forested Centre province. He bought almost 400 planks of illegal Gu Lau timber, an endangered species, in Quang Binh to supply his processing factory in Nam Dinh (Quang Binh Investigation Police Agency 2012).

Additionally, in Vietnam, there is a common form of timber processing business called "villages of traditional carpentry" or "wood craft villages" that have a well-established business in timber processing with the participation of the majority of the villagers (Nguyen and Phan 2014). Phu Khe Thuong village, Dong Ky ward and Huong Mac commune in Bac Ninh province are nationally well-known examples of these villages.[23] These workshops consume hundreds of thousands of m^3 of timber yearly, much of this is endangered species (Bac Ninh Environmental Police 2012a; VTV2 2013). However, a report by Vietnam Television reveals that among 300 timber-processing villages in Vietnam buying timber without receipts, which is a proof of the timber's legality, is rather common. Often, the processors do not ask timber traders for receipts since having receipts means paying more value-added tax, which increases the product's price and reduces competitiveness (VTV2 2013). This practice creates the conditions for illegal timber to enter the processing factories (VTV2 2013). In practice, the law enforcement agencies have already detected a number of incidents of transporting illicit timber to processing locations in Bac Ninh province (Bac Ninh Environmental Police 2012a; Tienphong 2011) (Fig. 4.7).

Fig. 4.7 Timber is sold in the Phu Khe Thuong timber market, Source: Own photos taken in Bac Ninh province in September 2013

Although the processing factories in lowland localities may use illegal timber, it is highly likely that they have a licence. This is because it is far easier for the law enforcement agencies to visit and inspect the processing factories in the lowlands. Furthermore, running such a factory requires a fixed location, several noisy machines and many employees, which makes the processing operation difficult to completely conceal from the anti-trafficking forces (4TT01, 37TT03).

In Brief: The Nature and Extent of Timber Trafficking in Vietnam

Taking into account all the findings presented in this chapter, first it can be claimed that over the five year period 2008–2012 there has been a generally decreasing trend in the number of cases of timber trafficking detected in Vietnam. Indeed, while nationally the number of detected cases of timber trafficking is dropping by nearly 2000 every year; locally, the crime is no longer an everyday problem in many provinces. Two decades ago, for instance, the practice of a significant proportion of villagers coming to nearby forests to participate in illegal logging would be widespread in most forested areas in Vietnam (Sunderlin and Huynh 2005); but this practice is seen as gradually declining.

However, irrespective of the decreasing prevalence, the operation of timber trafficking is likely to be increasing in sophistication and certainly it remains a concern in many parts of Vietnam. New sophisticated techniques have been employed in most stages of timber trafficking: forestry policies have been abused to carry out large-cale logging; advanced methods such as recycling documents in collaboration with corrupt officers have been used to transport timber; and various supporting activities are rife. In brief, timber trafficking still persists as a "pressing concern" in Vietnam (Forest Trends 2013: 1) and the control of the crime "has not fundamentally improved" (Department of Legal Affairs—MARD 2012: 24).

What happens in Vietnam appears to be consistent with the current evaluation by UNEP and Interpol of the worldwide trend of timber trafficking: that the reduction in the number of cases of timber trafficking worldwide is simply "temporary", and that "an apparent decline in illegal logging is due to more advanced laundering operations masking criminal activities, and not necessarily due to an overall decline in illegal logging" (Nellemann 2012: 7). Although the drops in the share of illegal timber are seen in major countries where timber trafficking occurs, such as Indonesia, Brazil, Cameroon, Ghana and Malaysia (Lawson and MacFaul 2010), it is a highly possible tendency that timber trafficking has become "more prominent, increasingly organised, sophisticated and transnational" (Stewart 2014: 241).

With regard to the illicit value of timber trafficking in Vietnam, it is suggested that it is annually worth at least £130 million—equivalent to the total of the salaries of all 12,000 forest protection officers in Vietnam for nine years. Although there is no way to claim that the estimate is perfectly accurate, the monetary gravity of timber trafficking in Vietnam tends to match the observation by Green et al. (2007: 117) that among internationally recognised green crimes, timber trafficking is "almost certainly the most economically significant".

Having looked into the current detailed typology of timber trafficking in Vietnam, this chapter proposes a comprehensive typology that is constructed from five interrelated components: harvesting, smuggling, trading, supporting and processing. Each component plays a different role in the trafficking chain as a whole, and is made up of further sub-components.

To begin with illegal harvesting, it is suggested that there are three distinctive forms of illegal timber harvesting in Vietnam: SSITH, MSITH and LSITH. With regard to SSITH, this is the illegal harvesting of small volumes of timber (e.g. less than 10 m^3 of plain timber in production forests), which is opportunistically or occasionally committed by individuals or small groups of poor forest-based residents, using rudimentary methods under the motive of subsistence demands. Fundamentally different from SSITH is MSITH that involves specialist loggers who illegally harvest relatively large volumes of timber (e.g. from 10 m^3 to 100 m^3 of plain timber in production forests) in a well-planned manner for the main purpose of gaining profits. Teamwork, good planning, resilience and strong support from scouts and violent protectors are typical features of MSITH.

The third form of illegal logging in Vietnam is LSITH, coordinated by commercial companies or timber barons who abuse forest-based policies and flout relevant legal requirements to harvest large volumes of timber for commercial purposes (typically more than 100 m^3 of plain timber in production forests). The three most abused policies are thought to be (1) forestland conversion into hydropower construction and rubber plantations, (2) planned logging and (3) housing policy for forest-based poor households. Each of these policies brings about different ways for unlawful timber to be acquired, mainly including misrepresenting rich forests as poor ones and high-quality timber species as plain ones, falsifying logging documents, flouting the approved areas and volumes, and exploiting the "legitimate outskirts" to collude with state officials.

It can be seen that whilst having a certain overlap, overall SSITH, MSITH and LSITH are fundamentally distinctive not only in terms of the volumes of illicit timber and the legal definitions of illegal logging, but more importantly in terms of the attributes of the harvesters, the degree of their involvement, their motivations, and the level of organisation of criminal commission. It is then suggested that illegal timber harvesting should be categorised into three types: SSITH, MSITH and LSITH. This is a fresh way to understand the dynamics of illegal timber harvesting.

The second component of timber trafficking in Vietnam is the step of smuggling. In order to smuggle larger timber volumes to long-distance destinations, counterfeit documents (either entirely forged, partly forged or re-cycled) in tandem with strong support from corrupt officers are often required for the smuggling to succeed. These are considered by this research as "advanced methods". Concurrently, for smaller timber volumes, often less than 1 m^3, that do not have any valid documents, that are carried over shorter distances such as inside the territory of a district or province, and where no relationship with officers is established, clandestine techniques are preferred, and this is named as "traditional methods".

In both methods, a myriad of specific tactics is used to facilitate the illicit transportation ranging from the manipulation of forged documents, the hoaxes directed at authorities, the use of various modified means of transport, the division of illegal timber to smaller portions, the combination of advanced and traditional tactics, to the consideration of temporal and spatial dimensions. Jointly, they reveal a truly diverse and complex dynamic of the transportation of illegal timber in Vietnam. However, there is little information on the transport operation presented in the existing literature on timber trafficking, which makes it difficult to compare and contrast this research's findings in Vietnam with those found in other countries. It is plausible that not all law enforcement officers are aware of the techniques revealed in this research, so it is hoped that this will enrich anti-trafficking officers' understanding of timber smuggling, thereby helping these officers to prepare more effective measures to control the smuggling.

With regards to the trading component, the act of selling and buying illegal timber for profit is a key part of timber trafficking. This trade occurs when profits are made and then distributed to different actors in the entire chain of timber trafficking. This distribution of profits serves as the chief, if not the only, motivation for almost all the actors. The traders may also work as coordinators of the whole trafficking chain, employing and assigning tasks to loggers, transporters, scouts and protectors. The traders are usually better-off, more competent at taking advantage of legal loopholes, and receive a much greater share of the profit than the rest. However, thanks to their close relationship with anti-trafficking officers, their ability to influence other associates, their strategy of indirect involvement, and particularly the simultaneous running of legal businesses, mean these traders feature disproportionally infrequently in the official list of offenders charged with timber trafficking.

The fifth element of timber trafficking in Vietnam is the supporting activities that are required to deal with the current disadvantages facing offenders, such as the loss of plentiful timber supplies and the improved performance of criminal justice agencies. Assistance from corrupt officers aside, the two types of supporter examined in this chapter are scouts and violent guardians. If the former is in charge of much less arduous tasks such as observing the operation of law enforcement agencies and then guiding the traffickers' operation, the latter embarks upon much more dubious acts, for instance violently protecting illegal timber and loggers, seizing the harvesting areas of other groups, and threatening other perpetrators.

As reported in Chap. 1, violence is an inherent and endemic aspect of timber trafficking, particularly in the world's largest tropical rainforests such as Brazil, Indonesia, Peru and Honduras (Boekhout van Solinge 2010b; CIP and EIA 2005; FAO 2007; Kaimowitz 2003; Nellemann 2012). In this sense, it is observed in Vietnam that such violence might have been used the past; recently, however, the scale of violence and intimidation in Vietnam seems to have decreased. At present, violence is a part of timber trafficking, but it tends to be unsystematic. Indeed, by and large, the professional violent actors are not necessarily a fundamental part of the timber trafficking chain. In most cases, the traffickers have their own territories, and many of them have relatively long-standing and trusting relationships, which means violence or threats of violence are not always essential. It is, therefore, argued that presently, the involvement of violence in timber trafficking in Vietnam is, on the whole, likely to be less severe and rampant than some other major locations of timber trafficking.

Finally, with regard to the pattern of illegal timber processing, Brack (2003) suggests three main types of activities involved in illegal timber processing worldwide: (1) processing of illicitly attained timber, (2) processing without licence or with fake licences and (3) processing with illicitly attained licences. While the evidence available in this research does not seem to be sufficient to examine the third type, it is suggested that the first two types do take place in Vietnam. Both forms (processing without a licence and processing of illegally harvested timber) happen more visibly in processing sawmills inside or close to forests. At the same time, the timber-processing factories and carpentry villages in lowland localities, in spite of having licences, are not immune from the illicit practice.

Conclusion

This chapter has revealed a comprehensive five-component typology of timber trafficking in Vietnam, which answers in depth the first main question: how is timber trafficking currently occurring in Vietnam? It can be concluded that illegal timber can be harvested via various techniques, by different types of loggers, and under varying motives. The illicit timber then flows through varying routes. It may either terminate in the villages or sawmills inside or right next to logged forests, or be sold to traders, then having been legalised, is amassed into greater volumes, before being transported to further markets. An overall typology of timber trafficking in Vietnam can be seen in Fig. 4.8.

Once the first main question of how timber trafficking in Vietnam occurs has been addressed, the next chapter looks for the answer of the second key question: how does timber trafficking affect various aspects of Vietnamese society?

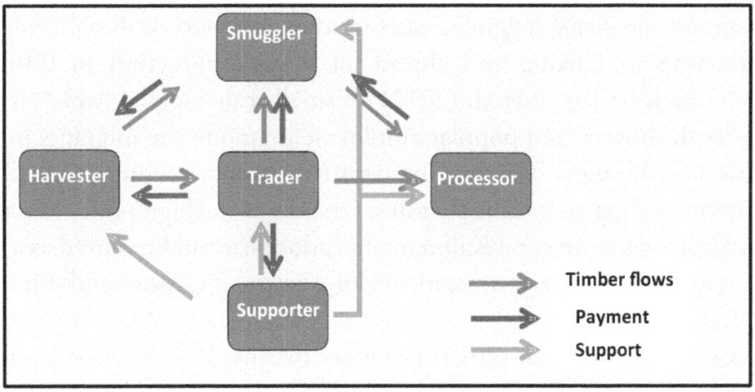

Fig. 4.8 An overall typology of timber trafficking in Vietnam

Notes

1. The most common forms of administrative sanction are monetary fines and confiscations of illegal forest products, equipment and vehicles used to carry out the violation. There are a number of violations that are detected by the law enforcement agencies, but not recorded as penalised cases. This is mainly because the anti-trafficking agencies cannot find the violators.
2. Apart from timber, there are small number of cases of illegal harvesting, transportation, processing and trade in other forest products most commonly orchids, mushrooms and bamboos.
3. The endangered timber species (Group IA and Group IIA) are specified in the Decree 32/2006/ND-CP; whereas all timber species in Vietnam are classified into eight groups (Group I to Group VIII) according to Decision 2198/CNR of the Ministry of Forestry dated 26 November 1977. Plain timber species, defined in this research, are those that are not listed in the Decree 32/2006/ND-CP.
4. The estimates of timber prices are based on the suggestions of the research interviewees (4TT01, 7FO1), other sources that note the values of timber prices (Kon Tum Investigation Police Agency 2011; Xuan Hoai 2014) and official documents specifying the prices of forest products, issued by provincial people's committees such as Dak Nong (Decision 21/2012/QD-UBNN), Quang Binh (Decision 749/QD-UBNN in 2010), Dak Lak (Decision 28/2013/QD-UBNN) and Lam Dong (Decision 2566/QD-UBNN in 2013).
5. A wood-balance analysis by Lawson and MacFaul (2010) suggests that small-scale illegal logging makes up three-quarters of illegal timber production in Ghana, and almost all illegal production in Cameroon. Research by Jagger et al. (2012) on small-scale logging, which is largely illegal, shows it is a popular employment among the migrants in southwestern Uganda. This logging does help reduce income inequality and improve their economic standing. Thanks to the high profits, small-scale illegal logging in rural Kalimantan, Indonesia "can be viewed as a lifeline supporting numerous segments of society" (Casson and Obidzinski 2007: 62).
6. A study by To et al. (2014) estimates that by 2012 while the total consumption of timber for house-building throughout Vietnam was 3.4 million m^3, the amount of timber harvested by all poor forest house-

holds (15% of Vietnamese household numbers) was 0.8 million m^3, accounting for 24% of the total timber amount used for house-building in the country.

7. Apart from the logging undertaken for house construction, in some areas indigenous people still maintain a tradition, in which they log several big planks of high-quality timber, and keep them in their house as valuable assets for either urgent situations or future generations (11IP02).
8. A police officer demonstrated this attribute in a recent event, during which over 20 officers, divided into two teams, conducted a raid on a harvesting scene that involved four loggers. However only one logger was arrested, the rest successfully escaped thanks to their familiarity with the forest terrain (20EP07).
9. The Joint Circular 19/2007 specifies, for instance, that for a criminal charge, the minimum volume of timber listed in Group IIA is 7 m^3 and for Group III is 20 m^3. Loggers then often harvest volumes that, after being divided between each logger, are less than these levels.
10. Some of the most notable locations for large-scale illegal logging are Cambodia (Le Billon 2000; Peters 2000), Russia (Pye-Smith 2006), Papua New Guinea (Laurance et al. 2011), Indonesia (Nellemann 2007; Telepak and EIA 2007; EIA and Telapak 2009), the Congo Basin (Ruiz Pérez et al. 2005), Honduras (CIP and EIA 2005), Madagascar (Global Witness and EIA 2009) and the Amazon (Laurance 1998; Boekhout van Solinge 2010a).
11. Key works in Vietnamese such as "các dự án thủy điện và cao su phá rừng" ("hydropower and rubber projects destroy forests") and "lợi dụng dự án cao su khai thác gỗ trái phép" ("taking advantage of rubber plantation projects to illegally harvest timber") yield a large number of relevant newspaper articles.
12. The conclusion by Lam Dong Province Inspection (2013) on the implementation of projects of forestland conversion into rubber plantation indicates that in three districts alone, there are 241 illegal cases of logging and forestland grabbing. These violations take place in 25 out of 44 companies inspected with almost 1800 m^3 of timber illegally harvested.
13. In Dak Nong province, for example, the Government Inspectorate (2014a) officially confirms that "the harvesting of timber and forest resources during the implementation of the projects of land and forest allocation and land lease shows many loopholes and shortcomings. This leads to the fact that a large amount of timber and forest resources are lost". In Gia Lai, in 2008 alone, over 30,000 m^3 of timber were lost

during the implementation of rubber plantation projects (Government Inspectorate 2014b).

14. Many projects of forestland conversion into rubber plantation in Lam Dong province, for example, actually harvest 1800 m^3 of timber from Groups II to IV, but the logging permits indicate that the timber species for the logging are in Groups V–VIII (Lam Dong Province Inspection 2013).
15. The policy is specified in the Forest Protection and Development Act 2004, Decree 23/2006/ND-CP, Decision 186/2006/QĐ-TTg and most specifically in Circular 35/2011/TT-BNNPTNT partly amended by Circular 70/2011/TT-BNNPTNT. It allows forest owners to annually harvest certain amounts of timber. In the case of natural forests or state-funded plantation forests, which account for the vast majority of forests in Vietnam, the logging must be permitted by different local authorities (MARD 2011).
16. There have been a number of cases detected of abusing planned logging activities that take place in different parts of Vietnam (Anh The 2013). A notorious example was the criminal case of the Mang Den SFE, which took place from 2002 to 2003 in Kon Tum and Gia Lai province and was judged as the most serious criminal case of illegal logging in the Central Highlands brought into court to date. The Mang Den SFE was permitted to log over 2300 m^3 of timber within the 61 hectares of forest that was managed by this enterprise. However, the enterprise, in close cooperation with five other logging firms, illegally harvested a total of 5500 m^3 from an area of 300 hectares in the Kon Ka Kinh National Park (My Ngoc and Nguyen Tran 2004). Due to the high level of criminal organisation, the provincial investigators appeared to be unable to gather sufficient legal evidence, and the case then had to be investigated by a central exclusive investigative task-force. A total of 27 people were prosecuted and mainly charged with either timber trafficking or corruption. The defendants were sentenced with a total of 123 years imprisonment (My Ngoc and Nguyen Tran 2004).
17. There are a number of cases found in different provinces such as Binh Dinh (Ngo 2008), Thua Thien Hue (Hoang Quan 2012); it appears to be particularly serious in the provinces of Quang Nam (Ha Vy 2011; Thanh Nien 2013), Nghe An (Khanh Hoan 2013; Phung, 2011; Tran Thu 2014) and Quang Tri (Dinh Thieu 2014; Phan 2011; Quang Tri Environmental Police 2012d).

18. Article 7 in Joint Circular No. 08/2009/TTLT-BXD-BTC-BKHĐT-BNNPTNT-NHNN on Guiding the Implementation of Decision No. 167/2008/QĐ-TTg states "The maximum timber volume is no more than 10 m3 of round timber for each household".
19. The criminal case in Bac Kan province, which is considered as a SSITH, is an example. In order to have timber to build a house, a villager asked four other villagers to come to Ba Be National Park and illegally cut down two Nghien trees, equivalent to 17.5 m^3 of timber (Ba Be Investigation Police Agency 2013a).
20. A police officer reveals some details of the technique. In checking stations, the entire timber volumes loaded in the trucks are confirmed to be consistent with the documents. So the documents will be stamped "Already Checked" and the trucks will pass through. To reuse the legal timber amount, the corrupt checkers will not write the specific checking times on both documents held by the timber owner and the supervision record managed by the checking station. After they pass the checkpoint, if they are unexpectedly inspected by mobile inspectors, they attempt to bribe the unwelcome inspectors. In many cases they succeed, but if for any reason, the bribe attempt fails, they will swiftly write the required specific time on the document. Then it becomes valid. They will then swiftly inform the previous checkers about this "tough" inspection, and the necessary information will also be added to the supervision records (20EP07).
21. In this technique, on the first three occasions, for example, the transporters carry small volumes of low-value timber, which they are willing to have seized. On the fourth occasion, however, if caught again, the carriers plead with officers for forgiveness, saying that previous haulages have been captured by the authorities, and their entire family relies on this journey. They then entreat the officers to give them and their family a chance to live. In this situation, Kiem Lam officers may reluctantly do them the favour of allowing the carriers to pass because the officers might fear that if the timber is seized again, these "nothing-to-lose" lawbreakers may take harsh revenge against the officers. This time, typically, the timber volume is highly valuable and can compensate for the preceding lost deliveries (05LR01). A similar tactic is to use intentionally dubious behaviours with a view to hoaxing the authorities (12EP04). In these cases, transporters do not carry any illegal timber on the first few occasions, but behave in a manner that makes the operations appear illegal. Having observed the suspicious behaviour, the authorities often request

to check the commodities. After checking on the first few occasions repeatedly indicates that nothing is illegal, the transporters strongly condemn the officers for constantly harassing their legal business, and in some cases, the officers are threatened with being sued for harassment. On the following occasions, the accusers may actually transport illegal timber, but with a greatly reduced probability of being inspected (12EP04).

22. The official statistics on timber trafficking consist of only four steps: illegal harvest, transport, trade and processing. The supporting activities (scouting, physical protection and patronage of officers) are considered as a component of timber trafficking in this research, but are not included in the official statistics on the crime.
23. The wood processing business in Dong Ky ward, for example, has been developed over decades, creating stable and relatively well-paid employment for more than 60% of the whole village population. The monthly salary for workers ranges from £100 to £270, depending on how skilful the workers are. The volume of timber that is annually processed in the village reaches over 100,000 m^3 (VTV2 2013). During my observation of the timber market in Phu Khe Thuong, close to Dong Ky ward, which is believed to be a unique timber market in Vietnam, large quantities of a variety of timber including highly valuable species were for sale. The market store was not spacious enough, so huge volumes of timber were openly loaded along streets (see Fig. 4.7).

References

Anh The. (2013). Công ty lâm nghiệp mở đường xuyên rừng để... phá rừng [The forestry company opens the road through forests to ... deforest]. *Dan Tri*, 17/10/2013. Retrieved January 4, 2015, from http://dantri.com.vn/kinh-doanh/cong-ty-lam-nghiep-mo-duong-xuyen-rung-de-pha-rung-790968.htm

ANTV. (2014). Tận thu gỗ lòng hồ thủy điện hay phá rừng? [The salvage of timber in hydropower reservoirs or deforestation?]. *ANTV*, 4/9/2014. Retrieved January 2, 2015, from http://antv.gov.vn/xahoi/quang-ngai-nguy-trang-tinh-vi-cho-xe-go-lau/63201.html

Ba Be Investigation Police Agency. (2013a). *Investigative conclusion of criminal case No.24/KLĐT dated 10/5/2013 by the Investigation Police Agency of the Ba*

Be District Public Security,on the criminal case of violating the regulations on forest exploitation and protection taking place on 25/11/2012 in 51 Zone, 104 Point, the Ba Be National Park. Ba Be, Bac Kan (in Vietnamese).

Bac Ninh Environmental Police. (2012a). Conference paper: The work of protecting, fighting and handling illicit activities in trading and transporting forest products in Bac Ninh province. In *Preventing and combatting violations against legislations on forest protection* (pp. 30–34). Hanoi: The Department of Preventing and Combating Environmental Crimes (in Vietnamese).

Bich Ngoc. (2013). Lộ rõ lợi ích nhóm 'phù phép' phá rừng trồng cao su [Revealing the group interests in mythologising to deforest for rubber plantation]. *Đất Việt*, 01/10/2013. Retrieved June 2, 2015, from http://www.baodatviet.vn/kinh-te/doanh-nghiep/lo-ro-loi-ich-nhom-phu-phep-pha-rung-trong-cao-su-2356205/

Bisschop, L. (2012). Out of the woods: The illegal trade in tropical timber and a European trade hub. *Global Crime, 13*(3), 191–212.

Boekhout van Solinge, T. (2010a). Deforestation crimes and conflicts in the Amazon. *Critical Criminology, 18*(4), 263–277.

Boekhout van Solinge, T. (2010b). Equatorial deforestation as harmful practice and criminological issue. In R. White (Ed.), *Global environmental harm.* Cullompton, UK: Willan.

Brack, D. (2003). Illegal logging and the illegal trade in forest and timber products. *International Forestry Review, 5*(3), 195–198.

Casson, A., & Obidzinski, K. (2007). From new order to regional autonomy: Shifting dynamics of illegal logging in Kalimantan, Indonesia. In L. Tacconi (Ed.), *Illegal logging, law inforcement, livelihoods and the timber trade.* London: Earthcan.

CIP & EIA. (2005). *The illegal logging crisis in Honduras: How U.S. and E.U. imports of illegal Honduran wood increase poverty, fuel corruption and devastate forests and communities.* Washington, DC: Center for International Policy & Environmental Investigation Agency.

Crow, M., Shelley, T. O. C., & Stretesky, P. (2014). Camouflage-collar crime: An examination of wildlife crime and characteristics of offenders in Florida. In T. Spapens, R. White, & M. Kluin (Eds.), *Environmental crime and its victims: Perspectives within green criminology* (pp. 185–204). Surrey: Ashgate.

Dak Nong Environmental Police. (2012b). Conference paper: The work of preventing and combatting violations of laws on forest protection in Dak Nong province. In *Preventing and combatting violations against legislations on forest protection* (pp. 158–166). Hanoi: The Department of Preventing and Combating Environmental Crimes (in Vietnamese).

Dang, T. K. (2013a). Doanh nghiệp buôn gỗ lậu bằng hồ só giả [Firms smuggling timber by fabricated documentations]. *Lao Dong*, 4/10/2013. Retrieved October 25, 2014, from http://laodong.com.vn/phap-luat/doanh-nghiep-buon-go-lau-bang-ho-so-gia-141181.bld

Dang, T. K. (2013b). Kiểm lâm "biến" gỗ lậu thành... gỗ hợp pháp [Kiem lam "mythecising" illegal timber into legal timber]. *Lao Dong*, 22/6/2013. Retrieved October 25, 2014, from http://laodong.com.vn/phap-luat/kiem-lam-bien-go-lau-thanh-go-hop-phap-122941.bld

Dang, T. K. (2014). Nhiều doanh nghiệp phá rừng, bán đất dự án [Many companies destroy forests, sell the project land]. *Lao Dong*, 11/12/2114. Retrieved January 1, 2015, from http://laodong.com.vn/xa-hoi/nhieu-doanh-nghiep-pha-rung-ban-dat-du-an-277157.bld

Department of Legal Affairs—Mard. (2012). Conference paper: The status of forest protection and solutions to enhance the work of forest protection. In *Preventing and combatting violations against legislations on forest protection* (pp. 18–29). Hanoi: The Department of Preventing and Combating Environmental Crimes (in Vietnamese).

Dinh Thieu. (2014). Quảng Trị: Lợi dụng chính sách giao rừng để...phá rừng [Quang Tri: Abusing forest allocation policy to deforesting]. *VOV—Radio Voice of Vietnam*, 26/02/2014. Retrieved June 2, 2015, from http://vov.vn/xa-hoi/moi-truong/loi-dung-chinh-sach-giao-rung-depha-rung-312450.vov

Duong, V. L. (2012). Conference paper: Current status of violations of laws on forest protection in the Central Highland and solutions to prevent and combat. In *Preventing and combatting violations against legislations on forest protection* (pp. 11–17). Hanoi: The Department of Preventing and Combating Environmental Crimes.

Duy Hau & Quang Tao. (2011). Lâm tặc "mở hội" ở Vườn quốc gia Yok Đôn [Lam tac "opening festival" inside the Yok Don National Park]. *DanViet*, 12/5/2011. Retrieved February 6, 2015, from http://danviet.vn/thoi-su/lam-tac-mo-hoi-o-vuon-quoc-gia-yok-don/42754p1c24.htm

EIA & TELAPAK. (2009). *Up for grabs: Deforestation and exploitation in Papua's plantations boom*. London and Bogor, Indonesia: Environmental Investigation Agency and Telapak. Retrieved from http://www.eia-international.org/wp-content/uploads/up-for-grabs.pdf

Environmental Police Department. (2012). Conference paper: Opening paper—The work of preventing and combating the violations on forest protection. In *Preventing and combatting violations against legislations on forest protection* (pp. 7–10). Hanoi: The Department of Preventing and Combating Environmental Crimes (in Vietnamese).

FAO. (2007). *State of the world's forests 2007*. Rome: Food and Agriculture Organization of the United Nations.

Forest Trends. (2013). *Small-scale illegal logging in Vietnam: Implications for FLEGT and REDD+*. The Forest Trends.

FPD. (2009). *Report on the operational performance in 2008 and the working programme in 2009*. Hanoi: The Forest Protection Department (in Vietnamese).

FPD. (2010). *Report on the operational performance in 2009 and the working programme in 2010*. Hanoi: The Forest Protection Department (in Vietnamese).

FPD. (2011). *Report on the operational performance in 2010 and the working programme in 2011*. Hanoi: The Forest Protection Department (in Vietnamese).

FPD. (2012). *Report on the operational performance in 2011 and the working programme in 2012*. Hanoi: The Forest Protection Department (in Vietnamese).

FPD. (2013). *Report on the operational performance in 2012 and the working programme in 2013*. Hanoi: The Forest Protection Department (in Vietnamese).

FPD. (2014). *Report on the operational performance in 2014 and the working programme in 2015*. Hanoi: The Forest Protection Department (in Vietnamese).

FPD Bac Kan. (2009). *Report on the operational performance in 2009 and the working programme in 2010*. Bac Kan: The Forest Protection Sub-Department (in Vietnamese).

FPD Bac Kan. (2011). *Report on the operational performance in 2011 and the working programme in 2012*. Bac Kan: The Forest Protection Sub-Department (in Vietnamese).

Glasbeek, H. (2002). *Wealth by stealth: Corporate crime, corporate law, and the perversion of democracy*. Between the Lines.

Global Witness & EIA. (2009). *Investigation into the illegal felling, transport and export of precious wood in SAVA Region Madagascar*. London: Global Witness and Environmental Investigation Agency.

Government Inspectorate. (2014a). *The declaration No.2490/TB-TTCP, 15/10/2014 of the inspection conclusion on the land management and utilisation with focus on land allocation and forest allocation in a number of projects in Dak Nong province in the 2004–2011 period*. Hanoi: Government Inspectorate of Vietnam (in Vietnamese).

Government Inspectorate. (2014b). *The declaration No.2834/TB-TTCP, 19/11/2014 of the inspection conclusion on the land management and utilisation and construction investment management in a number of projects in Gia Lai province*. Hanoi: Government Inspectorate of Vietnam (in Vietnamese).

Government Office. (2013). *Announcement No.159/TB-VPCP, dated 11/4/2013—The conclusion of the Deputy Prime Minister Hoang Trung Hai in the meeting on forest protection and development of the Central Highland provinces*. Hanoi: Officer of the Government (in Vietnamese).

Green, P., Ward, T., & McConnachie, K. (2007). Logging and legality: Environmental crime, civil society, and the state. *Social Justice, 34*(2), 94–110.

GSO. (2009). *Migration and urbanization in Vietnam: Situation, trends and differences*. Hanoi: The General Statistics Office.

Ha Thanh. (2014a). Giữ rừng ở Bạch Thông, thực trạng và giải pháp [Keeping the Bach Thong forest. Current situation and solutions]. *Bac Kan Online*, 22/9/2014. Retrieved June 2, 2015, from http://www.baobackan.org.vn/channel/2262/201409/giu-rung-o-bach-thong-thuc-trang-va-giai-phap-2339944/

Ha Tinh People's Council. (2012). *The Program: Sustainable protection and development of forests—The period 2012–2015 with orientation forwards 2020 (Issued together with Resolution No. 32/2012/NQ-HĐND dated 19.07.2012 of the Province People's Council)*. XVIth Ha Tinh People's Council—3rd Section, Ha Tinh (in Vietnamese).

Ha Vy. (2011). Một giám đốc "ăn" hơn 220m3 gỗ [A director "eats" over 220m3 of timber]. *Cong An Nhan Dan*, 1/10/2011. Retrieved May 29, 2015, from http://cand.com.vn/ANTT/Mot-giam-doc-an-hon-220m3-go-186617/

Hartley, R. D. (2008). *Corporate crime: A reference handbook*. Santa Barbara, CA: ABC-CLIO.

Hoai Nam. (2014a). Ngang nhiên vận chuyển gỗ lậu bằng xe máy ở Quảng Nam [Blatanly transporting illegal timber by motorbikes in Quang Nam]. *VOV*, 21/9/2014. Retrieved June 2, 2015, from http://vov.vn/doi-song/ngang-nhien-van-chuyen-go-lau-bang-xe-may-o-quang-nam-353231.vov

Hoang Nam. (2011). Khó ngăn phá rừng [Difficult to prevent timber trafficking]. *Tienphong*, 10/7/2011. Retrieved June 2, 2015, from http://www.tienphong.vn/phap-luat/kho-ngan-pha-rung-544555.tpo

Hoang Nam. (2014b). Lộ đường dây vận chuyển gỗ lậu quý hiếm bằng đường sắt [Revealing the network of transporting illegal timber by railway]. *Tien Phong*, 29/5/2014. Retrieved December 22, 2014, from http://www.tienphong.vn/xa-hoi/lo-duong-day-van-chuyen-go-lau-quy-hiem-bang-duong-sat-710266.tpo

Hoang Quan. (2012). Cán bộ chỉ đạo con cháu... phá rừng [Officials direct their descedants ... to deforest]. *Cong an Thanh Pho Ho Chi Minh*, 6/7/2012. Retrieved January 2, 2015, from http://www.congan.com.vn/?mod=detnews&catid=703&p=&id=474017

Hoang, T. D. (2012). Conference paper: The work of preventing and combatting violations of laws on forest protection in Thua Thien Hue province. In *Preventing and combatting violations against legislations on forest protection* (pp. 123–127). Hanoi: The Department of Preventing and Combating Environmental Crimes (in Vietnamese).

Hoang, T. T., & Nguyen, T. D. (2012). Conference paper: Violations against legislations on forest protection and arising issues over the work of education and training. In *Preventing and combatting violations against legislations on forest protection* (pp. 112–117). Hanoi: The Department of Preventing and Combating Environmental Crimes (in Vietnamese).

Investigation Security Agency. (2013b). *The investigative conclusion on the case of violation of regulation on forest exploitation and protection, taking place in Dak Nong, Binh Phuoc and Tien Giang in 2012*. Security Investigation Agency—The Ministry of Public Security, Ho Chi Minh City (in Vietnamese).

Jagger, P., Shively, G., & Arinaitwe, A. (2012). Circular migration, small-scale logging, and household livelihoods in Uganda. *Population and Environment, 34*(2), 235–256.

Kaimowitz, D. (2003). Forest law enforcement and rural livelihoods. *International Forestry Review, 5*(3), 199–210.

Khanh Hoan. (2013). Vụ lật xe chở gỗ lậu khiến 10 ngườ̀i chết: 8 bị cáo lãnh án tù [The case of flipped truck of illegal timber killing 10 people: 8 defendants are sentenced to prison]. *Laodong*, 24/1/2013. Retrieved June 2, 2015, from http://www.thanhnien.com.vn/pages/20130124/vu-lat-xe-cho-go-lau-khien-10-nguoi-chet-8-bi-cao-lanh-an-tu.aspx

Kon Tum Investigation Police Agency. (2009). *The investigative conclusion No.07/BC-PC15 dated 7/9/2009 on the case: "Violation of regulations on forest exploitation and protection", take place in Zone 101, protection forest, Dak Glei district, Kon Tum province*. Kon Tum: Investigating Police Agency, Kon Tum Public Security Department (in Vietnamese).

Kon Tum Investigation Police Agency. (2011). *The investigative conclusion No.02/KLDT-PC46 dated 30/9/2011 on the case: Violation of regulations on forest exploitation and protection, take place in Dak Uy special-used forest, Dak Ha district, Kon Tum province*. Kon Tum: Investigating Police Agency, Kon Tum Public Security Department (in Vietnamese).

Kramer, R. C., Michalowski, R. J., & Kauzlarich, D. (2002). The origins and development of the concept and theory of state-corporate crime. *Crime & Delinquency, 48*(2), 263–282.

Lam Dong Environmental Police. (2012c). Conference paper: Shortcomings and violations in implementing the forestland conversion policies in Lam Dong province. In *Preventing and combatting violations against legislations on forest protection* (pp. 137–140). Hanoi: The Department of Preventing and Combating Environmental Crimes (in Vietnamese).

Lam Dong Province Inspection. (2013). *Conclusion No.42 /KL-TTr, 11/3/2013 of the inspection on the projects of rubber plantation of the non-state companies in the districts of Da Teh, Da Huoai and Bao Lam.* Department of Inspection—Lam Dong Province People's Committee, Lam Dong (in Vietnamese).

Laurance, W. F. (2008). The need to cut China's illegal timber imports. *Science, 319*(5867), 1184–1185.

Laurance, W. F., Kakul, T., Keenan, R. J., Sayer, J., Passingan, S., Clements, G. R., et al. (2011). Predatory corporations, failing governance, and the fate of forests in Papua New Guinea. *Conservation Letters, 4*(2), 95–100.

Lawson, S., & MacFaul, L. (2010). *Illegal logging and related trade indicators of the global response.* The Royal Institute of International Affairs.

Le Billon, P. (2000). The political ecology of transition in Cambodia 1989–1999: War, peace and forest exploitation. *Development and Change, 31*(4), 785–805.

Le Minh. (2014). Lâm tặc là ai? [Who are Lam Tacs?]. *Tien Phong*, 2/12/2014. Retrieved December 21, 2014, from http://www.tienphong.vn/xa-hoi-chuyen-hom-nay/lam-tac-la-ai-790928.tpo

Lescuyer, G. (2007). Livelihood and the adaptive application of the law in the forests in Cameroon. In L. Tacconi (Ed.), *Illegal logging: Law enforcement, livelihoods and the timber trade* (pp. 167–190). London: Earthscan.

Liddick, D. (2011). *Crimes against nature: Illegal industries and the global environment.* Santa Barbara, CA: Praeger.

Luttrell, C., Obidzinski, K., Brockhaus, M., Muharrom, E., Petkova, E., Wardell, A., et al. (2011). *Lessons for REDD+ from measures to control illegal logging in Indonesia.* CIFOR.

MARD. (2011). *Circular No.35/2011/TT-BNNPTNT dated 20 May 2011: Guiding the implementation of harvesting and salvage of timber and NTFPs.* Hanoi: Minister of Agriculture and Rural Development (in Vietnamese).

MARD. (2014). *Official letter 1895/BNN-TCLN on strengthening the management of timber processing facilities.* Hanoi: Ministry of Agriculture and Rural Development (in Vietnamese).

MARD & FSSP. (2014). *Report on the development of the forestry sector in 2013*. Hanoi: Ministry of Agriculture and Rural Development & Forest Sector Support Program and Partnership. Retrieved from http://www.vietnamforestry.org.vn/mediastore/fsspco/2014/04/01/Bao_cao_PT_nganh_2013-ENGLISH.pdf.

McElwee, P. (2004). You say illegal, I say legal: The relationship between 'illegal' logging and land tenure, poverty, and forest use rights in Vietnam. *Journal of Sustainable Forestry, 19*(1-3), 97–135.

Michalowski, R. J., & Kramer, R. C. (2006). The critique of power. In R. J. Michalowski & R. C. Kramer (Eds.), *State-corporate crime: Wrongdoing at the intersection of business and government* (pp. 1–17). Piscataway, NJ: Rutgers University Press.

Minh Dat. (2014). Tăng cường công tác quản lý và bảo vệ rừng. *Cong An Kon Tum*, 6/11/2014. Retrieved January 2, 2015, from http://congankontum.gov.vn/hdccat/bao-dam-trat-tu-xh/52885-tang-cuong-cong-tac-quan-ly-va-bao-ve-rung.html

Minh Thi—Minh Trieu. (2014). Chuyện những người giữ Giáng hương [The stories on protectors of Giang Huong forest]. *Gia Lai Online*, 2/10/2014. Retrieved June 2, 2015, from http://baogialai.com.vn/channel/1622/201410/chuyen-nhung-nguoi-giu-giang-huong-2342405/

My Ngoc & Nguyen Tran. (2004). Xét xử vụ án phá rừng lớn nhất Tây Nguyên: Giám đốc lâm trường "bắt tay" doanh nghiệp để phá rừng [Adjudicating the biggest illegal logging case in the Central Highland: Forestry enterprise directors "handshake" firms to destroy forests]. *Thanh Nien*, 29/11/2013. Retrieved February 6, 2015, from http://www.thanhnien.com.vn/news1/pages/200447/102302.aspx

Nellemann, C. (2012). *Green carbon, black trade: Illegal logging, tax fraud and laundering in the worlds tropical forests. A rapid response assessment*. Birkeland Trykkeri AS, Norway: Interpol Environmental Crime Programme—United Nations.

Nellemann, C., Miles, L., Kaltenborn, B. P., Virtue, M., & Ahlenius, H. (2007). *The last stand of the Orangutan. State of emergency: Illegal logging, fire and palm oil in Indonesia's national parks*. Arendal: UNEP/GRID.

Ngo, H. N. (2008). Giao khoán bảo vệ rừng, thực hiện chính sách dân tộc và tác động đối với bảo vệ rừng ở Bình Định [The allocation of forest protection, the implementation of ethnic policies and the impacts on the work of forest protection in Binh Dinh]. *Kiem Lam Newsletter*. Retrieved June 2, 2015, from http://kiemlam.org.vn/Desktop.aspx/News/So_1_2_nam_2008/Giao_khoan_bao_ve_rung_thuc_hien_chinh_sach_dan_toc_va_tac_dong_doi_voi_bao_ve_rung_o_Binh_Dinh/

Nguyen, T. T. T., & Phan, M. T. (2014). *Wood processing industry—Activity code: No-5—Support business associations to prepare studies on development strategy for two selected industries*. Hanoi: EU-MUTRAP & Vietnam Chamber of Commerce and Industry. Retrieved from http://www.trungtamwto.vn/sites/default/files/publications/nghien_cuu_nganh_go.pdf

Nguyen, V. S. (2008). Wildlife trading in Vietnam situation, causes, and solutions. *The Journal of Environment & Development, 17*(2), 145–165.

Nhan Dan. (2005). Vụ phá rừng lớn nhất ở Quảng Ngãi: Trùm Hai Chiên đã đúợc dung túng? [The biggest case of illegal logging in Quang Ngai: Was Hai Chien magnate condoned?]. *Nhan Dan*, 25/10/2005. Retrieved June 2, 2015, from http://www.nhandan.com.vn/phapluat/thoi-su-phap-luat/item/4448702-.html

Nhien Di. (2013). Rừng Tây Nguyên teo tóp [The forests in Central Highland are atrophied]. 24/6/2013. Retrieved April 19, 2015, from http://nld.com.vn/thoi-su-trong-nuoc/rung-tay-nguyen-teo-top-20130624102358371.htm

Peters, J. L. (2000). Illegal trafficking of timber in Cambodia. *Colorado Journal of International Environmental Law and Policy, 11*, 102.

Pham, D. X. (2008). *Investigation delivered by the investigation police on crimes of economic management and official position over the violations against regulations on forest exploitation and protection*. The thesis of PhD in Law, The People's Police Academy (in Vietnamese).

Phan, T. B. (2011). Lợi dụng chính sách xóa nhà tạm để tàn phá rừng đại ngàn [Taking advantage of the policy of eliminating temporary houses to ... destry massive forests]. *Cong An Nhan Dan*, 15/2/2011. Retrieved January 2, 2015, from http://cand.com.vn/Xa-hoi/Loi-dung-chinh-sach-xoa-nha-tam-de-tan-pha-rung-dai-ngan-173905/

Phung, V. M. (2011). Để việc thực hiện chủ trúóng, chính sách của Đảng và Nhà núớc ở có sở không bị lợi dụng [To eliminate the grounds for abusing the guideline and policies of the Party and the State]. *Xay Dung Dang*, 25/10/2011. Retrieved January 2, 2015, from http://www.xaydungdang.org.vn/Home/dien-dan/2011/4332/De-viec-thuc-hien-chu-truong-chinh-sach-cua-Dang-va-Nha.aspx

PMV. (2003). *Directive No.12/2003/CT-TTg dated 16/05/2003—Regarding "implementing urgent solutions in forest protection and development"*. Hanoi: The Prime Minister of Vietnam.

Pye-Smith, C. (2006). *Logging in the Wild East: China and the forest crisis in the Russian Far East*. Forest Trends & Rights and Resources.

Quang Binh Investigation Police Agency. (2012). *The investigative conclusion No.25/KLDT-PC46 dated 12/1/2012 on the case: Duong Van Tanh offence on violation of regulations on forest exploitation and protection*. Investigating Police Agency, Quang Binh Public Security Department, Quang Binh (in Vietnamese).

Quang Thai. (2015). "Nóng" tình trạng khai thác lâm sản trái phép tại tỉnh Kon Tum ["Hot" situation of illegal harvest of forest products in Kon Tum]. 7/1/2015. Retrieved January 10, 2015, from http://www.vietnamplus.vn/nong-tinh-trang-khai-thac-lam-san-trai-phep-tai-tinh-kon-tum/300738.vnp

Quang Tri Environmental Police. (2012d). Conference paper: The Quang Tri police of preventing and combatting environmental crimes and the work of preventing and combating crimes and violations of forest protection and natural reservation. In *Preventing and combatting violations against legislations on forest protection* (pp. 149–153). Hanoi: The Department of Preventing and Combating Environmental Crimes (in Vietnamese).

Ruiz Pérez, M., Nguinguiri, J.-C., Nzala, D., Toirambe, B., Yalibanda, Y., Ezzine De Blas, D., et al. (2005). Logging in the Congo Basin: A multi-country characterization of timber companies. *Forest Ecology and Management, 214*(1), 221–236.

Saunders, J. (2014). *Trade in illegal timber: The response in Vietnam—A Chatham House Assessment*. Retrieved from http://www.illegal-logging.info/sites/files/chlogging/CHHJ2362_Vietnam_Logging_Research_Paper_FINAL.pdf

Schloenhardt, A. (2008). *The illegal trade in timber and timber products in the Asia-Pacific region—Research and Public Policy Series No. 89*. Canberra: Australian Institute of Criminology.

Sikor, T., & To, P. X. (2011). Illegal logging in Vietnam: Lam Tac (forest hijackers) in practice and talk. *Society & Natural Resources, 24*(7), 688–701.

Stahl, J. (2010). The rents of illegal logging: The mechanisms behind the rush on forest resources in Southeast Albania. *Conservation and Society, 8*(2), 140.

Stewart, D. (2014). Project leaf, and interpol's work on illegal logging and forest crime. In T. Spapens, R. White, & M. Kluin (Eds.), *Environmental crime and its victims: Perspectives within green criminology* (pp. 236–247). Surrey: Ashgate.

Sunderlin, W. D., & Huynh, T. B. (2005). *Poverty alleviation and forests in Vietnam*. Subur Printing, Indonesia: Center for International Forestry Research—CIFOR.

Tacconi, L. (2007). The problem of illegal logging. In L. Tacconi (Ed.), *Illegal logging: Law enforcement, livelihoods and the timber trade* (pp. 1–16). London: Earthscan.

Tan Vu. (2014a). Tan nát rừng đầu nguồn [Smashed watershed forests]. *Tuoi Tre*, 12/12/2014. Retrieved January 4, 2015, from http://tuoitre.vn/tin/ban-doc/20141212/tan-nat-rung-dau-nguon/683775.html

Telapak & EIA. (2007). *The thousand-headed snake: Forest crimes, corruption and injustice in Indonesia*. Bogor, Indonesia.

Thanh Nguyen. (2014). Hung thần gỗ lậu [Deadly elfs of illegal timber]. *Thanh Nien*, 6/10/2014. Retrieved June 3, 2015, from http://www.thanhnien.com.vn/pages/20141005/hung-than-go-lau.aspx

Thanh Nien. (2013). Khai thác trái phép gần 1.000 m3 gỗ dựng nhà tái định cư [Illegally harvesting almost 1,000 m^3 of timber for building resettlement houses]. *Con Nguoi & Thien Nhien—Thiennhien.Net*, 4/2/2013. Retrieved June 9, 2015, from http://www.thiennhien.net/2013/02/05/khai-thac-trai-phep-gan-1-000-m3-go-dung-nha-tai-dinh-cu/

Thien Nhan. (2013). Hãi hùng xe máy chở gỗ lậu [Horrificly motorcycles carrying illegal timber]. *Thanh Nien*, 02/12/2013. Retrieved June 3, 2012, from http://www.thanhnien.com.vn/pages/20131201/hai-hung-xe-may-cho-go-lau.aspx

Tienphong. (2011). Bắt 15 toa tàu chở gỗ lậu quý hiếm [Arresting 15 train wagons illegally transporting rare and endanger timber]. *Tienphong*, 15/12/2011. Retrieved June 3, 2015, from http://www.tienphong.vn/Phap-Luat/561202/Bat-15-toa-tau-cho-go-lau-quy-hiem-tpp.html

To, P. X., Tran, L. H., & Nguyen, T. Q. (2014). *The overview of the Vietnamese timber sector: Current status and trends*. Forest Trends/VIFORES/FPA Binh Dinh.

To, X. P. (2012). Natural forests and its suffering from timber need [Rừng tự nhiên chịu sức ép từ nhu cầu sử dụng gỗ]. *Con Nguoi & Thien Nhien*, 6/11/2012. Retrieved June 3, 2015, from http://www.thiennhien.net/2012/11/06/rung-tu-nhien-chiu-suc-ep-tu-nhu-cau-su-dung-go/

To, X. P., & Tran, H. N. (2014). *Rubber expansion and forest protection in Vietnam*. Hue, Viet Nam. Retrieved from http://www.forest-trends.org/documents/files/doc_4671.pdf

Tombs, S., & Whyte, D. (2009). The state and corporate crime. In R. Coleman, J. Sim, S. Tombs, & D. Whyte (Eds.), *State, power, crime* (pp. 103–115). London: Sage.

Tombs, S., & Whyte, D. (2015). *The corporate criminal: Why corporations must be abolished*. Abingdon, Oxon; New York, NY: Routledge.

Tran Thu. (2014). Lợi dụng xóa nhà tạm để... phá rừng? [Taking advantage of the elimination of temporary houses to ... deforest]. *Dan Viet*, 7/3/2014. Retrieved January 2, 2015, from http://danviet.vn/dien-dan-ban-doc/loi-dung-xoa-nha-tam-de-pha-rung-196903.html

Tuong Vy. (2009). 9 "quan" hầu tòa vụ phá rừng Khe Diên [9 "leaders" brought into the court in an illegal deforestation case]. *Dan Tri*, 28/07/2009. Retrieved

June 3, 2015, from http://dantri.com.vn/phap-luat/9-quan-hau-toa-vu-pha-rung-khe-dien-340122.htm

USAID. (2013). *Vietnam tropical forest and biodiversity assessment.* Retrieved from http://www.usaid.gov/sites/default/files/documents/1861/Vietnam118_119FinalDraft%2015Oct2013%20(Public%20Version).pdf

Uyen Thu. (2013a). *Vì sao hơn 130,000 ha rừng Tây Nguyên bị bức tử: Kỳ 1 "Cơn lốc" cao su "tàn sát" rừng [Why more than 130,000 ha of forests in Central Highlands are forced to suicide: Vol 1: The rubber "cyclones" massacre the forests].* Gia Lai, p. 3, released 16/9/2013: Gia Lai Newspaper.

Uyen Thu. (2013b). *Vì sao hơn 130,000 ha rừng Tây Nguyên bị bức tử: Kỳ 2: Nhiều bất cập trong quản lý rừng [Why more than 130,000 ha of forests in Central Highlands are forced to suicide: Vol 2: Many shortcomings in forest management]. Gia Lai.* Gia Lai, p. 3, released 17/9/2013: Gia Lai Newspaper.

Van Thanh. (2014b). Rừng bảo tồn thiên nhiên Ea Sô bị xẻ thịt [The reservation forest Ea So is being slaughtered] *Cong An Nhan Dan*, 19/12/2014. Retrieved December 19, 2014, from http://cand.com.vn/Ban-doc-cand/Rung-bao-ton-thien-nhien-ea-So-bi-xe-thit-335128/

Viet Bao. (2005). Những kiểm lâm "ăn lương" của lâm tặc [Kiem Lam officers get salaries from Lam Tac]. *Viet Bao*, 10/11/2005. Retrieved June 3, 2015, from http://vietbao.vn/An-ninh-Phap-luat/Nhung-kiem-lam-an-luong-cua-lam-tac/55089059/302/

VTV2. (2013). *Timber villages in Vietnam in the context of international integration. TV Programme of Green Vietnam.* Education Science Board, Vietnam Television Broadcast.

VTV. (2014). Xưởng mộc—Nơi ẩn náu và tiếp tay cho lâm tặc [The carpentry workshops that shelter and abailt illegal loggers]. *Dai Truyen Hinh Viet Nam*, 27/10/2014. Retrieved December 22, 2014, from http://vtv.vn/trong-nuoc/xuong-moc-noi-an-nau-va-tiep-tay-cho-lam-tac-20141027161722054.htm

Vu, D. T. (2014b). Khai thác rừng trồng phá luôn rừng phòng hộ [Harvesting production forests, the destroy protection forests]. *Nong Nghiep Viet Nam*, 6/11/2014. Retrieved January 4, 2015, from http://nongnghiep.vn/khai-thac-rung-trong-pha-luon-rung-phong-ho-post134185.html

Wyatt, T. (2013a). The Russian Far East's illegal timber trade: An organized crime? *Crime Law and Social Change, 1*(61), 15–35.

Wyatt, T. (2013b). *Wildlife trafficking—A deconstruction of the crime, the victims and the offenders.* Basingstoke, UK: Palgrave Macmillan.

Xuan Dung—Trong Loi. (2014). Giữ bình yên cho những cánh rừng giáp ranh [Keep the peacefulness for the adjacent forests]. *Binh Dinh Online*,

26/10/2014. Retrieved January 21, 2015, from http://baobinhdinh.com.vn/viewer.aspx?macm=5&macmp=5&mabb=30307

Xuan Hoai. (2014). Về vụ khai thác lâm *sản* trái phép *tại* Quảng Nam: Khởi tố bị can trong tháng này [The case of illegal logging in Quang Nam—Initiating the accused with criminal procedures in this month]. *Cong an THANH PHO HO CHI MINH*, 6/10/2014. Retrieved June 3, 2015, from http://trangtin.congan.com.vn/?mod=detnews&catid=1101&p=&id=526393

Xuan Huy & An Bang. (2011). Nỗi lo "hung thần" chở gỗ lậu [The concern of deadly elfs transporting illegal timber]. *Phu Yen Online*, 18/8/2011. Retrieved January 11, 2015, from http://www.baophuyen.com.vn/89/68101/noi-lo--hung-than--cho-go-lau.html

Xuan Tho. (2014). Rừng bảo tồn Mường La đang bị "rút ruột" [Muong La Reservation forest is "being gutted"]. *VOV—Dai Tieng Noi Viet Nam*, 24/11/2014. Retrieved December 23, 2014, from http://vov.vn/xa-hoi/rung-bao-ton-muong-la-dang-bi-rut-ruot-366763.vov

Yokoyama, M. (2007). Environmental pollution by corporations in Japan. In H. N. Pontell & G. Geis (Eds.), *International handbook of white-collar and corporate crime* (pp. 327–346). New York: Springer.

5

Impacts of Timber Trafficking on Human Security

This chapter reveals the victimisation from timber trafficking in the context of Vietnam. Victimisation can be seen in all seven interconnected dimensions of human security: economic, food, health, environmental, personal, community and political security. The data analysis shows that some pairs of these dimensions prove difficult to separate, thus they are grouped into four broad sections: (1) economic and food security, (2) health and environmental security, (3) personal security and (4) community and political security. In each dimension, this research looks for various insights into who, how and how much it is impacted by timber trafficking.

Economic and Food Security

Economic security requires "an assured basic income—usually from productive and remunerative work or in the last resort from some publicly financed safety net", whereas food security calls for all people at all times to "have both physical and economic access to basic food" (UNDP 1994: 25, 27). The economic and food security dimensions are briefly recapped by the Commission on Human Security (2003: 73):

N.A. Cao, *Timber Trafficking in Vietnam*, Palgrave Studies in Green Criminology, DOI 10.1007/978-3-319-64280-2_5

> When people's livelihoods are deeply compromised—when people are uncertain where the next meal will come from, when their life savings suddenly plummet in value, when their crops fail and they have no savings—human security contracts. People eat less and some starve. They pull their children out of school. They cannot afford clothing, heating or health care. Repeated crises further increase the vulnerability of people in absolute or extreme poverty.

Under this conceptual framework, it is suggested that in the context of Vietnam, timber trafficking and its inevitable consequence—the loss of forests—generates severe and diverse threats not only to the livelihoods of forest-based communities, but also to the employment of other population groups that make use of forest resources.

Impacts on the Livelihoods of Forest Dwellers

As introduced in Chap. 3, food, livelihoods and the incomes of 24–30 million Vietnamese are still either entirely or partly reliant on forests, and the dependence is more intensified among ethnic minorities living in highland areas (FSIV and FAO 2009; Nguyen et al. 2007; RECOFTC 2014; Xuan Minh 2014). In the forested areas surveyed in this research, although reliance on forest resources may be less than in the past, it seems that the forest resources, especially NTFPs, still play a major role in the livelihood of forest-dwelling communities. In some mountainous villages in Thanh Hoa province, for instance, a forest protection officer estimates that there are still two-thirds of villagers whose livelihoods are carried out inside forests (2FO01). Similarly, an environmental police officer believes that most people in the forested areas of Quang Binh province have to rely heavily on the forests because "all of their expenses ranging from schooling to buying everyday items are dependent on their livelihoods inside forests" (20EP07). As examined in Chap. 3, this forest dependency of forest-based inhabitants is explained by many reasons such as lack of cultivable land and alternative jobs, isolated geographical location, poor infrastructure and poor education (ICEM 2003; Sunderlin and Huynh 2005; Tran et al. 2010; World Bank 2009). These reasons are confirmed by the research respondents. Furthermore, some respondents also emphasise that many forest dwellers truly enjoy this tradition of "living with forests", and prefer forest-based jobs to others such as rice farming (17EP05, 29IP04, 30LR04). A local resident says that "climbing up to the forests is easier, cleaner and more enjoyable than jumping down into watery rice fields" (30LR04).

As indicated by many interviewees, the indigenous come to forests to collect timber and NTFPs used for various purposes including food stuffs such as animals, bamboo shoots, vegetation, leaves, fruits, grains, spices, honey, swallow nests and edible insects; for medicines such as medicinal plants and aromatic substances; and for other common usages in rural communities such as bamboo, rattan, resins and oils, small wild animals, birds and insects. These products are used by indigenous people for their own everyday consumption as well as to sell for money. They play a vital part in ensuring the economic and food security of the forest dwellers, particularly ethnic minorities. As asserted by a local authority in the Central Highlands: "the ethnic minorities have truly intimate links with forests, so if the forests are lost, they lose everything. This is terrifying" (33LA03).

As the occurrence of illegal logging causes forest loss, this threatens the economic and food resources of the forest dwellers. It is also evident that the more dependent on forest resources the forest-based people are, the more severe the threats to their economic and food security will be. This means that the larger scale and the less sustainable the logging is, the greater the threat is. Thus, it can be seen that different forms of timber harvesting generate varying degrees of impacts on the economic and food security of local people. Indeed, SSITH entails cutting down a small number of trees and this is done infrequently, using rudimentary methods of logging. This makes less substantial detrimental impacts on the forest resources than MSITH, and especially LSITH, which cause much greater loss of forests (Chap. 4).

It is thus relatively clear to see that the detrimental impacts of timber trafficking on livelihoods of the forest-based Vietnamese are similar to those found in other countries (Chan 2010; FAO 2007; Global Witness 2001; Pye-Smith 2006). Nevertheless, the literature tends to overlook the impacts to non-forestry livelihoods that are not necessarily delivered inside forests or by forest-based communities. These impacts will be examined in the next sub-section.

Non-forestry Employment

The first and probably most apparent non-forestry employment negatively affected by timber trafficking in Vietnam is farming, such as growing fruit, rice or corn, that is carried out in the areas close to damaged forests. Accidents from heavily logged forests such as flooding, land erosion and landslides during the rainy season and drought during the dry season regularly happen, affecting, or even sweeping away, entire crops (3EP02, 12EP04, Minh Tuan—Huy Toan 2014) (see Fig. 5.1).

Fig. 5.1 Forests have been wholly "shaved", causing landsides and land erosions, Source: Own photos taken in Bac Kan and the Central Highlands in August and September 2013

A forestry official in Son La province observes that in the early 2000s, on average, one hectare of farm land could yield more than 10 tons of sweet corn without using much fertiliser; whereas now using a large amount of fertiliser, the highland farmers only grow 5–6 tons (35FO07). The main reason for this is due to the nutrient depletion of the farmlands caused by the heavy constant water flows from the upstream forests damaged by illegal logging (35FO07). During the fieldwork conducted in the rainy season in the mountainous areas in Thanh Hoa and Bac Kan provinces, it was clear that landslides and avalanches of big rocks from harshly logged forests, which lay on top of many rice plots belonging to local people, completely damaged the plots. Similarly in Gia Lai province, when asked about the impacts of illegal logging on farming, an ethnic minority villager described:

> Illegal logging makes plants and tree stumps in the upstream forests decay. When it is raining, it brings this rubbish down to our plots of land. This has a very bad effect on our plots. Still, in our village there are some incidents where Lam Tacs bring many buffalos to carry logs. Because of the inattention of the Lam Tacs, the buffalos eat all the rice plants on large spaces of our field. This means we will harvest no rice in the next six months. (30LR04)

In addition to farming, those who work on the production of firewood, charcoal and handicrafts are also affected by forest loss. While forests in Vietnam annually provide 24.5 million tonnes of firewood for 75% of the Vietnamese population (FSIV and FAO 2009), forest degradation will reduce employment for the firewood makers and negatively affect their daily life. In villages inside Ba Be National Park, firewood collection is a common job among the local residents. On average, every day each seven-member household consumes about 30–40 kg of firewood for heating and cooking food for the family and for cattle (Tran et al. 2010).

Producing charcoal from small branches of the forest trees is also a popular occupation in some localities in Vietnam, but as a consequence of forest exhaustion, recently, this livelihood has been negatively affected (5LR01). What is more, the handicraft businesses whose materials are

NTFPs also play an important role for rural people in Vietnam. Currently, over 400 villages with nearly 100,000 villagers in Vietnam engage in handicraft production that creates total earnings of £25 million for these communities (FSIV and FAO 2009). Forest loss, which decreases the supply and pushes up the price of these materials, will affect the business as well.

Along with farming, firewood, charcoal and handicraft employment, the jobs in legitimate timber-related companies may also be threatened by timber trafficking. Indeed, once timber trafficking has reached an uncontrollable scale, in order to protect the remaining forests, the Vietnamese government at both central and local levels have to establish exceedingly strict policies including closing forest gates, logging bans and imposing tremendously complex procedures on timber harvesting and trading. Such policies, as addressed by a forest company director, tend to be extreme and sometimes counterproductive, generating severe impacts on the normal operation of legitimate companies in the forestry sector and wood industry. This negatively affects the economic security of company workers (28TT02):

> I think harvesting timber in a sustainable way would not generate significant harm to the forests, but yield substantial sources of jobs and salary for a large number of employees in forest companies like us. The extreme logging ban, of course, removes, or at least extensively cuts, such sources. This for sure badly influences the income and jobs of my employees as well as myself. (28FF02)

Indirect Economic Impacts

In addition to employment opportunities, the reduction of timber capacity and other forest resources due to timber trafficking generates indirect economic effects. Perhaps, the most visible impact is on the material supply for house building. As mentioned in Chap. 3, every year, the mountain households in Vietnam need a total of 800,000 m^3 of timber for building their houses. A senior NGO interviewee observes that three decades ago, almost all indigenous people living near forests in Vietnam found it easy to get timber for building their houses but now, in many areas, these forests are no longer rich enough to supply their needs (39NG05). With no hard timber available, the local people have to use culms and bamboo trees or other material from less durable trees to build their houses (6FO02, 39NG05).

What is more, illegal logging of endangered timber species often takes place in national parks and natural reserves, some of which are famous tourist attractions. These locations often provide the local people with employment related to tourist goods and services. Since illegal logging damages the attractiveness of the natural landscape, it will lead to negative impacts on these jobs. Ba Be National Park, for instance, creates jobs in the guest house, boat sightseeing and souvenir shops for a number of local people (13LA01). However, in addition to the forest damage caused by illegal logging, as a result of forestland erosion from the damaged forest areas, every year the Ba Be Lake, the heart of the Ba Be National Park, shrinks by several meters (12EP04). This has pernicious effects on the natural environment and the beauty of the park, which will potentially attract less visitors to the park, and so provide less employment opportunities for the local people (12EP04) (see Fig. 5.2).

As discussed in Chap. 1, the literature on timber trafficking also documents the worrying loss of government revenue. There are considerably varying estimates on the annual lost revenue, ranging from several

Fig. 5.2 Local people in Ba Be National Park sell fish and run boat sightseeing trips for visitors; and the Ba Be Lake has shrunk as a result of land erosion in the logged forests, Source: Own photos taken in August 2013

million pounds in Cameroon (SGS Trade Assurance Services 2002), tens of millions in Mozambique (EIA 2013), hundreds of millions in the Bolivian, Brazilian and Peruvian Amazon (Gutierrez-Velez and MacDicken 2008) to over a few billion in Indonesia (Chan 2010; UNODC 2010). This loss should interest scholars of human security since reduced government revenue means reduced government spending on essential social services such as law enforcement, health, education and sustainable development, which has an important security implication[1] (Wyatt 2013).

In the context of Vietnam, as estimated in Chap. 4, the annual value of timber trafficking is at least £130 million. According to the Act of Added Value Tax 2008, specified in Decree 219/2013/BTC by the Ministry of Finance, the tax rate imposed on the timber trade is 10%. This means that every year timber trafficking deprives the Vietnamese government of at least £13 million in taxation revenue. In fact, since many types of timber illegally traded in Vietnam are exceedingly valuable, the revenue obtained from authorised activities would be significant if the trade was legal with the timber legally sold. In the criminal case of trafficking Sua timber in Phong Nha Ke Bang National Park in 2012, for example, the investigation conclusion by Quang Binh Public Security (2013) indicates that the act of illegally harvesting three Sua old trees by 12 offenders in this case had caused a total monetary loss to the state of over VND70.9 billion (£2.26 million). It can be seen that the illegal harvesting of only three trees in one single criminal case already costs the state millions of pounds, indicating that the monetary loss caused by timber trafficking in reality would be enormous.

Nevertheless, the tax loss is only a part of the financial cost, the Vietnamese government perhaps loses more when it comes to the immense spending on all governmental agencies entrusted in combating timber trafficking and fixing its consequences. Interestingly, in the context of Vietnam, a police officer notes that spending on the policing of timber trafficking may be higher than the cost of other offences (32IP05). Indeed, to cope with timber smuggling, a larger number of law enforcement agencies have to take part in the battle mainly Kiem Lam, environmental police, investigative police, local authorities, market inspectors, customs and the border army, which collectively is very costly for the state budget (32IP05). The multiple costs to the state budget limit national expenditure on improving social welfare for the Vietnamese.

This research has so far examined the impacts of timber trafficking on material aspects of human security consisting of the economic and food security of different groups in Vietnamese society. The findings appear to correspond with the overall observation worldwide that environmental problems, such as deforestation and biodiversity loss, are a source of security problems for some individuals and social groups because they reduce the access to, and the quality of, natural resources that are important to sustain them (Barnett and Adger 2010). Likewise, climate change, which is also widely seen as a major security threat, harmfully impacts crop productivity, food availability and household and individual incomes, which may "potentially interrupt progress toward a world without hunger" (Wheeler and von Braun 2013: 508).

However, as Thomas (2001: 161) rightly agues "material sufficiency is a necessary, but not sufficient, condition of human security that entails more than physical survival". In the field of green victimisation, Hall (2013a: 34) also argues that "there is no clear line between economic and non-economic impacts. Consequently, to focus exclusively on the monetary calculation of harm is likely to underestimate the true extent of environmental victimisation". That said, there are other major non-material consequences of green crime and timber trafficking. The next section will look into impacts of the crime on health and environmental security.

Health and Environmental Security

Good health is both essential and instrumental to achieve human security. It is essential because at the very heart of security is protecting human lives. Health security is, therefore, at the vital core of human security, and illness, disability and avoidable death are "critical pervasive threats" to human security (Commission on Human Security 2003: 96). This research evidence shows that illicit timber traffickers, especially harvesters and transporters, in Vietnam have to work in dangerous conditions and face numerous physical health hazards.

Physical Health

It is true that logging is a physically onerous and risky job, particularly so when it comes to illegal logging that often takes place in remote jungles at night without personal protective equipment. Although there are no official statistics for Vietnam on the number of loggers and carriers killed or severely wounded as a result of accidents occurring from logging and carrying heavy logs, the research participants show their concern about the potential and reality of unfortunate outcomes from the clandestine act of logging and transporting. Concern is well demonstrated by various, albeit somewhat anecdotal, evidence. A forest resident in the Central Highlands who used to partake in illicit logging describes:

> It was extremely onerous work. Before cutting, we always had to think very carefully about the forest terrain, guess the direction where the trees would fall. Otherwise, the trees would easily hit us, and then we could drop off our corpses in the middle of jungles. (25LR03)

In "the logging commune" of Son Hong in Quang Tri province, where the majority of the commune's population is involved in illegal logging, the "profession" gives the community some income, but it also causes really lamentable consequences for its residents' health (Phan 2011). Over the past few years at least three commune loggers have been killed by being crushed by large trees, while the number of wounded loggers was "uncountable" (Phan 2011). In Ba Be National Park in 2012,a local logger illegally cut a big Nghien tree at night. When the tree fell down, it hit a big rock. The rock then suddenly rolled and fell down, bashing and killing one of the loggers (15FO04).

Additionally, there are many cases where loggers and even law enforcement officers are bitten by poisonous snakes, contract malaria and severe illnesses, and encounter natural disasters due to the extreme working conditions when loggers and officers stay inside the forest for days, or even months, with a poor diet and in unhygienic conditions (15FO04, 26EP09; Quang Nhat 2013). Statistics by the FPD show that over the past 40 years, hundreds of Kiem Lam officers have died as a consequence of malignant malaria and encounters with bombs and mines in forests (FPD 2013).

Impacts on the health security of traffickers are not confined to inside forests, but also take place outside forests during illegal timber smuggling. Again, there is no systematic figure for accidents due to timber transportation but anecdotal evidence indicates that such accidents are not rare. An interviewee witnessed a timber truck overturning and killing the driver when it passed his village in 2012 (30LR04). As shown in Chap. 4, in many parts of Vietnam, the motorbikes used to carry heavy volumes of illegal timber by reckless drivers, driving over 60km/hour at night on notoriously dangerous roads, cause many deaths and serious injuries (Hoai Nam 2014; Xuan Huy and An Bang 2011; Thanh Nguyen 2014b).

The most tragic accident was the flipping over of a timber truck in Quy Hop district, Nghe An province at 3.30 a.m. in December 2011. In this case, as reported in a number of media articles (Nguyen Duy 2013; Xuan Hoa 2011; Khanh Hoan 2013), the illegal timber in the truck was owned by the chief of Kiem Lam Agency in Pu Huong Nature Reserve who employed 14 local porters and asked three subordinate Kiem Lam officers to escort the truck. During transport, the porters sat on top of the cargo space which was fully loaded with large timber planks while the Kiem Lam officers were in the cab of the truck. The truck suddenly overturned when attempting to climb a slope. The accident instantly killed 10 porters and critically wounded the rest, while two Kiem Lam officers suffered minor injuries.

At the trial, eight offenders in this case including four Kiem Lam officers received a total of 206 months' imprisonment. Sadly, these local porters were poor young villagers; most of them were relatives of each other, and all of them were the main breadwinners in their families. This means that such a tragic accident not only caused the deaths of the unlucky porters but also impacted the emotional well-being and economic security of their family members. Additionally, before this accident, exactly in this road section, named by the local people as "cung đường gỗ lậu" (the curve of illegal timber route), 14 other timber porters were killed and dozens were severely wounded (Xuan Hoa 2011).

Clearly, in the context of Vietnam the lives and physical health of the people who get involved in timber trafficking is severely threatened by their involvement. This situation is similar to that found in other countries where the illicit loggers face extremely harsh and risky working conditions (Casson and Obidzinski 2007; Chan 2010; OECD 2012). Most workers in illegal sawmills in Kotawaringin Timur, Indonesia, for example, can only endure the work for six months up to a maximum of one year because of its tremendously strenuous and dangerous nature with many losing hands due to the use of band saws (Casson and Obidzinski 2007). On several occasions, the illicit harvesters are shot by police agencies (OECD 2012).

However, in tandem with the offender's health security, the safety of those who are not a part of the illicit business is also negatively impacted by the crime. The next sub-section will examine other impacts of timber trafficking that fall between health security and environmental security.

Suffering from Natural Disasters

According to statistics from the Ministry of Natural Resources and the Environment, in the past 10 years, over 9500 Vietnamese were killed or went missing because of natural disasters such as hurricanes, floods and landslides (Phuong Anh 2014). While thousands of hectares of forests in Vietnam have been destroyed every year as a result of illegal logging (Chap. 4), some interviewees (12EP04, 26EP09, 35FO07) blame the forest loss as being one of the key contributors to natural disasters, particularly flooding in Vietnam. They also believe that this kind of impact is the harshest consequence of timber trafficking in Vietnam, emphasising the seriousness of flooding via a Vietnamese proverb: "nhất thủy nhì hỏa" (first flooding, second fire), which means that in Vietnam flooding is the most severe type of natural disaster. Notably, there is a correlation in Vietnam that over the past decade, some of the locations in which timber harvesting has been the most severe tend also to be the locations of the harshest natural disasters, and vice versa.

The continuing floods, landslides and rock avalanches happening in Van Chan district, Yen Bai province—a hotspot of timber trafficking in the North—is a prominent example. Among the incidents is one in 2005 that resulted in 50 people being killed or reported missing, some 200 houses washed away, tens of bridges collapsed, and hundreds of hectares of rice fields were completely destroyed, becoming uncultivable for years (Tran

Thuong 2009). One of the key conditions for this disaster, as shown by scientists, is severe harvesting of the large forests that contain valuable Pơmu timber. This is because the forests are located in upstream areas in the district and serve the function of curbing water speed (Tran Thuong 2009).

Ha Tinh, another major location of timber trafficking in the Centre, has frequently experienced a phenomenon of "flooding escalates flooding" (lũ chồng lên lũ). Often, while this province is still heavily suffering from a first flooding, subsequent flooding disasters arrive. Again, the scientists argue that one of the explanations for such flooding is the destruction of protected forests in this province located in the watershed areas (Phan 2011). Conversely, in some localities in Bac Kan province where the forests have been substantially enriched in the past five years, the problem of landslides and flash floods have been notably reduced (12EP04).

Thus, suffering caused by natural disasters as a consequence of timber trafficking in Vietnam is vivid and extensive. This does not seem to be much different from other forested parts of the world. As mentioned in Chap. 1, globally over 200 million people have been affected with 1.5 million killed annually by natural disasters in the past two decades, in which 90% of the people exposed are from the developing world (UNEP 2007). Timber trafficking is one of the major causes of such disasters, particularly forest fires, flash flooding, landslides and soil erosion.[2] There are further substantial health and environmental impacts that are observed in Vietnam, but not examined much in the literature on timber trafficking. These impacts are related to the water supply for forest-based dwellers.

Water Supply is Polluted and Altered

Because it is clear that "without water, survival, human or otherwise, is impossible", water is one of the "special issues" in human security (Commission on Human Security 2003: 15), and a severe shortage of drinking water and sanitation is a "major threat" to human security (Nsiah-Gyabaah 2010: 246). Every year, approximately 3.4 million people worldwide die due to water-related diseases (DFID 2002). It is noted in this research that illegal timber harvesting directly impacts the water supplies of many forest inhabitants. Indeed, in some woodland parts of

Vietnam, as emphasised by a number of the research participants, the impact of illegal logging that is most apparent to aboriginal forest inhabitants is probably the pollution and alteration of water supplies both for drinking and cultivation purposes (3EP02, 6FO02, 24LA02, 41IP06). These respondents provide a straightforward explanation that the forests are often located in the watershed and upstream areas that provide the primary, if not the only, source of domestic and irrigation water for the residents living in downstream areas. In the dry season, since forests are lost, their function of storing water is reduced; many downstream communities suffer from a shortage of water. In the rainy season, heavy rains in barren forests bring mud and dirty material into the water sources of the communities, thereby heavily polluting the drinking water and risking the health of those downstream.

This impact on water sources is intensified in some communities whose residents raise livestock such as buffaloes, cows and goats whose faeces contaminate the water flow (3EP02) or in upstream areas where pesticides and herbicides are used for plantations in the upland forests (24LA02). A commune civil servant describes an incident that is currently occurring in his commune:

> After carrying out the shady act of clear felling of 20 hectare of rich forest, the district SFE uses a large amount of herbicides to clear the remaining plants in this forest. The commune people collected over 80 bottles of herbicide there. The problem is that the logged forest is the watershed one. It is very steep (up to 50 degrees), very close to the streams, and just 10 km away from the local community in the lower land. It causes serious consequences for the water source of the community. (24LA02)

Alternatively, in some cases the water supply is severely altered when the illegal logging is conducted via destructive means such as the use of explosives and quarries to either harvest stumps of high-value timber or construct roads to carry logs out of forests, which profoundly alters the natural flow of water. This practice is seen in Quang Binh (Huong Giang 2014) and Lang Son provinces (VOV 2014) where explosives are used to harvest stumps of Sua timber. This evidence would support the observation by Stewart (2014: 239) that to avoid detection, illegal logging virtually ignores best-practice logging techniques to reduce environmental impacts, but rather employs "the most destructive forms of wood extractions". The environmentally destructive behaviour of illegal logging links to its impact on the ecosystems of harvested forests, which will be examined in the next sub-section.

Damaged Forest Ecosystems

Illegal timber harvesting has numerous salient environmental impacts, most obviously being the loss of forests. As reported by the Environmental Police Department (2012), from 2005 to 2010 in Vietnam, around 5400 hectares of forests have annually been lost as a consequence of illegal logging. Also, it is highly likely that there is a significant area of forests cleared by LSITH that has not been captured by the official statistics. For example, in the past years over 45,000 hectares of forest every year have been cleared for projects of forestland conversion (FPD 2014). As indicated in Chap. 4, a large number of the projects may not fully comply with the relevant legal provisions, meaning that each year at least another several thousand hectares of forests have been illegally cleared, but not addressed by law enforcement agencies. A police officer suspects another way that illegal logging may cause the forest loss:

> Although such cases have not been officially detected yet, it is possible that after illegal logging, the harvesters intentionally burn the entire zones of logged forest, so that nobody can discover their logging. If this is the case, everything in the forests is completely destroyed. (17EP05)[3]

Severe forest degradation was clearly observed during fieldwork in the different regions of Vietnam. These regions were supposed to have very rich forests, but it was sadly observed that apart from some preservation zones and national parks, the current state of the forests is mostly poor. In the Central Highlands, for instance, it is a long-lasting image among the Vietnamese that the region is abundant in rich forests. This is, however, no longer true. There are different drivers of deforestation in Vietnam,[4] but illegal logging (defined in this research as comprising

SSITH, MSITH and LSITH) should be considered as one of the most important. There is a common expression gloomily echoed by many interviewees in this research: "thanks to illegal logging, the work of destroying forests has basically been achieved in Vietnam". This sentence sums up well the destructive outcome of illegal logging on Vietnamese forests in the past decades.

In addition to forest loss, timber trafficking also poses a severe threat to forest biodiversity. Forests are known throughout the world as the largest terrestrial ecosystem and the biggest reservoir of plants, animals and biodiversity on land (FAO 2007). Within tropical forests alone, although they may only cover 3% of the Earth's surface, they house over half of all known plant and animal species on the planet (Boekhout van Solinge 2008). Being home to around 10% of the world's species, Vietnam is recognised as the 16th richest biodiverse countries and one of the richest agro-biodiverse countries in the world (USAid 2013). While biodiversity degradation, particularly in endemic and endangered species, has remained a critical issue, illegal logging has been one of the proximate causes of the problem (USAid 2013; World Bank 2005).[5]

This impact is largely confirmed by many interviewees in this research. A senior officer observes that in the 1980s, Ben En National Park in Thanh Hoa province had as many deer as cows. "Every afternoon large herds of deer went to the forest outskirts eating and drinking and herds of elephants came to the town center. But it is now hard to see them even inside the park" (3EP02). In Ba Be National Park in Bac Kan province, over the past 10–15 years while the wealth of 300-year-old Nghien timber has been substantially reduced, some other tree species such as De, Doi and Dinh no longer exist in the park (8EP03). In addition, Sua timber, ebony timber and valuable orchids are now hardly found in Phong Nha Ke Bang National Park in Quang Binh province (23FO05). The worldwide population of Thuy tung trees (*Glyptostrobus pensilis*) is 162 individuals. All of them now grow only in Dak Lak province of Vietnam, and are on the critical brink of extinction (Van Thanh 2014c).

All of the cited sources point out that illegal logging noticeably intensifies the problem of biodiversity degradation. An environmental police officer who has extensive understanding of biodiversity spells out one of possible outcomes:

> There are animals such as buffalos, cows, dogs, pigs and chickens that can stay with human beings, receive good living conditions and may become better. But many wild animals cannot reside, and become worse, when humans encroach on their lives. With the encroachment of loggers, tigers and bears, for instance, cannot breed. The large human invasions into the forests for illicit logging, in combination with the destructive methods of cutting and carrying the logs, very much disturb and damage the ideal natural environment of animals, forcing them to run away. (3EP02)

A further observation emerged in this research: the impacts of timber trafficking on the forest ecosystems are of particular concern in the context of Vietnam. Indeed, driven by the traditional demand for endangered timber (Chap. 3), much of the timber trafficking in Vietnam particularly targets rare, precious and endangered timber such as Sua, Trac, Nghien, Lim and Thuy Tung, which takes hundreds of years to grow to their normal size for logging (3EP02, 8EP03, 12EP04). Official statistics indicate that the majority of 5400 hectares of forest annually destroyed by illegal logging are old-growth forests and watershed protected forests located in national parks and natural reservation zones (Environmental Police Department 2012). These forests, whilst having tremendously high ecological value, account for only a minimal share of the total forest area in Vietnam (USAid 2013). The old-growth forests, for example, equate to only 8% of the total forest area in the country (MNRE 2010). Furthermore, an overall feature of Vietnamese forests is their heterogeneity, which means most of the forests in Vietnam often have many different layers of fauna and flora containing various species (17EP05). That said, once a big tree falls down, lots of other small trees, vegetation zones and animals are adversely affected; so too when big trees are pulled out of the forests (17EP05). This damage is even more severe in the Northern forests of Vietnam that have very steep terrains (3EP02).

To recapitulate: timber trafficking harmfully affect the health and environmental security of various types of victim, and there are four main groups of affected victims, namely (1) illegal harvesters and transporters killed and wounded by logging and transportation accidents, (2) innocent people in logging hotspots traumatised by natural disasters, (3) indigenous forest-based inhabitants victimised by altered and polluted water supplies and (4) the forest ecosystems severely damaged by destructive logging activities and disturbed by the intensive intrusion of illegal loggers. There are other ways in which timber trafficking severely affects human health security, which will be looked at in the next section on the impacts on personal security.

Personal Security

Personal security is chiefly concerned with physical safety from, and psychological wellbeing against, violence particularly from violent crime (Commission on Human Security 2003; UNDP 1994). In the context of timber trafficking in Vietnam, as briefly indicated in Chap. 4, the violence actually involves the crime, in which professional traders employ violent actors to support their illicit business. Under the framework of human security, this section will investigate in more detail the mechanism of how the use of violence affects the personal security of a variety of stakeholders including offenders themselves, entrusted officers, journalists, whistle-blowers, and ordinary people.

Personal Security of Offenders

Violence against other offenders is often associated with the idea of "outsider" or "intruders". If the traffickers, loggers, traders or smugglers know each other well, violence is unlikely. In SSITH, for example, violence is not common because virtually all the indigenous loggers know each other well and they respect each other's subsistence need for timber. However, the arrival of new traffickers to the "territory" or "manor" of others is likely to cause violent conflicts. Violence in MSITH seems to be far more likely because it involves both local actors and alien migrants[6] (Chap. 4).

Likewise, in the trading step, each forest area is considered as a "manor" that is normally controlled by a timber trader who buys all the timber illegally sourced in the manor. In the case of another trader breaking the hidden rule, coming to the village to buy the timber and usually offering higher prices, a violent clash between the traders is almost unavoidable. As a local resident describes:

> The situation of violence, on the one hand, is more intensified if the "traditional trader" has operated in this area for a long time especially when he has already invested some money for the local loggers. In this case, the long-standing trader can bring his violent squad to teach the new trader's team a costly lesson. The violent acts can take place right at the location of timber loading or along transporting streets. After that, he often also informs the police about the illegal acts of the new trader. However, on the other hand, the situation of violence would be soothed if the new trader discusses with, and gives something to, the "traditional trader" in advance. (5LR01)

Concerning conflicts during illicit buying and selling, the research participants basically agree that the illicit trade in Vietnam is largely based on the principle of "thuận mua, vừa bán" (happy buying, satisfied selling). However, occasionally, traders have minor clashes when the price of timber is still valued based on market rules, but the traders coerce local loggers not to sell the timber to other traders, but only to them (20IP07). The worse scenario, though it is rare, is that a trader may coerce local loggers into selling the illegal timber for a lower price than it is actually worth or the local loggers may be harassed by violent traffickers (33LA03). Thus, it can be seen that violent threats by professional traffickers to local loggers do not seem, in general, to be tremendously harsh.

Nonetheless, on several occasions, the personal security of local loggers, especially those who harvest highly valuable timber, is severely threatened by brutal timber robbers. The most shocking incident that shows the danger that timber harvesters face in Vietnam is the case of the murder of five people looking for Agarwood timber in Quang Tri province in March 2013. In this case, a group of three people, equipped with an AK rifle intended to kidnap people who came to the forest to seek Agarwood—a highly valuable timber. After failing to kidnap three timber hunters on 22 March inside a forest location close to the border between Laos and Vietnam, on the next day, the group abducted seven other timber loggers and demanded a ransom of £500 for each hunter.

One of the hunters was released to return to his village and collect the ransom. However, at midnight on that day, when the released hunter did not come back, five abducted loggers were killed one by one. One of the hostages successfully escaped while the massacre was taking place. The slaughterers tied the victims' hands, asked them to kneel down, battered their heads with the AK butt until they died, and finally buried all of them in a hole (Huu Thanh 2014b; Thanh Nien 2013).[7] This is not an isolated case. In Minh Tien village in Quang Binh province where half of the village population are non-professional Agarwood hunters, in the first three months of 2013 alone, these timber hunters experienced three cases of being robbed by armed groups (Quang Nhat 2013).

This research's findings on the fatal impacts on health security and personal security of those who partake in timber trafficking provide firm evidence for the observation by Hall (2011, 2013a, b, 2014, b), Hall and Farrall (2013), and Pemberton (2014) that there is a substantial overlap between victims and the offenders of green crime when some of those most harshly affected by green crime might actually be green offenders themselves; and that becoming a green offender increases the risk of being a green victim.

Personal Security of Officers

It is predictable that the personal security of offenders is threatened as a consequence of their own criminal involvement. However, the personal security of officers who combat the crime is also violently threatened. Given this topic, the overall evaluation by the majority of the research respondents is that in most provinces in Vietnam, instances of severe violence and intimidation towards officers have recently been decreasing. That is to say, during the detection of, and arrest for, a violation, as largely agreed among the interviewees, the most common acts jeopardising personal security of on-duty officers are the ones of slander, humiliation and intimidation of the officers, sometimes in combination with using stones, knifes or cudgels.

Serious incidents of using firearms and/or grenades to attack officers are now not very common. If guns are used, they are often homemade and rudimentary (17EP05). A commonly used intimidation tactic used against an on-duty official is saying something like "be careful because we know where you are living and where your children are studying" (6FO02). This is more common when offenders outnumber the officers; though when more law enforcers come to the scene, the offenders stop the intimidation (2FO01). Similarly, when the offenders are drunk, they become more aggressive; but afterwards they obey the statutory procedures (27FO06).

Officers are sometimes wounded when they use motorbikes to chase vans carrying illegal timber, with the drivers zig-zagging and then eventually offloading the logs onto the road once the officers get too close (2FO01, 11IP02, Xuan Huy and An Bang 2011).[8] The violent threats continue after detection. Subsequently, some violators keep intimidating officers by coming to Kiem Lam stations to disturb them and send offensive mobile phone messages to officers to menace them and their families (6FO02). However, in most cases, these acts are confined to spoken intimidation, rather than physical acts.

While most instances of violence towards officers are at the lower end of the scale, this does not necessarily mean that there are no extremely serious cases of violence against the officers. Statistics by the FPD show that from 2008 to 2012, in all of Vietnam there were 274 cases of attacking on-duty officers (about 55 cases every year), including some serious cases that killed four and wounded 182 officials. The FPD is concerned that in recent years, the modus operandi of aggressiveness towards on-duty officials in a number of provinces of Vietnam still remains severe, exacerbated by some serious cases conducted in a well-planned manner (FPD 2013).

There are plenty of vivid examples to demonstrate the problem. The situation of timber trafficking in the Dak Ang Protection Forest in Kon Tum province is still happening quite often, but the in situ forest rangers are not active in fighting the traffickers who have carried out many assaults against the forest staff (Hoang 2012). The associate head of the forest Management Boards worries:

> Lam Tacs are harvesting almost every evening but it is difficult to arrest them. If we have four or five members, we do not dare to come to the forest. Sometimes, we detect them, then they chase back, attack and even kidnap us. (Hoang 2012)

Similarly, in the case of Khe Giua SFE in Quang Binh province, the enterprise director complains:

> If we arrest Lam Tacs, they will definitely take revenge. There was a logger who cut off the ear of a forest ranger, but he has not been prosecuted yet. We sometimes see him roaming around the forests, continuing to illegally log. Forest rangers know he is logging, but simply report to leaders and do not dare to get close to him. (Hoang Nam 2011)

Analysing different sources of research data suggests that the severe violent threat to the personal security of officers may be explained by the coexistence of four main conditions: work location, inherently violent timber traffickers, lack of staffing and inappropriate methods of handling the potentially violent incidents. To begin with, forestry staff often work in locations that are either remote, where traffickers can easily escape after beating up the officials, or they are very close to the traffickers' communities where they know the forest terrain well and get support from their families and villagers to intimidate the staff. Second, professional timber traffickers are typically rash and aggressive; they feel free to take severe revenge against other offenders as well as officers who "destroy" their careers (5LR01, 20EP07; Nguyen Cau 2014; Viet Quoc 2015). An interviewee who used to take part in the unlawful industry asserts:

> Generally speaking, the circle of specialist Lam Tacs is basically reckless and fierce. They are always ready to use a variety of means of violence if they think it is necessary to maintain their business. There is a common saying among these smugglers that "if I die [i.e. if I'm arrested], I would not leave you [officers] living normally". (5LR01)

Third, there is a lack of staff and equipment to sufficiently protect the personal security of the forestry forces. As will be seen in Chap. 6, by and large, the forestry forces need many more staff and operational tools to deliver effective enforcement, and the dominance of the offenders over the officers in terms of manpower and equipment is one of the main reasons why traffickers are able to threaten the forces. A Kiem Lam officer complains in a pitiful tone "the operational equipment and personnel of Kiem Lam are still very insufficient, obviously far less than those of Lam

Tac" (23FO05). At times a forest protection team has three or four staff equipped with only machetes while a specialist gang of timber harvesters and porters may have 10–20 members who are potentially violent, carry machetes, axes, cudgels and even homemade guns and cans of pepper spray, which help the gang go unchallenged (21IP03, 23FO05; Minh Thi—Minh Trieu 2014; Nong nghiep Vietnam 2011; Viet Quoc 2015). Even when Kiem Lam officers are equipped with guns, quite often they hesitate to use them because of the exceedingly complex requirements when reporting shooting events (6FO02, 9FO03, 15FO04). A Kiem Lam officer states "because of the intricate proceedings of gun administration in our force, guns are carried mostly for showing off. We have to think very carefully before firing any single bullet" (9FO03).

A further contributor to the violent responses of timber smugglers towards anti-trafficking officers lies in the methods used by officers in dealing with criminal incidents. There are, in this sense, quite a few cases of officer misconduct or at least unprofessionalism that prompt the vehemence of the offenders. First is the situation of "unclean officers". An interviewee suspects that, periodically, a Kiem Lam officer is attacked not because he is always upright, but rather he has already received bribes and has either tried to deceive the traffickers or made no effort to protect them (5LR01, 25LR03). An interviewee describes:

> Sometimes, corrupt Kiem Lam officers treat bribing Lam Tacs very badly and they should take the blame for this. For example, they bargain with Lam Tacs over the amount of the bribe in an unashamed manner. In other cases, after receiving a bribe from a transporter, the corrupt officer may inform other staff, who are working inside forests, about the van with illegal timber. These officers, subsequently, can get outside the forest, stop the van, ask for a bribe and come back to continue their work inside the forest. Such blatant behaviour often motivates hatred and attacks against officers. (5LR01)

This situation suggests that while, on the one hand, the increasing collaboration between traffickers and officers bound by bribery generally helps lessen the occurrence of violence (Chap. 4), corruption, on the other hand, is at times also a trigger for violent reactions against the officers' personal security.

Another scenario is the domineering style of officers in handling potentially violent events. The head of the division of environmental police in Quang Binh province observes that:

> In some cases, officers handle complex violations in an unprofessional and clumsy manner, showing a lack of communication skills when explaining wrongdoings to the offenders. More serious may be the bossy and authoritarian acts coupled with the priority of their own personal gains while handling the cases, which prompts anger among the offenders and then triggers violence against the officers. (Duong 2012: 106)

This authoritarian manner is possibly one of the main explanations for an interesting observation by a police officer:

> Both the Central Highlands and the Centre are hotspots of timber trafficking in Vietnam, but severe attacks on officers seem to be less common and serious in the Central Highlands. I think the underlying reason comes partly from the methods of the officers used in handling incidents. The operational methods of staff in the Central Highlands seem to be generally more cordial and accommodating than ones in the Centre. (32IP05)

This observation is supported by the information that in two Central Highlands provinces of Kon Tum and Gia Lai, for example, although the provinces are hotspots of timber trafficking, in the past years there were no serious cases of attacking officers resulting in death or serious wounds (26EP09, 29IP04, 32IP05, 40EP11). Conversely, in some localities in the Centre, severely violent reactions against forestry forces are very common and alarming (20EP07, 21IP03, 23FO05, 24LA02). Indeed, the Lam Tac attacks against on-duty staff in the Centre province of Quang Binh is frequently reported in the Vietnamese media (Hoang Nam 2011; Nong nghiep Vietnam 2011). An interviewee stresses sadly "forest rangers working for Khe Giua SFE in Le Thuy district are *regularly* beaten by illegal loggers" (21IP03). In Khe Giua SFE alone, in the first half of 2011, there were four cases where Lam Tacs attacked forest protection officers and forest rangers.

Notable was a 2009 case when a group of timber traffickers in the Le Thuy district fractured the legs of a forest ranger and then used a machete to cut off his ear (Hoang Nam 2011). The worrying use of violence is also confirmed by the Quang Binh environmental police:

> Over the last years, the situation of assaulting on-duty officers has been taking place in a complex and formidable fashion. Perpetrators are increasingly heartless, forceful and reckless, using serious weapons such as firearms and daggers to blatantly attack and kidnap officers, gathering a large number of people to vandalise official cars and stations. (Duong 2012: 105)

Thus, it can be seen that there are various factors that contribute to the violent threats to personal security of law enforcers. The factors can be external (e.g. isolated work locations and inherently violent offenders) and internal (lack of staff and inappropriate handling methods). The internal factors are probably more significant in explaining the degree of aggression varying from region to region as well as from force to force. To be more specific, it is plausible that because of the lower level of manpower capacity and professionalism, the safety of Kiem Lam officers and forest rangers tends to be more heavily affected than police officers. The same applies to the different regions where the fierce reactions against the Centre staff tend to be more common and severe than the Central Highlands staff.

The threats to the personal security of timber traffickers and anti-trafficking officers have been explored in this sub-section. The research evidence also shows that timber trafficking threatens the personal security of ordinary people who are neither criminals nor members of law enforcement agencies. In the next sub-section, the other cohorts of victims of violence are investigated, including forest-based residents and those who serve as whistle-blowers.

Personal Security of Other People

In regard to the personal security of forest-based dwellers, a particular concern is that professional traffickers may violently force indigenous people to partake in the illicit industry, and then ove-rexploit their labour. This happens, for example, in Brazil where in cattle ranching and sugarcane plantations that serve as part of the frontline of illegal deforestation,

25,000 slave labourers are kept in debt bondage and prevented from escaping by armed guards, irrespective of government efforts to free thousands every year (Boekhout van Solinge 2010). Likewise, 33,000 loggers are under forced labour conditions in the Peruvian Amazon (EIA 2007). However, there is common agreement among the vast majority of interviewees that there is no one who is really forced to take part in trafficking activities in Vietnam. The illegal timber business predominantly relies on the principle of consent; which means all actors have the freedom to make their own decisions on whether or not to partake in smuggling activities.

Although logging and timber transporting are commonly known to be onerous, risky and unfairly paid jobs, getting involved or not is the local people's choice. A timber trader confidently assesses: "when joining the logging, the local loggers are paid much better than in other normal jobs; as such there is neither coerciveness nor overexploitation at all" (28TT02). This opinion is agreed with by a commune authority officer:

> Because the residents get good earnings, they participate in the work. There is no force. Among all the occupations available for the commune people now, Nghien logging is the most profitable. Cutting only one Nghien board can yield up to VND180,000–200,000 [£6–7], which is enough for their living. (13LA01)

A common explanation is that using coercive force on the local people is unwise. Timber traffickers are now aware that these people are the most capable and cheapest source of labour, so instead of threatening they should "flatter" them by paying in advance, aiding them in harsh times or paying higher wages than other jobs so that the local people remain interested in their illicit businesses (17EP05, 29IP04, 25LR03, 36LR06, Van Thanh 2014d). A resident in the Central Highlands who used to illegally log says:

> To be honest with you, I used to do illegal logging for over ten years, but there was no way anyone could force us to log. If traders wanted to buy timber from us, after getting the deal, they had to pay us in advance; then we logged for them. Otherwise, we frankly told them to go away. (25LR03)

As to whether Lam Tacs intimidate and abuse the local people, a police officer quickly asserts:

> No, no, never! But rather, in order to commit their crime, Lam Tacs have to rely on, and have to be supported by, the local people. Simply because they need the villagers to log, carry and hide the timber and they want to avoid snitching from the villagers. If the local people don't like Lam Tacs, they can even stop the timber trucks, and then Lam Tacs will be powerless, particularly when the trucks damage their village roads. Moreover the traffickers know that the poor ethnic minorities often receive privileges from the ethnic policy of the Party and the State. These privileges can be abused for the illicit activities by, for example, asking the ethnic minorities to store, hide or to be the owner of the illegal timber. (29IP04)

Nevertheless, in contrast to the evidence that local villagers are not coerced, a villager shows his anger with Lam Tacs. He says in a discontented tone:

> Lam Tacs cause much damage to our village. They ruin the roads, cause traffic accidents and damage the farms. We ask them to pay for these damages but they don't care. Because Lam Tacs already pay for Kiem Lam and forestry companies, they don't care about anyone and nobody can stop them. We talk to them through but it goes in one ear and out the other. They sometimes spook us and bully us. They warn us not to tell anyone about their logging. (31LR05)

This statement shows that, on some occasions, Lam Tacs can intimidate the personal security of the villagers who disclose their illegitimate acts. It is believed that this action is not groundless albeit perhaps infrequent. A forest protection officer suspects that, in some cases, brutal timber offenders generate fear among local people, emphasising that "in many parts, local people often are more frightened of the traffickers than the law" (6FO02). An interviewee comments:

> In fact, professional Lam Tacs may not be able to attack the local people in their village, but when the villagers get outside their community, they can be easily intimidated. Because these Lam Tacs often live in towns, where

villagers have to come to buy food or sell their crops, the Lam Tacs can easily find them to stop and teach them a lesson. Even if the traffickers are imprisoned, their families may take revenge for them. (6FO02)

There are already a number of serious cases revealed where local whistle-blowers have been attacked by violent timber traffickers. In Binh Thuan in 2014, on the basis of suspecting a villager to be an informant for Kiem Lam, more than 10 Lam Tacs came to the villager's home, burned his wood house, and threatened him with further consequences if he kept informing Kiem Lam (Thuy Nguyen 2014c). Apart from the villagers, as addressed in the media, journalists who attempt to unveil illegal operations may also be the victims of violent timber traffickers. The journalist who investigated the illegal logging inside the Khe May forest in Quang Nam province was warned that if he continued to visit the illegal logging site he "would have no way to leave the forest" (*Tuoitre News* 2014). In the case of timber trafficking in Yok Don National Park, the head of Public Service Broadcasting of Buon Don District who decided to film and publicly condemn the criminal activities was severely wounded by timber offenders who used big bricks to beat his head and face until he was unconscious (Do 2011).

For analytical purposes, each cohort of victim (e.g. offenders, on-duty officers, forest inhabitants and journalists) whose personal security is jeopardised by timber trafficking in Vietnam is examined, but in reality there are cases where timber traffickers simultaneously use violence against the different groups of victims. The Hai Chi case in Binh Thuan province is possibly the most notable illustration. During the 10-year period of operation before being sentenced in 2007, the gang committed some 60 violent incidents against ordinary people, other timber traders, Kiem Lam officers, commune authorities and police officers. The most notable were assaulting the commune police chief in 2002 and torturing a patrol police officer in 2003[9] (Nguoi Lao Dong 2007).

It can be summed up that, in the context of Vietnam, the health impacts caused by timber trafficking on various kinds of victims are clear and worrying. These consequences, whilst echoing the concern over the involvement of violence addressed in previous research,[10] are similar to the health impacts caused by other forms of green crime found in other countries. These effects are, for example, generated by climate variability and change in the USA (Gubler et al. 2001), exposure to toxic waste, pesticides and dioxin mostly in the USA and the UK (Lynch and Stretesky 2001), air pollution in the UK (Walters 2010b), genetically modified food worldwide (Walters 2010a), the oil spill in the Gulf of Mexico (Lee and Blanchard 2010; Spencer and Fitzgerald 2013), and e-waste importation, transport and dumping in Ghana (Bisschop and Walle 2013). The severe and tangible health impacts found in this research also appear to support the argument by Hall (2013a: 27) that the impacts on the health of human beings generated by a variety of forms of environmental harm and crime are among the "clearest and most immediately worrying".

So far the impacts of timber trafficking on various security aspects at the individual level have been presented. Some of the interviewees were also concerned that in addition to individuals, collectives as a whole can be injured by timber trafficking, though such impacts are more difficult to recognise. In the framework of human security, these impacts are considered as threats to community and political security.

Community and Political Security

Community security is chiefly concerned (1) with membership of a community that "can provide a cultural identity and a reassuring set of values", and (2) with being safe from oppressive community practices and ethnic conflict (UNDP 1994: 31). To be more operational, community security can be considered as the freedom from threats to key community values particularly "cultural norms, rules, regulations and behaviours" (Bryant 1995: 6). It is revealed in this research that four community values are jeopardised both tangibly and intangibly by timber trafficking in Vietnam. Interestingly, some of these values are closely connected with the aspect of political security in the human security framework.

Eroding Sacred Spirituality and Traditional Norms

In parallel with economic, social and environmental significance, it is widely recognised that forests have aesthetic, religious and cultural values particularly for long-established forest-based communities (Buttoud

2000; Daniel et al. 2012; Trigger and Mulcock 2005). Because it is clear that "their cultures are as rich as the forest ecosystems" and that "when the forest goes their culture goes too", on-going forest exploitation has deep impacts on the cultural values of indigenous tribal residents (Grainger 1993: 109). In this sense, the Vietnamese forests are no exception. While the Vietnamese forest-based communities have resided inside, and relied on, forests for centuries, the forests play an important role in the cultural and spiritual life of these communities (01EP01, 14LR01, 29IP04). The forests are not simply a material resource, but also a spiritual place (LandNet 2013); they house the spirits of gods, demons and/or ancestors (ICEM 2003).

LandNet (2013), Quang Long (2014) and Tran (2013) provide intriguing insights into the cultural and spiritual merits of forests in some ethnic minority communities in Vietnam; and more importantly how forest degradation endangers these values. For Ha Nhi inhabitants in Lao Cai province, for example, forests are their "Deity" who has protected their community for generations (Tran 2013). A statement at the recent scientific conference on "Solutions to preserve and develop the culture of the Central Highlands" emphasises that forest protection is the most important and core solution (Nguyen 2014a).

In addition to the suffering caused by threats to forest-inspired sacred values, the forest-based communities are also distressed by the encroachment of outside timber traffickers who may disturb the moral values of the communities (1EP01, 6FO02, 20EP07, 26EP09). Engaging in timber trafficking brings good earnings for community dwellers; some of them, mostly young people, spend their earnings on excessive drinking, gambling and particularly buying dugs and sexual services that all undermine the traditional norms and create potential disorder for their community (1EP01, 6FO02, 8EP03, 21IP03). Additionally, payments and share-outs of the profits from timber trafficking are on occasion not timely or fair, which also creates disputes and clashes among the villagers involved (20EP07). Of more concern is that some young village loggers involved in drug use then become junkies, which is a threat to family cohesion and a source of risk to the community's order and safety[11] (6FO02, 8EP03).

Generating Ethnic Tension between Communities

It is revealed in this research that timber trafficking may trigger conflicts between different ethnic communities. Indeed, over the past few decades in Vietnam, there have been a large number of immigrants who come from the Northern parts to the Southern areas particularly the Central Highlands and the South East where they believe that there are rich forest resources and cultivable forestlands (Chap. 3). During the past several years, although the local authorities have helped 75% of immigrants settle down, there are still over 90,000 migrants residing inside the forested areas across Vietnam (Department of Legal Affairs—MARD 2012).[12]

As a means of livelihood, sizable numbers of these migrants carry out illegal logging (Pham 2008; Quynh Anh 2014). Indeed, from 1997 to 2006, among a total of over 1800 criminal offenders charged with timber trafficking in Vietnam, 273 defendants were migrants (15%) who had moved from the North to the Central Highlands and the South-East (Pham 2008). Also, during this period, in three provinces in the Central Highlands (Gia Lai, Kon Tum and Dak Lak) alone, tens of thousands of hectares of forests were damaged as a result of illegal logging and forest-land appropriation carried out by these migrants (Pham 2008).

Unfortunately, many of the affected forests have been residential areas as well being the primary sources of livelihood for indigenous ethnic minorities such as Jrai, Ede and Bahnar for generations. In some cases, the migrant loggers cooperate with the in situ forest occupants to conduct illicit activities such as cutting, carrying, storing and hiding the illegal timber (29IP04; Nguyen Cau 2014; Xuan Dung—Trong Loi 2014). However, in many cases, the migrants ignore the indigenous people, carry out these activities alone, and even occupy or damage the forests previously claimed by the aboriginals, which has generated disputes and tension between the original communities and the migrant groups especially the Kinh majority people (30LR04, 33LA03, Nguyen Luan 2014).

Going beyond ethnic tensions, timber trafficking may nurture the ideology of ethnic discrimination. It is perceived by indigenous people that the scale of timber harvesting conducted by Kinh majority groups is far greater, more frequent and more blameworthy than the logging by indigenous ethnic minorities; yet seemingly, more residents of ethnic minorities are arrested and sentenced to prison than their Kinh counterparts (30LR04, 33LA03). Official statistics appear to concur with this concern. According to statistics from the Supreme People's Court of Vietnam, in the five-year period from 2005 to 2009, 62% of timber traffickers brought to criminal trial were ethnic minorities, whereas these minorities account for only 14.2% of the Vietnamese population[13] (Vu 2010).

The high share of ethnic minorities charged in criminal courts can, in part, be explained by the fact that most ethnic minority regions are dominated by forests (World Bank 2009), thus their inhabitants are more vulnerable to participation in illegal logging. Even so, this disproportionality may raise the politically sensitive issue of state discrimination against ethnic minorities (33LA03). Indeed, a minority of residents regularly observe that they cut down just a few trees, but encounter multiple inspections from state agencies including commune committees, Kiem Lam, police and forest rangers; whereas the Kinh traders both legally and illegally log and carry huge amounts of timber with no obstruction (33LA03, Nguyen Luan 2014).

Even worse, in some forested areas, while the aboriginal people, who were born and raised inside the forests, have made every effort to protect these forests for centuries, external harvesters under authority patronage come to harvest and then appropriate the forests (Minh Tuan—Huy Toan 2014; Nguyen Luan 2014). This practice leads indigenous people to not only become frustrated (Minh Tuan—Huy Toan 2014; Nguyen Luan 2014), but also to assume that the state authorities heavily discriminate against ethnic minorities (33LA03). An ethnic minority resident complains that:

> It is true that some households in our village have five to seven children, we need a bigger or a new house to live. When we want to have some timber to build the house, Kiem Lams ask us to do a lot of paper work. This is very complex for us. Meanwhile, Lam Tacs cut a lot of timber but don't need any paper work....Lam Tacs are the well-off Kinh people. They are very clever. We are scared of them and run away from them. (31LR05)

This problem is compounded by the abuse of large-scale projects of forestland conversion, which has been found in many localities in the Central Highlands. In Gia Lai and Dak Nong provinces for example, two in-depth investigations by the Government Inspectorate (2014a, b) conclude that many forestland conversion projects in these provinces have appropriated large areas of forestlands that have long been the residential and cultivable lands of ethnic minorities, but these minorities have been offered no proper resettlement measures, leading to long-lasting, intense disputes and complaints. A local authority employee is deeply worried:

> The perception of state discrimination against ethnic minorities is further exacerbated by the propaganda crusaded by some racial antagonists. The racist idea spread by these antagonists is that the primary purpose of Kinh people coming to the Central Highlands is to steal the forests and lands from the ethnic minorities. This is a growing worrying in the context where the ethnic minorities are facing an increasing population and shrinking rich forests. (33LA03)

What is more, these apparent incidents of discrimination may stimulate the ethnic minorities either to (1) illegally log before the forests are completely damaged or appropriated by state companies or other communities or (2) unlawfully grab forestland as much as possible for the purpose of affirming their forestland ownership and then claiming compensation if the state allows forestland conversion projects in the occupied lands (33LA03, Dak Nong Environmental Police 2012). If these unlawful acts are detected and prevented by state agencies, the violators gather a large number of indigenous people to vehemently oppose the agencies[14] (Dak Nong Environmental Police 2012; Duong 2012).

In short, illegal timber harvesting, especially MSITH and LSITH, is a salient threat to community security via the generation of ethnic tension and particularly a perception of state discrimination against ethnic minorities. In the framework of human security, state discrimination against ethnic minorities becomes a form of threat to the political security of these residents (UNDP 1994). The evidence presented in this subsection tends to support the argument in the existing literature on timber trafficking that the crime may generate severe conflict between local forest communities and the incomers becoming involved in the illicit business (Banks and Rice 2008; FAO 2007). In a broader sense, the evidence also underpins the claim that unequal access to environmental resources such as forests, water and land is among the potentially important drivers of conflict between communities (UNEP 2007).

Impacts on Future Generations

Although previous studies on timber trafficking virtually ignore the impacts of the crime on future generations, the research data suggest that timber trafficking in Vietnam would have long-term pernicious effects on future generations. In addition to the impacts on young local loggers who, as mentioned earlier, participate in activities such as the use of prostitutes and excessive gambling, drinking and drug use, timber trafficking also damages their education and career opportunities. A police officer is profoundly concerned that timber trafficking cultivates the thinking among the young people in forest communities that since logging is enough for the future, there is neither a need for schooling nor for learning other skills (20EP07). These young villagers, who either take part in the illicit business or see their parents logging, believe that "in the future, going to forests will bring a decent living; schooling and other employment trainings are, therefore, unnecessary and unworthy" (20EP07).

A report by Duy Hau and Quang Tao (2011) on the alarming problem of timber trafficking in the Krong Na commune, Dak Lak province found that irrespective of being a poor commune, most of the commune children have mobile phones. The reason is because the children are provided with the phones and instructed by Lam Tacs to use them to inform their employers if they see the law enforcement forces. Due to the money received, some of the children do not go to school but rather work inside the forest taking care of cows and assisting the Lam Tacs (Duy Hau and Quang Tao 2011).

Likewise, in Nghe An province where a timber truck overturned and killed 10 local people, the children are trained to carry logs from the ages of six or seven. These children, despite being unable to speak and write fluently, use buffalos to carry illegal logs out of forests (Xuan Hoa 2011)

Fig. 5.3 A child knows how to use buffalos to carry logs since childhood, Source: Xuan Hoa (2011)

(see Fig. 5.3). A villager says that "the villagers have been used to taking part in logging work since childhood. Growing up, we only know a livelihood that relies on logs. Generation to generation, this livelihood remains unchanged" (Xuan Hoa 2011).

As emphasised by the Commission on Human Security (2003: 114), a human security perspective underscores the importance of basic education that "can give people freedom to promote their human security and that of others". It highlights that "in addition to the human security benefits stemming from education, schools can act as delivery points for other human security interventions" (Commission on Human Security 2003: 115). In this sense, timber trafficking that distorts the younger generation's awareness of education is a security threat.

Another distinctive and perhaps more fundamental way that timber trafficking can adversely affect future generations of forest communities is that illegal logging currently conducted by community members undermines efforts to boost the sustainable economic development that is beneficial for future generations of these communities. The Vietnamese government and international institutions have made strenuous efforts to improve forest management within forest communities, which will greatly benefit future generations (19NG03). However, it seems that these communities current involvement in illegal logging appreciably erodes such efforts, thus impacting their descendants in an adverse albeit intangible fashion. This is well demonstrated by a senior NGO official:

> Basically that [sustainable forest development] means reducing consumption now, so that in the future there will be increased benefits. Basically, it is limiting exploitation however you look at it. That is the message. That message won't work if while, on the one hand, NGOs or governments are limiting or moving production of timber to a more sustainable footing here; but around the villages, there are a lot of illegal activities going on. What it does is it weakens; it fatally weakens any initiatives that try to introduce or improve forest management within the village areas because they know on the other side of the mountain there are chainsaws…The logical response is to get in and cut to make much money today and don't worry about tomorrow. (19NG03)

Thus, there are at least two ways in which future generations would be negatively affected by the current illegal logging in their communities: their disengagement with education and the loss of benefits from sustainable forest management. If this is the case, the effects of timber trafficking on the younger generations found in this research contribute empirical evidence to the developing proposition that future generations would be a category of victims of environmental harm and crime[15] (Skinnider 2011).

This sub-section has revealed four main values: forest-inspired sacred beliefs, moral norms, inter-ethnic solidarity and priority for future generations who are the subjects of victimisation from timber trafficking in the context of Vietnam. The results from this sub-section again corroborate the argument widely addressed by green criminologists that green crimes victimise communities as a whole and threaten the cultural values of communities. Brack (2004: A81), for example, postulates that green crimes, unlike most other types of crime, victimise not just individuals but "society as a whole", whereas Skinnider (2011) argues that victims of environmental crime are usually victimised collectively via the loss of culture and traditions.[16] Additionally, the multidimensional impacts of timber trafficking on community security provide an account to underpin the argument by Stretesky et al. (2014) that ecological disorganisation and destruction to ecosystems may create various forms of social disorganisation, in which social problems such as unemployment, conflict, crime and violence often originate from "exploitative relationships between the ecology and the economy" (Stretesky et al. 2014: 92). Illegal logging as a form of environmental exploitation results in not only environmental consequences, but also severe economic, social and political issues.

Positive Aspects

In conjunction with harmful consequences, this chapter is also interested in the possible positive impacts generated by timber trafficking in the context of Vietnam. Nonetheless, when asked about this, most interviewees show a cynical attitude. A police officer firmly asserts "there are no positive impacts at all, but rather, lots of very detrimental and long-term consequences" (17EP05). Yet, there is some evidence challenging this assertion, which suggests there are some benefits stemming from timber trafficking.

The first and perhaps most observable benefit can be seen in the form of SSITH that helps poor forest people meet their subsistence needs, especially for house building from the illegally sourced timber. As indicated in Chap. 4, in many highland areas, it is often extremely expensive to build houses using bricks and cement because of the high cost of transporting such materials from the lowlands. Without trees to build houses, poor people living in special-use forests, where most logging activities are criminalised, may not have houses to live in.

Although Project 167 and Project 134 on housing policy for poor households allow some poor forest-based households to log for the purpose of house building, Article 51 of the Forest Protection and Development Act stipulates that in all special-use forests such as national parks and natural conservation zones, any logging activities are illegal, except for collecting dead or wrecked trees and selective harvesting for scientific purposes. While 85% of the protected forests in Vietnam are located in residential areas of "medium" and "high" poverty (Pham et al. 2012), it is extremely tricky for the needy families in these poor areas to build their houses without illegally cutting some trees. It is believed that a large portion of the total 0.8 million m^3 of logs used to build the houses of mountain households in Vietnam every year is unlawfully harvested from the nearby forests (3EP02, 12EP04, Ha Thanh 2014a; MARD and FSSP 2014; Xuan Tho 2014). The fieldwork observation indicates that almost all houses in the core of Ba Be National Park are made by Nghien timber harvested in the park although by law timber harvesting is not allowed.

Furthermore, the earnings from unlawful timber harvesting partly support local loggers. Although their earnings are small compared to the total profits of the criminal operations, these earnings help the local actors improve their living conditions particularly in the slack periods of farming (13LA01; 23FO05; 24LR02; 25LR03; Hoai Nam 2014; Van Thanh 2014d). After cutting or transporting only one Nghien plank, the loggers and porters in Bac Kan province can get at least £2–3. With only one plank a day, their monthly earning could be £60–70, which is already double their annual total income from farming (9FO03, 12EP04, 13LA01). That explains the fact that in some forest-based villages, a sizable number of the villagers have partaken in timber trafficking, considering it a "traditional occupation". This is certainly the case in Son Hong commune in Ha Tinh province where illegal logging has become a popular long-term job with the commune having 232 illegal loggers, 248 chainsaws, 11 sawmills and 14 out of 15 villages in the commune engaging in illegal logging (Duy Tuan 2012).

Additionally, thanks to illegal logging in combination with land grabbing of the logged forestland, the loggers, particularly the Northern migrants, who first came to forests in the Central Highlands with "empty hands", now have residential and arable land which is essential for stable settlement (25LR03). Illicit logging also brings small shops and stalls (26EP09). What is more, thanks to the bribes in timber trafficking, the corrupt officers, who are struggling with low wages, have opportunities to improve their families' living conditions; in return the corrupt officers help the illegal businesses operate more quickly and smoothly (4TT01).

Conversely, it is a point of view of many interviewees that the abovementioned benefits are temporary and insignificant, compared to the negative impacts that are long-term and far more severe. Given the involvement of the local loggers, for instance, an NGO officer affirms that it is "a very bad deal" for the local people to get involved in illegal logging because they receive only marginal payments from traders, whilst facing enormous risks to their health (19NG03).

In a similar vein, an officer anticipates that if the local communities give up illegal logging and participate in other works, they would not only have a more stable and decent income, but also reduce the detrimental consequences caused by illegal logging for many other communities (17EP05). This awareness is recognised in some areas. In the area of Ben En National Park, for instance, almost all of the local people used to participate in illegal logging, but recently very few people do so and only sporadically (2FO01, 3EP02). The vast majority of the local people now realise that carrying out other jobs such as planting acacia trees, logging the trees in production forests or collecting bamboo can earn at least a hundred thousand VND (£3) a day, which is an acceptable income. At the same time, they are also aware that illegal logging is now much more arduous and they are more likely to be arrested and imposed with stiff fines (2FO01, 3EP02).

In Brief: How Timber Trafficking is Affecting Vietnamese Society?

Placing all the findings presented throughout this chapter into the broad framework of human security, it can generally be argued that timber trafficking in Vietnam affects almost every aspect of human security in varying, but substantial, degrees. These impacts can be generated by every element of timber trafficking: harvesting, smuggling, trading, supporting and processing. However, the first element of harvesting brings about the most extensive consequences. Indeed, as presented throughout this chapter, illegal timber harvesting is a substantial threat to almost all human security aspects.

Nonetheless, it is believed that different forms of illicit harvesting precipitate varying degrees and dimensions of security implications. That is to say, SSITH characterised by small amounts of illegal timber, infrequent crime commission, rudimentary logging methods and subsistence-based motives is the least harmful in all elements of human security for any kind of victim whether logger, officer or other indirect victim. It even creates certain material benefits such as helping the forest households to build their houses and to obtain residential and cultivable land, even though such benefits are the subject of controversy. MSITH, meanwhile, defined by the involvement of professional harvesters, traders and even violent actors would be a major threat to personal security because an overwhelming part of the involvement of violence in timber trafficking in Vietnam originates from MSITH actors. At the same time, LSITH that abuses the legal platform to clear the largest forest areas may be the most worrying threat to various security dimensions, especially economic and environmental security.

Is Timber Trafficking in Vietnam a form of Environmental Racism?

This research's findings on instances of LSITH characterised by the involvement of rich Kinh majorities and the diverse security impacts on forest-based indigenous communities generated by LSITH lead to the question of whether the LSITH found in this research is a form of environmental racism. As discussed in Chap. 2, there are differing approaches to conceptualise environmental racism, and the involvement of intentional racial discrimination is the key point that differentiates these approaches (Holifield 2001).

It appears that the evidence presented in this research largely fails to support the suggestion that the Vietnamese forestry state agencies and large-scale logging companies deliberately target forest-based minority communities as logging locations due to deliberate racial discrimination or hostility. Instead, it is likely that the primary motivation of the Kinh majorities is to harvest large volumes of timber whenever and wherever they can, and they do so not with the intention of targeting ethnic minorities, but instead with the intention of targeting rich forests that are weakly governed. The fact that the majority of the 10-million-plus Vietnamese ethnic minorities, comprising around 14% of the population, reside in upland areas that form the habitat for most of the rich forests in Vietnam is likely to make these locations more vulnerable to be targeted for logging.

In a broader context, reports by international observers show no signs of significant intentional discrimination against ethnic minorities in Vietnam. Instead, it is observed that "minorities have not been ignored in policy and practice in Vietnam" (World Bank 2009: 1) and that "Vietnam has a large number of policies and programs specifically designed to assist ethnic minority development…these policies are generally well-understood and have been systematically implemented at all levels of government" (Baulch et al. 2008: 6).

Further, as seen in Chap. 3, the Vietnamese government at both national and local levels has been keen to initiated policies to provide ethnic minorities with privileges by, for example, providing them with residential and cultivable land as designated in Programme 167 and Programme 134. Additionally, the indigenous minorities are known as the cheapest and most effective loggers; thus, harvesting could be more profitable if logging sites are close to indigenous villages (17EP05, 29IP04, 25LR03, 36LR06). Thus, whatever considerations are taken into account, there is little evidence to support the claim that there is "deliberate targeting" under racial discrimination; rather, the cost-benefit advantages appear to determine the choice of logging locations.

However, as observed earlier in this chapter, the indigenous ethnic minorities are those who suffer disproportionately from the timber trafficking operations. Thus, if intent is not a fundamental element of environmental racism as proposed by Pulido (2000), LSITH could be considered as a form of environmental racism because it "burdens minority groups disproportionately" (Holifield 2001: 83).

What are the Common Characteristics of Victimisation from Green Crime?

Through the synthesis of existing discussions on green criminology and timber trafficking in combination with empirical evidence pertaining to the security impacts of timber trafficking in Vietnam, this book suggests three defining attributes of green victimisation which are (1) a suffering hierarchy, (2) victim-offender overlap, and (3) multidimensionality. This threefold framework is clarified as follows.

To begin with, corresponding to the concept of environmental justice (Bryant 1995; Julian 2004; Schlosberg 2007; Walker 2012), green criminologists argue that there is a hierarchy of green victimisation, in which disadvantaged victim groups are more likely to bear the brunt of the victimisation (Carrabine et al. 2009; Sollund 2013; South 2010; White 2008, 2011). From the account of various victims of timber trafficking in Vietnam, as detailed in this chapter, these victims can be broadly categorised into four main groups: (1) poor ethnic minorities who reside in or near to the illegally logged forests; (2) local loggers, porters, smugglers, violent supporters and other offenders; (3) Kiem Lam, forest rangers and police officers; and (4) various indirect victims such as farmers, workers using NTFPs, employees of legitimate timber companies, journalists, future generations and the Vietnamese people as a whole.

It can be clearly seen that the first cohort of victim—the needy forest dwellers—who already face more significant economic and social disadvantages[17] are also those who are far more likely to be victimised the most severely and extensively by timber trafficking. The indigent forest dwellers are those whose livelihood, supplies of food and water, physical health and cultural values are most at risk as a result of timber trafficking. In other words, unlike other types of victims, who only suffer from certain aspects of victimisation, for the local poor residents, all of their security elements are likely to be threatened, and their degree of suffering may also be the highest. A similar risk can be also seen in the poor indigenous loggers and porters who invariably receive the smallest share of illicit profits, but encounter the greatest likelihood of fatal accidents and arrest during their extremely arduous work.[18] This means that the most disadvantaged group of victim is likely to encounter the most intensive and extensive suffering from timber trafficking.

The second notable observation in green victimisation initiated by Hall (2011, 2013a, b, 2014, b), Hall and Farrall (2013) and Pemberton (2014) is the overlap between victims and offenders of green crime. It is posited that some of those most severely affected by green victimisation might actually be green offenders. This can be seen in the victims and offenders of timber trafficking in Vietnam, particularly when it comes to forest-based residents. These people are an important part of timber trafficking, serving as the main loggers and porters. Simultaneously, as addressed above, they are also a major source of victims of the crime, bearing the greatest risks to their physical health for example. In other words, these indigenous actors directly suffer severe consequences from their own unlawful engagement. However, the overlap is less clear when it comes to better-off offenders such as traders and LSITH coordinators. These actors play an important role in the trafficking chain, but all elements of their security are threatened much less intensively compared to those of local loggers and porters.

Finally, the findings on the diverse and interlinked impacts of timber trafficking on all seven aspects of human security in this research reveal a further typical feature of green victimisation, that of multidimensionality. There has been a range of green research that has already revealed the impacts of green harm and crime on a variety of dimensions such as economic consequences (Barnett and Adger 2010; Wheeler and von Braun 2013), human health (Bisschop and Walle 2013; Gubler et al. 2001; Lee and Blanchard 2010; Walters 2010b), cultural values of communities (Hall 2013a; Ruggiero and South 2013; Skinnider 2011; Wheatley 1997) and future generations (Agnew 2012; Skinnider 2011). It is this research's emphasis that while green harm and crime affect various types of victims at both individual and collective levels, the suffering of each victim category is simultaneously multifaceted.

As a result of timber trafficking, the forest dwellers, for example, suffer not only from environmental insecurity, but at the same time their economic, food, health, personal and community security are also endangered by the crime. In other words, the environmental crime of timber trafficking is likely to generate not only environmental, but parallel economic, social and cultural problems, the most notable of which are murder, kidnapping, violence, inter-class inequality, ethnic conflicts and erosion of public confidence in the rule of law. The multidimensionality of green victimisation is in line with one of the major arguments in the broad perspective of human security that "a threat to one element of human security is likely to travel—like an angry typhoon—to all forms of human security" (UNDP 1994: 33).

In short, the three distinctive traits of green victimisation proposed in this research are the victimisation hierarchy, victim-offender intersection and multidimensionality. The hope is that this proposal, while contributing to the newly established sub-discipline of green victimology, will be tested and developed by further research on victimisation from green crime. This framework may be particularly relevant to the "green" types of environmental harm and crime such as illegal wildlife poaching, deforestation, fishing and other illegal forms of destruction of flora and fauna habitats that take place in developing countries. However, this is not to say that this framework is of little relevance to other forms of criminality occurring in other socio-legal settings. The degree of relevance should be tested in further research.

The outcomes of the investigation into the security implications of timber trafficking in the context of Vietnam also provide a piece of empirical evidence confirming the assumption set up at the beginning of this research that it is achievable and fruitful to integrate a security perspective into green criminology in the course of examining green victimisation. Furthermore, this research also supports the call by Shearing (2015) and South (2015) to closely connect green criminology with security studies.

Conclusion

This chapter examines a wide array of impacts of timber trafficking in the context of Vietnam. Briefly, these impacts are diverse and profound. Timber trafficking threatens economic and food security by depriving forest-dwelling inhabitants who are reliant on forest resources of food, income and employment. Work that is carried out outside forests such as farming, the production of firewood, charcoal and handicrafts, tourism and that of legitimate forestry companies, are also adversely impacted by timber trafficking. While timber trafficking results in no tax being paid to the Vietnamese authorities, the economic cost that the government has to pay to combat this crime often reaches high levels compared to that spent on combating other offences. The financial loss to the authorities reduces the amount of money available to boost economic and social welfare for the Vietnamese.

Compared to other impacts of timber trafficking, impacts on health and environmental security tend to be more tangible. Many of those who take part in the illicit logging and transporting activities are killed or severely wounded. Additionally, timber trafficking is believed to be a major factor that contributes to natural disasters such as forest fires, flash floods and landslides that cause the death and disappearance of almost a thousand Vietnamese every year. It is observed that in Vietnam some of the locations where timber harvesting is most severe are also the zones where natural disasters are harshest. Still, indigenous forest-based inhabitants are affected by water supplies that are heavily altered and polluted by the loss of forests located upstream of their villages.

With several thousand hectares of forests cleared by illegal timber harvesting every year, the crime poses a significant threat to the forests' ecosystems and biodiversity. In all national parks surveyed in this research, there is an evident reduction in wildlife, especially in high-value species, many species are in danger of extinction and some have already permanently disappeared. Severe violence as a chief source of threat to personal security is, on several occasions, also found in the operation of timber trafficking in Vietnam. Although, on the whole, the violence seems to have dropped in most provinces, there are still many serious cases of using violence to threaten the personal security of various groups: timber traffickers, anti-trafficking officers and whistle-blowers who are villagers and journalists.

Additionally, since forests are the origin of some of the major cultural and spiritual values of many Vietnamese forest-dwelling communities, forest loss is an appreciable danger to the cultural values of these communities. Timber traffickers as outsiders also encourage drug addiction, gambling and the use of sex workers in forest communities. More seriously, the encroachment of migrant groups of illegal loggers may create disputes and conflicts between the migrants and the indigenous communities, which is intensified by racist antagonists spreading the idea that Kinh majority groups are stealing forests from local people. This leads the locals to assume that the authorities discriminate against their community. Finally, in some timber trafficking hotspots, since the crime gives young people a job and income, it keeps the young generation from attending school and building a legitimate career. In future, the younger generations will also be largely excluded from the benefits of sustainable forest governance programmes because these now cannot be effectively

Fig. 5.4 Multidimensional victimisation from timber trafficking in Vietnam

implemented as a result of illegal logging. The multifaceted victimisation from timber trafficking is demonstrated in Fig. 5.4.

Chapters 4 and 5 have respectively answered two principal questions of this research: (1) how is timber trafficking in Vietnam taking place and (2) how is it affecting various aspects of Vietnamese society? The next chapter deals with the third question of how to curb the crime and thus its detrimental impacts.

Notes

1. Every year, timber trafficking in Indonesia, for example, costs the Indonesian government $4 billion that is about five times higher than the health budget of the country by 2004 (UNODC 2010). Similarly, the annual loss caused by timber trafficking is sufficient to employ about 2500 schoolteachers in Nicaragua (Richards et al. 2003).

2. Previous research has shown a large number of serious incidents of natural disaster resulting from massive illegal logging. Some of these incidents were the flooding in China in 1998 that killed over 2500 people and rendered another 19 million homeless (Pye-Smith 2006), the extensive flooding in the Philippines in 2004 that resulted in over 1000 people dead or missing (Brack 2005), and the severe drought in Honduras (CIP and EIA 2005).
3. This suspicion seems to coincide with the observation in Yok Don National Park, Dak Lak province where in 2011 the remaining stumps of hundreds of high-value sandalwood trees, which were left after serious events of illicit logging, were burned. The burning had been carried out before the inspectors of the General Forestry Department arrived to check the problem of illegal logging in the park (Do 2011; Duy Hau and Quang Tao 2011; Tran 2011).
4. A study by Pham et al. (2012), which synthesises the findings from various research articles, highlights the three most important causes of deforestation in Vietnam: (1) land conversion for agriculture and infrastructure; (2) unsustainable logging (both illegal and legal logging); and (3) forest fires.
5. This assessment of the biodiversity in Vietnam by the International Union for Conservation of Nature (IUCN) indicates that 512 out of 3990 assessed species (13%) are threatened with extinction, placing Vietnam as one of the countries with the highest rates of threatened species worldwide (see USAid 2013). The World Bank (2005) estimates that 28% of mammals, 10% of birds and 21% of reptiles and land amphibians in Vietnam are threatened or recently extinct.
6. A notable example of the use of violence in MSITH was a case in Quang Binh province in 2012. In this case, once the information on the discovery of Sua trees in the Phong Nha Ke Bang National Park was circulated, hundreds of villagers living near the park, large number of professional timber traffickers and tens of violent gangs from different districts and provinces, equipped with weapons, came into the park in an attempt to gain parts of the trees. The situation then became uncontrollable with the involved traffickers vying for, forcing each other to sell, and stealing every slab of timber. There were at least seven people injured and taken to hospital for emergency treatment (Quang Binh Public Security 2013; Tienphong 2012).
7. These murderers, including a Laotian, were arrested three days after the event and then prosecuted for four different offences: murder, robbery,

kidnapping in order to appropriate property and illegal use of military weapons (Huu Thanh 2014b; Thanh Nien 2013).

8. A case taking place in Khanh Hoa province in 2013 was an example. When driving a van loaded with 63 large planks of illegal timber at midnight, which was being closely chased by Khanh Vinh district police, the transporters criss-crossed, crashed the van into the police motorbikes and then suddenly offloaded the timber planks onto the streets. Two police officers in the team were severely wounded in this incident (Nguyen Dung 2013).
9. In this case, between 2003 and 2004 alone, Kiem Lam forces detected 12 incidents of illegal timber transportation carried out by this gang. In six out of 12 incidents, the Kiem Lam forces had to fire shots in response to the gang's violent activities (TTXVN 2005). Because of the serious criminal acts, Hai Chi was sentenced to 19 years' imprisonment for three different offences, including six years' imprisonment for the offence of "Inflicting Grievous Bodily Harm with Intent" (Minh Thuan and Que Ha 2007).
10. Research by Boekhout van Solinge and Kuijpers (2013), for example, show that from 1971 to 2004, there were 772 human rights and forest activists murdered and hundreds facing the threat of murder in the Para state of Brazil. Similarly, during illegal logging, the incidents which cause deaths, injuries and property destruction occur twice a day in Indonesia (Reboredo 2013).
11. The interviewees also mention different situations in which local loggers can become drug users. Some of the young people are first provided with free drugs by timber traders. Then once they become addicted, they find the best way to earn money for drugs is to take part in timber harvesting and smuggling. They are also encouraged to recruit more local loggers, thus their earnings will be higher (6FO02, 8EP03). Alternatively, some timber traders sell cheap drugs so that local timber porters can be physically stronger to carry much more timber; consequently, these porters also become drug addicts (28TT02).
12. In Dak Lak—a Central Highlands province—alone, by 2014, there were over 6500 households with 32,600 inhabitants who were temporarily living inside the forests; the vast majority of these being poor and ill-educated (Quynh Anh 2014).
13. Similarly, between 2001 and 2006, of the 80 people criminally accused of illegal logging and illegal deforestation in Gia Lai province, 56.8% were ethnic minorities such as Jrai and Bahnar (Pham 2006).
14. In April 2011, for example, when the People's Committee of Tuy Duc district, Dak Nong province seized a forest zone that had been illegally

grabbed and harvested by indigenous households, 60 local people gathered together, fought against the law enforcement forces, and vandalised two bulldozers, seven excavators and one emergency car (Dak Nong Environmental Police 2012). Another serious dispute over forest-land ownership occurred between 76 local households and local authorities in Dak Mil district. The dispute has already lasted 25 years but still continues (Government Inspectorate 2014a).

15. Climate change, for example, will generate durable effects on subsequent generations (Skinnider 2011). One of the possible explanations is that "climate change will disrupt families, as some family members die, become ill, or migrate. Further, the increased poverty and strain associated with climate change will have a negative impact on parenting practices" (Agnew 2012: 32).
16. Likewise, Hall (2013a: 35) believes that social and cultural impacts generated by green victimisation can have "very real consequences (some more practical and other less so) and indeed cause harm of a systemic and long-term nature". In a similar vein, Ruggiero and South (2013: 364) posit that victims of environmental crime "include not only individuals and their physical or emotional health, but the social, cultural and economic life of entire communities". Social and cultural impacts of green crimes on native communities have been seen in some studies such as Wheatley (1997) who looks into the social and cultural impacts of mercury pollution on aboriginal people in Canada, and Brook (1998) who is concerned about dangers to Native American sovereignty triggered by the industrial dumping of toxic wastes.
17. Although 53 ethnic minority groups in Vietnam account for less than 15% of the national population, in 2010 they made up 47% of the poor in the country (World Bank 2012). At the same time, 66.3% of ethnic minorities were poor, compared to only 12.9% of the Kinh majority population (World Bank 2012). There are also other significant discrepancies in economic development, social welfare and healthcare for the poor ethnic minorities in Vietnam (Baulch et al. 2007; SRV 2012; World Bank 2009, 2012).
18. This is also documented in many other areas worldwide (Human Rights Watch 2009, Richards et al. 2003). In Indonesia, for example, between 2005 and 2008, 156 out of 205 (76%) cases of illegal logging prosecuted were against low-level labourers; whereas 35 out of 49 (71%) cases, which were against government officials or high-level businessmen, were acquitted (Human Rights Watch 2009).

References

Agnew, R. (2012). Dire forecast: A theoretical model of the impact of climate change on crime. *Theoretical Criminology, 16*(1), 21–42.

Banks, D., & Rice, M. (2008). *Environmental crime: A threat to our future.* London: Environmental Investigation Agency.

Barnett, J., & Adger, W. N. (2010). Environmental change, human security, and violent conflict. In R. A. Matthew, J. Barnett, B. McDonald, & K. L. O'Brien (Eds.), *Global environmental change and human security* (pp. 119–136). Cambridge, MA: MIT Press.

Baulch, B., Chuyen, T. T. K., Haughton, D., & Haughton, J. (2007). Ethnic minority development in Vietnam. *The Journal of Development Studies, 43*(7), 1151–1176.

Baulch, B., Pham, H. T., & Phuong, N. T. T. (2008). *The Economic development of ethnic minorities in Vietnam.* Institute of Development Studies and the Department of Economics—University of Sussex and Centre for Analysis and Forecasting of Vietnam Academy of Social Sciences. Retrieved from http://www.ids.ac.uk/files/policybrief-final-Eng.pdf

Bisschop, L., & Walle, G. V. (2013). Environmental victimisation and conflict resolution: A case study of e-Waste. In R. Walters, D. Westerhuis, & T. Wyatt (Eds.), *Emerging issues in green criminology: Exploring power, justice and harm* (pp. 34–54). Houndmills, Basingstoke, UK: Palgrave Macmillan.

Boekhout van Solinge, T. (2008). Eco-crime: The tropical timber trade. In D. Siegel & H. Nelen (Eds.), *Organized crime: Culture, markets and policies* (pp. 97–111). New York: Springer.

Boekhout van Solinge, T. (2010). Equatorial deforestation as harmful practice and criminological issue. In R. White (Ed.), *Global environmental harm.* Cullompton, UK: Willan.

Boekhout van Solinge, T., & Kuijpers, K. (2013). The Amazon rainforest—A green criminological perspective. In N. South & A. Brisman (Eds.), *The Routledge international handbook of green criminology*. London: Routledge.

Brack, D. (2004). The growth and control of international environmental crime. *Environmental Health Perspectives, 112*(2), A80–A81.

Brack, D. (2005). Controlling illegal logging and the trade in illegally harvested timber: The EU's Forest Law Enforcement, Governance and Trade initiative. *Review of European Community & International Environmental Law, 14*(1), 28–38.

Brook, D. (1998). Environmental genocide: Native Americans and toxic waste. *American Journal of Economics and Sociology, 57*(1), 105–113.

Bryant, B. I. (1995). *Environmental justice: Issues, policies, and solutions*. Washington, DC: Island Press.

Buttoud, G. (2000). How can policy take into consideration the "full value" of forests? *Land Use Policy, 17*(3), 169–175.

Carrabine, E., Cox, P., Lee, M., Plummer, K., & South, N. (2009). *Criminology: A sociological introduction*. New York, NY: Routledge.

Casson, A., & Obidzinski, K. (2007). From new order to regional autonomy: Shifting dynamics of illegal logging in Kalimantan, Indonesia. In L. Tacconi (Ed.), *Illegal logging, law inforcement, livelihoods and the timber trade*. London: Earthcan.

Chan, A. (2010). *Illegal logging in Indonesia: The environmental, economic and social costs*. Minneapolis: BlueGreen Alliance.

CIP & EIA. (2005). *The illegal logging crisis in Honduras: How U.S. and E.U. imports of illegal Honduran wood increase poverty, fuel corruption and devastate forests and communities*. Washington, DC: Center for International Policy & Environmental Investigation Agency.

Commission on Human Security. (2003). *Human security now*. New York: United Nations.

Dak Nong Environmental Police. (2012). Conference paper: The work of preventing and combatting violations of laws on forest protection in Dak Nong province. In *Preventing and combatting violations against legislations on forest protection* (pp. 158–166). Hanoi: The Department of Preventing and Combating Environmental Crimes (in Vietnamese).

Daniel, T. C., Muhar, A., Arnberger, A., Aznar, O., Boyd, J. W., Chan, K. M., et al. (2012). Contributions of cultural services to the ecosystem services agenda. *Proceedings of the National Academy of Sciences, 109*(23), 8812–8819.

Department of Legal Affairs—Mard. (2012). Conference paper: The status of forest protection and solutions to enhance the work of forest protection. In *Preventing and combatting violations against legislations on forest protection* (pp. 18–29). Hanoi: The Department of Preventing and Combating Environmental Crimes (in Vietnamese).

DFID. (2002). *Poverty and environment*. London: Department for International Development.

Do, D. H. (2011). Yok Đôn, hổ thẹn với rừng! [Yok Don, what a shame with forests]. *Laodong*, 30/07/2011. Retrieved June 2, 2015, from http://laodong.com.vn/lao-dong-cuoi-tuan/yok-don-ho-then-voi-rung-44208.bld

Duong, D. V. (2012). Conference paper: Fighting against on-duty officers in the work of forest protection in Quang Binh—Current status and solutions. In *Preventing and combatting violations against legislations on forest protection* (pp. 118–122). Hanoi: The Department of Preventing and Combating Environmental Crimes (in Vietnamese).

Duy Hau & Quang Tao. (2011). Lâm tặc "mở hội" ở Vườn quốc gia Yok Đôn [Lam tac "opening festival" inside the Yok Don National Park]. *DanViet*, 12/5/2011. Retrieved February 6, 2015, from http://danviet.vn/thoi-su/lam-tac-mo-hoi-o-vuon-quoc-gia-yok-don/42754p1c24.htm

Duy Tuan. (2012). 'Xã lâm tặc' và những khu rừng bị chảy máu ['Lam Tac commune' and the bleeding forests]. 09/03/2012. Retrieved February 6, 2015, from http://vietnamnet.vn/vn/xa-hoi/63333/-xa-lam-tac--va-nhung-khu-rung-bi-chay-mau.html

EIA. (2007). *Attention Wal-Mart shoppers: How Wal-Mart's sourcing practices encourage illegal logging and threaten endangered species*. McKinleyville, CA: Environmental Investigation Agency.

EIA. (2013). *First class connecions—Log smuggling, illegal logging, and corruption in Mozambique*. London: Environmental Investigation Agency.

Environmental Police Department. (2012). Conference paper: Opening paper—The work of preventing and combating the violations on forest protection. In *Preventing and combatting violations against legislations on forest protection* (pp. 7–10). Hanoi: The Department of Preventing and Combating Environmental Crimes (in Vietnamese).

FAO. (2007). *State of the world's forests 2007*. Rome: Food and Agriculture Organization of the United Nations.

FPD. (2013). *Forest protection force of Vietnam: Forty years of construction and development (1973–2013)*. Hanoi: The Forest Protection Department (in Vietnamese).

FPD. (2014). *Report on the operational performance in 2014 and the working programme in 2015*. Hanoi: The Forest Protection Department (in Vietnamese).

FSIV & FAO. (2009). *Vietnam forestry outlook study*. Bangkok: Forest Science Institute of Vietnam and FAO Regional Office for Asia and the Pacific.

Global Witness. (2001). *The logs of war, the timber trade and armed conflict*. London: Global Witness.

Government Inspectorate. (2014a). *The declaration No.2490/TB-TTCP, 15/10/2014 of the inspection conclusion on the land management and utilisation with focus on land allocation and forest allocation in a number of projects in Dak Nong province in the 2004–2011 period*. Hanoi: Government Inspectorate of Vietnam (in Vietnamese).

Government Inspectorate. (2014b). *The declaration No.2834/TB-TTCP, 19/11/2014 of the inspection conclusion on the land management and utilisation and construction investment management in a number of projects in Gia Lai province*. Hanoi: Government Inspectorate of Vietnam (in Vietnamese).

Grainger, A. (1993). *Controlling tropical deforestation*. London: Earthscan Publications.

Gubler, D. J., Reiter, P., Ebi, K. L., Yap, W., Nasci, R., & Patz, J. A. (2001). Climate variability and change in the United States: Potential impacts on vector- and rodent-borne diseases. *Environmental Health Perspectives, 109*(2), 223–233.

Gutierrez-Velez, V. H., & MacDicken, K. (2008). Quantifying the direct social and governmental costs of illegal logging in the Bolivian, Brazilian, and Peruvian Amazon. *Forest Policy and Economics, 10*(4), 248–256.

Ha Thanh. (2014a). Giữ rừng ở Bạch Thông, thực trạng và giải pháp [Keeping the Bach Thong forest. Current situation and solutions]. *Bac Kan Online*, 22/9/2014. Retrieved June 2, 2015, from http://www.baobackan.org.vn/channel/2262/201409/giu-rung-o-bach-thong-thuc-trang-va-giai-phap-2339944/

Hall, M. (2011). Environmental victims: Challenges for criminology and victimology in the 21st century. *Varstvoslovje: Journal of Criminal Justice & Security, 13*(4), 371–391.

Hall, M. (2013a). *Victims of environmental harm: Rights, recognition and redress under national and international law*. London: Routledge.

Hall, M. (2013b). Victims of environmental harms and their role in national and international justice. In R. Walters, D. Westerhuis, & T. Wyatt (Eds.), *Emerging issues in green criminology: Exploring power, justice and harm* (pp. 218–241). Houndmills, Basingstoke, UK: Palgrave Macmillan.

Hall, M. (2014). Environmental harm and environmental victims: Scoping out a 'green victimology'. *International Review of Victimology, 20*(1), 129–143.

Hall, M., & Farrall, S. (2013). The criminogenic consequences of climate change—Blurring the boundaries between offenders and victims. In N. South & A. Brisman (Eds.), *The Routledge international handbook of green criminology* (pp. 120–133). London: Routledge.

Hoai Nam. (2014). Ngang nhiên vận chuyển gỗ lậu bằng xe máy ở Quảng Nam [Blatanly transporting illegal timber by motorbikes in Quang Nam]. *VOV*, 21/9/2014. Retrieved June 2, 2015, from http://vov.vn/doi-song/ngang-nhien-van-chuyen-go-lau-bang-xe-may-o-quang-nam-353231.vov

Hoang, C. N. (2012). Kon Tum: Lâm tặc táo tợn bắt và tra tấn chủ rừng [Kon Tum, Lam tac audaciously arrest and torture forest owners]. *Vietnam+*, 24/2/2012. Retrieved March 2, 2014. http://www.vietnamplus.vn/kontum-lam-tac-tao-ton-bat-va-tra-tan-chu-rung/130869.vnp

Hoang Nam. (2011). Khó ngăn phá rừng [Difficult to prevent timber trafficking]. *Tienphong*, 10/7/2011. Retrieved June 2, 2015, from http://www.tienphong.vn/phap-luat/kho-ngan-pha-rung-544555.tpo

Holifield, R. (2001). Defining environmental justice and environmental racism. *Urban Geography, 22*(1), 78–90.

Human Rights Watch. (2009). *"Wild Money": The human rights consequences of illegal logging and corruption in Indonesia's forestry sector*. New York: Human Rights Watch. Retrieved from http://www.hrw.org/sites/default/files/reports/indonesia1209web_0.pdf

Huong Giang. (2014). Bắt giữ nhiều vụ dùng thuốc nổ khai thác gỗ sưa [Arresting many cases using explosives to harvest Sua timber]. *Nhan Dan Online*, 5/5/2014. Retrieved January 19, 2015, from http://www.nhandan.com.vn/mobile/_mobile_phapluat/_mobile_thoisu/item/23104402.html

Huu Thanh. (2014b). Vụ thảm sát 5 phu trầm: Các bị can bị truy tố 4 tội danh [The massacre of five agarwood hunters: The accused are prosecuted with four offences]. *Tienphong*, 12/02/2014. Retrieved June 2, 2015, from http://www.tienphong.vn/Phap-Luat/vu-tham-sat-5-phu-tram-cac-bi-can-bi-truy-to-4-toi-danh-677867.tpo

ICEM. (2003). *Vietnam national report on protected areas and development—Review of protected areas and development in the lower Mekong River region*. Queensland, Australia: International Center for Environmental Management. Retrieved from http://cmsdata.iucn.org/downloads/vietnam_pad.pdf

Julian, P. (2004). Inequality, social differences and environmental resources. In R. D. White (Ed.), *Controversies in environmental sociology*. Cambridge: Cambridge University Press.

Khanh Hoan. (2013). Vụ lật xe chở gỗ lậu khiến 10 người chết: 8 bị cáo lãnh án tù [The case of flipped truck of illegal timber killing 10 people: 8 defendants are sentenced to prison]. *Laodong*, 24/1/2013. Retrieved June 2, 2015, from http://www.thanhnien.com.vn/pages/20130124/vu-lat-xe-cho-go-lau-khien-10-nguoi-chet-8-bi-cao-lanh-an-tu.aspx

LandNet. (2013). Không để Rừng Tây Nguyên trở thành hoài niệm! [Don't let the Central Highland forests became nostalgia]. *LandNet*, 25/9/2013. Retrieved January 21, 2015, from http://land.net.vn/info/Tin-tuc/Khong-de-Rung-Tay-Nguyen-tro-thanh-hoai-niem!-61-317.html

Lee, M. R., & Blanchard, T. C. (2010). *Health impacts of deepwater horizon oil disaster on coastal Louisiana residents*. Unpublished manuscript, Department of Sociology, Louisiana State University, Baton Rouge, Louisiana. Retrieved from http://www.lsu.edu/pa/mediacenter/tipsheets/spill/publichealthreport_2.pdf

Lynch, M. J., & Stretesky, P. (2001). Toxic crimes: Examining corporate victimization of the general public employing medical and epidemiological evidence. *Critical Criminology, 10*(3), 153–172.

MARD & FSSP. (2014). *Report on the development of the forestry sector in 2013*. Hanoi: Ministry of Agriculture and Rural Development & Forest Sector Support Program and Partnership. Retrieved from http://www.vietnamforestry.org.vn/mediastore/fsspco/2014/04/01/Bao_cao_PT_nganh_2013-ENGLISH.pdf.

Minh Thi—Minh Trieu. (2014). Chuyện những người giữ Giáng hương [The stories on protectors of Giang Huong forest]. *Gia Lai Online*, 2/10/2014. Retrieved June 2, 2015, from http://baogialai.com.vn/channel/1622/201410/chuyen-nhung-nguoi-giu-giang-huong-2342405/

Minh Thuan & Que Ha. (2007). Kết thúc phiên só thẩm xét xử băng nhóm "Đồi Hoa Mai": Hai Chi bị tuyên phạt 19 năm tù [The end of the trial of "Doi Hoa Mai" gang: Hai Chi was sentenced to 19 years in prison]. *Thanhnien Online*, 4/10/2007. Retrieved February 6, 2015, from http://www.thanhnien.com.vn/news1/pages/200740/211212.aspx

Minh Tuan—Huy Toan. (2014). Bài 2: Tàn phá hàng trăm hecta rừng [Article 2: Destroying hundreds of ha of forests]. *Kinh Te Nong Thon*, 25/11/2014. Retrieved January 18, 2015, from http://www.kinhtenongthon.com.vn/Bai-2-Tan-pha-hang-tram-hecta-rung-140-48347.html

MNRE. (2010). *National Environmental Report—2010—Overview of the environmental in Vietnam*. Hanoi: Ministry of Natural Resources and Environment (in Vietnamese).

Nguoi Lao Dong. (2007). Trùm xã hội đen Hai Chi hầu tòa [Gangster boss Hai Chi in court]. *Nguoi Lao Dong*, 03/10/2007. Retrieved June 2, 2015, from http://nld.com.vn/phap-luat/trum-xa-hoi-den-hai-chi-hau-toa-203966.htm

Nguyen Cau. (2014). Rừng đặc dụng Bà Nà—Núi Chúa bị tàn phá do đâu [Why is the special use forest of Ba Na—Nui Chua destroyed?]. *Cong An Nhan Dan Online*, 31/10/2014. Retrieved January 20, 2015, from http://cand.com.vn/Xa-hoi/Rung-dac-dung-Ba-Na---Nui-Chua-bi-tan-pha-do-dau-276935/

Nguyen, D. Q. (2014a). Bảo tồn văn hóa Tây Nguyên, phải giữ rừng [Preserving the culture of Central Highland, must maintain forests]. *Tien Phong*, 25/5/2014. Retrieved January 21, 2015, from http://www.tienphong.vn/xa-hoi/bao-ton-van-hoa-tay-nguyen-phai-giu-rung-708775.tpo

Nguyen Dung. (2013). Khánh Hòa: Cuộc truy đuổi kinh hoàng xe lâm tặc [Khanh Hoa: The terrifying chase of a Lam Tac van]. *Dan Tri*, 14/07/2013. Retrieved April 23, 2014. http://dantri.com.vn/xa-hoi/cuoc-truy-duoi-kinhhoang-xe-lam-tac-1374228200.htm

Nguyen Duy. (2013). Vụ lật xe chở gỗ đè chết 10 người: Các bị cáo lĩnh 206 tháng tù [The case of flipped truck of illegal timber killing 10 people: The defendants get 206 month in prison]. *Dantri*, 25/1/2013. Retrieved June 2,

2015, from http://dantri.com.vn/phap-luat/vu-lat-xe-cho-go-de-chet-10-nguoi-cac-bi-cao-linh-206-thang-tu-689243.htm

Nguyen Luan. (2014). Phan Dũng: "Nóng" chuyện lâm tặc nơi khác đến phá rừng [Phan Dung: "Hot" issue of illegal loggers elsewhere coming to deforest]. *Binh Thuan Online*, 30/10/2014. Retrieved January 20, 2015, from http://www.baobinhthuan.com.vn/phap-luat/phan-dung-nong-chuyen-lam-tac-noi-khac-den-pha-rungbr-70734.html

Nguyen, M. H., Vu, V. D., Nguyen, V. S., Hoang, V. T., Nguyen, H. D., Pham, N. T., et al. (2007). *Report on the review of Vietnam's wildlife trade policy*. Hanoi: CRES/FPD/UNEP/CITES/IUED.

Nong Nghiep Vietnam. (2011). Quảng Bình: Bảo vệ rừng liên tục bị hành hung [Quang Binh: Forest rangers are regularly assaulted]. *Bao Moi*, 25/2/2011. Retrieved June 2, 2015, from http://www.baomoi.com/Quang-Binh-Bao-ve-rung-lien-tuc-bi-hanh-hung/141/5763937.epi

Nsiah-Gyabaah, K. (2010). Human security as a prerequisite for development. In R. A. Matthew, J. Barnett, B. McDonald, & K. L. O'Brien (Eds.), *Global environmental change and human security* (pp. 237–259). Cambridge, MA: MIT Press.

OECD. (2012). *Illegal trade in environmentally sensitive goods*. Paris: Organisation for Economic Co-operation and Development.

Pemberton, A. (2014). Environmental victims and criminal justice: Proceed with caution. In T. Spapens, R. White, & M. Kluin (Eds.), *Environmental crime and its victims: Perspectives within green criminology* (pp. 63–86). Surrey: Ashgate.

Pham, D. X. (2008). *Investigation delivered by the investigation police on crimes of economic management and official position over the violations against regulations on forest exploitation and protection*. The thesis of PhD in Law, The People's Police Academy (in Vietnamese).

Pham, T. L. (2006). *Characteristics of the criminal offense of violating the regulations on the exploitation and protection of forests in the province of Gia Lai*. Master dissertation, The People's Police Academy (in Vietnamese).

Pham, T. T., Moeliono, M., Nguyen, T. H., Nguyen, H. T., & Vu, T. H. (2012). *The context of REDD+ in Vietnam—Drivers, agents and institutions*. Occasional Paper 75, CIFOR, Bogor, Indonesia. Retrieved from http://www.cifor.org/publications/pdf_files/OccPapers/OP-75.pdf

Phan, T. B. (2011). Lợi dụng chính sách xóa nhà tạm để tàn phá rừng đại ngàn [Taking advantage of the policy of eliminating temporary houses to ... destry massive forests]. *Cong An Nhan Dan*, 15/2/2011. Retrieved January 2, 2015, from http://cand.com.vn/Xa-hoi/Loi-dung-chinh-sach-xoa-nha-tam-de-tan-pha-rung-dai-ngan-173905/

Phuong Anh. (2014) (03/06/2014). Hưởng ứng Ngày Môi trường thế giới 5/6/2014: Hành động để ngăn nước biển dâng [Responding to the World Environment Day 05/06/2014: Action to prevent rising sea]. *Natural Resources and the Environment.* Retrieved from http://tainguyenmoitruong.com.vn/huong-ung-ngay-moi-truong-the-gioi-562014-hanh-dong-de-ngan-nuoc-bien-dang.html

Pulido, L. (2000). Rethinking environmental racism: White privilege and urban development in southern California. *Annals of the Association of American Geographers, 90*(1), 12–40.

Pye-Smith, C. (2006). *Logging in the Wild East: China and the forest crisis in the Russian Far East.* Forest Trends & Rights and Resources.

Quang Binh Public Security. (2013). *The investigative conclusion of the criminal case No.13/KLDT-PC46 dated 15/01/2013: Violation of he regulations on forest exploitation and protection taking place in the Phong Nha—Ke Bang National Park.* Investigating Police Agency, Quang Binh Public Security Department, Quang Binh (in Vietnamese).

Quang Long. (2014). "Báu vật" rừng già [The "treasure" of old forests]. *Tien Phong,* 4/5/2014. Retrieved January 21, 2015, from http://www.tienphong.vn/xa-hoi/bau-vat-rung-gia-700950.tpo

Quang Nhat. (2013). Bi kịch đời phu trầm [The life tragedy of agarwood hunters]. *Nguoi Lao Dong,* 30/03/2013. Retrieved June 2, 2015, from http://dantri.com.vn/xa-hoi/bi-kich-doi-phu-tram-713495.htm

Quynh Anh. (2014). Kỳ II: Áp lực từ "cơn bão" di cư tự do" di cư tự do [Volume II: Pressure from "storms" of free migration]. *Thanh Tra,* 25/11/2014. Retrieved January20,2015,fromhttp://thanhtra.com.vn/xa-hoi/doi-song/ky-ii-ap-luc-tu-con-bao-di-cu-tu-do_t114c1159n82558

Reboredo, F. (2013). Socio-economic, environmental, and governance impacts of illegal logging. *Environment Systems and Decisions, 33*(2), 295–304.

RECOFTC. (2014). *Vietnam's forestry reforms: A brief overview of forest reform in the country.* The Center for People and Forest. Retrieved December 1, 2014, from http://www.recoftc.org/country/vietnam/basic-page/vietnam%E2%80%99s-forestry-reforms

Richards, M., Wells, A., Gatto, F. D., Contreras-Hermosilla, A., & Pommier, D. (2003). Impacts of illegal ity and barriers to legality: A diagnostic analysis of illegal logging in Honduras and Nicaragua. *International Forestry Review, 5*(3), 282–292.

Ruggiero, V., & South, N. (2013). Green criminology and crimes of the economy: Theory, research and praxis. *Critical Criminology, 21*(3), 359–373.

Schlosberg, D. (2007). *Defining environmental justice: Theories, movements, and nature*. Oxford: Oxford University Press.

SGS Trade Assurance Services. (2002). *Forest law assessment in selected African countries*. Geneva: World Bank & WWF.

Shearing, C. (2015). Criminology and the Anthropocene. *Criminology and Criminal Justice, 15*(3), 255–269.

Skinnider, E. (2011). *Victims of environmental crime: Mapping the issues*. International Centre for Criminal Law Reform and Criminal Justice Policy.

Sollund, R. (2013). The victimization of women, children and non-human species through traffcking and trade—Crime understood through an ecofeminist perspective. In N. South & A. Brisman (Eds.), *The Routledge international handbook of green criminology* (pp. 317–330). London: Routledge.

South, N. (2010). The ecocidal tendencies of the late modernity: Transnational crime, social exclusion, victims and rights. In R. White (Ed.), *Global environmental harm: Criminological perspective*. Cullompton, UK: Willan.

South, N. (2015). Anticipating the Anthropocene and greening criminology. *Criminology and Criminal Justice, 15*(3), 270–276.

Spencer, D. C., & Fitzgerald, A. (2013). Three ecologies, transversality and victimization: The case of the British Petroleum oil spill. *Crime, Law and Social Change, 59*(2), 209–223.

SRV. (2012). *Implementation of sustainable development in Vietnam—National report at the United Nations Conference on Sustainable Development (RIO+20)*. Socialist Republic of Vietnam, Hanoi.

Stewart, D. (2014). Project leaf, and interpol's work on illegal logging and forest crime. In T. Spapens, R. White, & M. Kluin (Eds.), *Environmental crime and its victims: Perspectives within green criminology* (pp. 236–247). Surrey: Ashgate.

Stretesky, P. B., Long, M. A., & Lynch, M. J. (2014). *The treadmill of crime: Political economy and green criminology*. Oxon and New York: Taylor and Francis.

Sunderlin, W. D., & Huynh, T. B. (2005). *Poverty alleviation and forests in Vietnam*. Subur Printing, Indonesia: Center for International Forestry Research—CIFOR.

Thanh Nguyen. (2014b). Hung thần gỗ lậu [Deadly elfs of illegal timber]. *Thanh Nien*, 6/10/2014. Retrieved June 3, 2015, from http://www.thanhnien.com.vn/pages/20141005/hung-than-go-lau.aspx

Thanh Nien. (2013). 3 arrested for brutally killing loggers in Vietnam. *Thanh Nien News*, 5/4/2013. Retrieved June 3, 2015, from http://www.thanhniennews.com/society/3-arrested-for-brutally-killing-loggers-in-vietnam-2982.html

Thomas, C. (2001). Global governance, development and human security: Exploring the links. *Third World Quarterly, 22*(2), 159–175.

Thuy Nguyen. (2014c). Nghi báo kiểm lâm bắt gỗ, nhóm lâm tặc đốt nhà dân [Suspecting of informing Kiem Lam to arrest timber, Lam Tac burn the people house]. *Phap Luat & Doi Song*, 17/7/2014. Retrieved June 3, 2015, from http://www.doisongphapluat.com/dia-phuong/mien-nam/nghi-bao-kiem-lam-bat-go-nhom-lam-tac-dot-nha-dan-a41416.html

Tienphong. (2012). Nghe lâm tặc kể chuyện chặt sưa nghìn tỷ [Listening to Lam Tac telling stories of cutting down Sua trees worth of thoundans of billion]. *TienPhong*. Retrieved from http://www.tienphong.vn/xa-hoi/nghe-lam-tac-ke-chuyen-chat-sua-nghin-ty-579702.tpo

To, X. P. (2012). Natural forests and its suffering from timber need [Rừng tự nhiên chịu sức ép từ nhu cầu sử dụng gỗ]. *Con Nguoi & Thien Nhien*, 6/11/2012. Retrieved June 3, 2015, from http://www.thiennhien.net/2012/11/06/rung-tu-nhien-chiu-suc-ep-tu-nhu-cau-su-dung-go/

Tran, C. D., Duong, T. N., Bui, V. Q., & Ly, V. T. (2010). Examining impacts of eco-social factors on forest resources of the Ba Be National Park—Ba Be distric, Bac Kan province. *The Ba Be National Park*, 199–210. Bac Kan: The Management Board of Ba Be National Park (in Vietnamese).

Tran, N. Q. (2011). Tàn sát Vườn quốc gia Yók Đôn [Massacring the Yok Don National Park]. *Thanh Nien*, 18/04/2011. Retrieved June 3, 2015, from http://www.thanhnien.com.vn/pages/20110418/tan-sat-vuon-quoc-gia-yok-don.aspx

Tran Thuong. (2009) (01/06/2009). Qua vụ lũ quét ở Yên Bái: Đâu là nguyên nhân gây thiệt hại? [From the flooding in Yen Bai: What are the causes?]. *Dân tộc và Phát triển*. Retrieved from http://cema.gov.vn/modules.php?name=Content&op=details&mid=2704

Tran, V. V. (2013). "Rừng thiêng" của người Hà Nhì ["Holy forest" of Ha Nhi people]. *Nong Nghiep Viet Nam*, 25/9/2013. Retrieved January 21, 2015, from http://land.net.vn/info/Tin-tuc/Khong-de-Rung-Tay-Nguyen-tro-thanh-hoai-niem!-61-317.html

Trigger, D., & Mulcock, J. (2005). Forests as spiritually significant places: Nature, culture and belonging in Australia. *The Australian Journal of Anthropology, 16*(3), 306–320.

TTXVN. (2005). Băng nhóm XHĐ Hai Chi tham gia phá rừng như thế nào? [How does Hai Chi gang takes part in timber trafficking?]. *Viet Bao*, 10/7/2005.

Retrieved June 3, 2015, from http://vietbao.vn/An-ninh-Phap-luat/Bang-nhom-XHD-Hai-Chi-tham-gia-pha-rung-nhu-the-nao/55073230/302/

Tuoitre News. (2014). Infiltrating 'deforestation factory' in Vietnam jungle. *Tuoi Tre News*, 22/9/2014. Retrieved June 3, 2015, from http://tuoitrenews.vn/features/22636/infiltrating-a-deforestation-factory-in-a-vietnam-jungle

UNDP. (1994). *Human Development Report 1994*. New York: United Nations Development Program.

UNEP. (2007). *Global Environment Outllook, GEO 4—Environment for development*. Valletta, Malta: United Nations Environment Programme.

UNODC. (2010). *The globalization of crime—A transnational organized crime threat assessment*. United Nations Office on Drugs and Crime.

USAID. (2013). *Vietnam tropical forest and biodiversity assessment*. Retrieved from http://www.usaid.gov/sites/default/files/documents/1861/Vietnam118_119FinalDraft%2015Oct2013%20(Public%20Version).pdf

Van Thanh. (2014c). Nguy có tuyệt chủng cây thủy tùng trên đất Tây Nguyên [The risk of extinction of thuy tung trees in the Central Highlands]. *Cong An Nhan Dan Online*, 12/12/2014. Retrieved January 19, 2015, from http://cand.com.vn/Ban-doc-cand/Nguy-co-tuyet-chung-cay-thuy-tung-tren-dat-Tay-Nguyen-334264/

Van Thanh. (2014d). Rừng bảo tồn thiên nhiên Ea Sô bị xẻ thịt [The reservation forest Ea So is being slaughtered] *Cong An Nhan Dan*, 19/12/2014. Retrieved December 19, 2014, from http://cand.com.vn/Ban-doc-cand/Rung-bao-ton-thien-nhien-ea-So-bi-xe-thit-335128/

Viet Quoc. (2015). Bình Thuận khó giữ được rừng giáp ranh [It is hard for Binh Thuan to protect the adjacent forests]. *VOV*, 7/1/2015. Retrieved January 30, 2015, from http://vov.vn/phap-luat/binh-thuan-kho-giu-duoc-rung-giap-ranh-376264.vov

VOV. (2014). Cảnh báo tình trạng phá đá "săn" Hoàng Đàn [Warning the situations of quarrying to "hunt" Hoang Dan timber]. *Lang Son Television and Radio*, 31/10/2014. Retrieved January 19, 2015, from http://langsontv.vn/node/46079

Vu, T. H. (2010). *The offence of violating regulations on forest exploitation and protection in the Vietnamese criminal law*. Master dissertation, School of Law, Hanoi National University (in Vietnamese).

Walker, G. P. (2012). *Environmental justice: Concepts, evidence and politics*. London: Routledge.

Walters, R. (2010a). *Crime, political economy and genetically modified food* (1st ed.). Hoboken: Taylor and Francis.

Walters, R. (2010b). Toxic atmospheres air pollution, trade and the politics of regulation. *Critical Criminology, 18*(4), 307–323.
Wheatley, M. (1997). Social and cultural impacts of mercury pollution on aboriginal peoples in Canada. *Water, Air, and Soil Pollution, 97*(1–2), 85–90.
Wheeler, T., & von Braun, J. (2013). Climate change impacts on global food security. *Science, 341*(6145), 508–513.
White, R. (2008). *Crimes against nature: Environmental criminolgy and ecological justice*. Cullompton, UK: Willan.
White, R. (2011). *Transnational environmental crime: Toward an eco-global criminology*. London: Routledge.
World Bank. (2005). *Vietnam environment monitor 2005—Biodiversity*. Hanoi: The Vietnamese Ministry of Environment and Natural Resources, the World Bank, and the Swedish International Development Cooperation Agency. Retrieved from http://siteresources.worldbank.org/INTEAPREGTOPENVIRONMENT/Resources/VN_Env_Monitor_05.pdf
World Bank. (2009). *Country social analysis: Ethnicity and development in Vietnam—summary report*. Washington, DC: World Bank. Retrieved from http://siteresources.worldbank.org/INTEAPREGTOPSOCDEV/Resources/499760ESW0Whit1C10VietnamSummary1LR.pdf
World Bank. (2012). *Well begun, not yet done: Vietnam's remarkable progress on poverty reduction and the emerging challenges*. Hanoi: World Bank. Retrieved from http://www.worldbank.org/en/news/feature/2013/01/24/poverty-reduction-in-vietnam-remarkable-progress-emerging-challenges
Wyatt, T. (2013). The security implications of the illegal wildlife trade. *The Journal of Social Criminology*, August, 130–158.
Xuan Dung—Trong Loi. (2014). Giữ bình yên cho những cánh rừng giáp ranh [Keep the peacefulness for the adjacent forests]. *Binh Dinh Online*, 26/10/2014. Retrieved January 21, 2015, from http://baobinhdinh.com.vn/viewer.aspx?macm=5&macmp=5&mabb=30307
Xuan Hoa. (2011). Vụ lật xe: Hé lộ bí ẩn kinh hoàng trên cung đường gỗ lậu [The case of flipped truck: Revealing the horrifying secrets on the curve of illegal timber road]. *Giao duc Viet Nam*, 14/12/2011. Retrieved June 3, 2015, from http://giaoduc.net.vn/Xa-hoi/Vu-lat-xe-He-lo-bi-an-kinh-hoang-tren-cung-duong-go-lau-post26592.gd
Xuan Huy & An Bang. (2011). Nỗi lo "hung thần" chở gỗ lậu [The concern of deadly elfs transporting illegal timber]. *Phu Yen Online*, 18/8/2011. Retrieved January 11, 2015, from http://www.baophuyen.com.vn/89/68101/noi-lo--hung-than--cho-go-lau.html

Xuan Minh. (2014). Khai thác cây lâm sản ngoài gỗ: Gắn phát triển với bảo tồn [Harvesting non-timber forest products: Linking development with conservation]. *Vietnam+*, 22/11/2014. Retrieved June 3, 2015, from http://www.vietnamplus.vn/khai-thac-cay-lam-san-ngoai-go-gan-phat-trien-voi-bao-ton/292572.vnp

Xuan Tho. (2014). Rừng bảo tồn Mường La đang bị "rút ruột" [Muong La Reservation forest is "being gutted"]. *VOV—Dai Tieng Noi Viet Nam*, 24/11/2014. Retrieved December 23, 2014, from http://vov.vn/xa-hoi/rung-bao-ton-muong-la-dang-bi-rut-ruot-366763.vov

6

Solutions to Curtailing Timber Trafficking in Vietnam

The identification of key causes and conditions of crime, and suggested appropriate solutions to eliminate these factors, thereby curtailing crime are an important task in criminological research. This is particularly meaningful in the field of environmental crime where practical solutions to control the crime have not yet received much attention (Chap. 2). As White and Heckenberg (2014: 19) rightly argue "knowing about the damage and about criminality is one thing. But in the end, it is how groups, organisations, institutions and societies respond to environmental harm that ultimately counts". This chapter considers the key factors that drive timber trafficking in the context of Vietnam and then suggests corresponding solutions to tackle these root causes. Two types of solutions are identified and explained in this chapter: policy framework and law enforcement.

Policy Framework

There is little dispute that "appropriate law reform to clarify and harmonise forestry laws should be a priority, and will significantly help law enforcement efforts" (Stewart 2014: 242). In order to obtain a more

N.A. Cao, *Timber Trafficking in Vietnam*, Palgrave Studies in Green Criminology,
DOI 10.1007/978-3-319-64280-2_6

appropriate policy framework for the control of timber trafficking in Vietnam, this section proposes three areas of policy reform.

Priority of Forest-based Households in Forest Governance

White (2011: 113) points out the importance of a justice-based approach in dealing with green issues, suggesting that any environmental destruction "is best remedied by social justice initiatives rather than criminal justice interventions". In the field of illegal logging, globally, it is frequently argued that an unjust policy framework in forest governance that marginalises forest-based residents from the harvesting of forest resources serves as an essential factor that pushes illegal logging into a higher level of pervasiveness and seriousness (Kaimowits 2007; Wells et al. 2007).

In the context of Vietnam, a centralised mechanism of forestland ownership that fully ignored forest communities was in place before the 1990s (Chap. 3). It is currently observed that while showing some signs of improvement (Chap. 3), the current legal regime on forest governance still has fundamental limitations in pursuing a justice-based approach, particularly in recognising the legitimate interests of forest-based people. More specifically, at present, there are two main problematic scenarios taking place simultaneously in Vietnam.

First, state actors still maintain the ownership rights to large areas of forests especially high-quality ones. According to Decision 3322/QD-BNN-TCLN on Declaration of National Forest Status by the MARD, as of 31 December 2013, state agencies (e.g. state forest companies,[1] management boards, commune people's committees and armed forces) still remain in charge of about 9.2 million of hectares, which accounts for over 65.6% of Vietnam's 14 million hectares of forests (MARD 2014). Almost all rich protected and special-use forests are currently managed by these state actors (To and Tran 2014). Meanwhile, despite the previous experience of exceedingly ineffective management, these state forest companies still carry out excessive forestland clearance (To and Tran 2014), log unsustainably (Pham et al. 2012), lack clarity on institutional mandates (USAid 2013), have an outdated organisational structure and insufficient operational capability (Government Inspectorate 2014), and collude with illicit timber traffickers (Chap. 4). A local authority officer pointedly states:

> I don't mean to completely rule out the role of SFEs, but I think they are more and more incapable of protecting and making the effective use of the forests. So that it is urgent to think again about their operation. (24LA02)

Alongside state forest companies' inadequate management of forests is the dilemma of forest allocation to commune people's committees. Indeed, the commune authorities are the owners of over 2.3 million hectares of forest (16% of Vietnamese forests) (MARD 2014), but they critically lack the skills, personnel and funding to monitor their forests (40IP11, FPD 2012; Pham et al. 2012).

The second and more profound problem is that the rights and the sustainable harvests of indigenous communities have not been adequately considered. While the majority (66%) of forests in Vietnam have not been allocated to forest households,[2] the forests that have been already distributed to the indigenous inhabitants, are frequently far smaller, more degraded and isolated than ones owned by state actors (3EP02, 7NG01, 40IP11, Ngoc Linh 2014; To and Tran 2014; VAF 2014). In fact, about 70% of the allocated forests are in poor quality and each household receives only about 2–3 hectares (To and Tran 2014), whereas, as a commune authority suggests, a mountainous household needs about 20 hectares of forestland in order to carry out the forestry work that provides a stable and good income (24LR02).

Additionally, at present, the majority of those who have been allocated forestland have not officially been granted certificates of land-use right, meaning that their legal status of forest ownership has not been formalised.[3] This leads to the fact that forestland receivers feel insecure about their forest ownership and hesitate to enter into economic transactions relating to their forest lands (Pham et al. 2012; To and Canby 2011; VAF 2014).

At the same time, for those who are employed by SFEs and management boards to undertake the task of protecting nearby forests, the payment for their work is very minimal, and is far from sufficient to maintain their daily life (13LA01, 24LA02, VAF 2014; World Bank 2009). A study by the World Bank (2009: 36) points out that "the very low rates of payment for forest protection mean this sector makes an almost negligible contribution to household livelihoods for most minorities". Some local households who are employed to protect the forests in Ba Be National Park receive only VND100,000 (£3.30) per hectare per year. This means that the earnings for successfully protecting one hectare of forest in an entire year, in contrast to bringing one small plank of illegal Nghien timber out of the forest, which requires only half a day, are almost the same (13LA01).

In any scenario, disregard for the legitimate interests of local communities has immense implications for the control of illegal logging. All participants in this research firmly agree that the marginalisation of forest-based inhabitants from forest benefits is one of the fundamental drivers of illegal timber harvesting in Vietnam, and that prioritising indigenous people in forestry policy would be one of the most effective solutions to reduce the occurrence of the crime. Overall, the existing framework of forest governance brings no meaningful benefits for the local inhabitants, which leads to these inhabitants having no motivation to protect the forests, instead taking part in unlawful logging. The facts have shown that in the past three years, all of the most serious cases of illegal timber harvesting in Vietnam took place in forests that are owned and managed by state organisations.[4]

Thus it is suggested that it is imperative for the Vietnamese policy makers to employ a more radical approach that fundamentally prioritises forest households. This can be achieved in three main ways. First, forest ownership should be comprehensively transferred into the hands of local people. The forests, particularly the rich ones, should be allocated to, and owned by, the forest dwellers as much as possible. This approach would create employment for millions of forest households in Vietnam, and would offer them the opportunity to receive meaningful and tangible benefits from the forests, thereby helping to alleviate their severe poverty. The underlying foundation is that when local people's livelihoods have not been ensured and poverty has not been reduced, the fight against illegal logging is pointless (9FO03, 21IP02, 23FO05, 40EP11), like "bắt cóc bỏ đĩa" (arrest toads, but put them on plates)[5] (N. Hung 2014).

Moreover, this approach also eliminates the embedded thinking that forests belong to others (25LR03), thereby transforming local people from forest demolishers into forest protectors and developers. The interviewees in this research stress that rather than state actors such as forest companies, forest management boards, Kiem Lam or police forces, the indigenous inhabitants who consistently reside in, and depend upon, forests could be the best protectors of those forests.

A local resident decisively asserts:

> If we still keep the old policy, the hungry keep getting hungrier, and the rich keep getting richer. Indeed, this policy enriches others, just small groups of people only. To be honest with you, Kiem Lam can't protect the forests. Only the local people can do this. They are the best forest protectors. Allocating forests to them, say 15 hectares for each family, teaching them to plant, to protect and to harvest the forest resources, giving them some loan to carry out the work. That is the best solution. You see lots of money has already been spent, but it's simply feeding someone else who protects nothing. If Kiem Lam protects the forests, they will definitely log. In contrast the local people respect themselves. Nobody dares to steal the local people's forests. (25LR03)

Second, also in agreement with the approach of prioritising forest communities, some interviewees warn that in Vietnam not all forests should be locally owned by forest dwellers, and that at the time, not all forest households are capable of being forest owners. Specifically, border forests that are important for national security and forests with extremely high conservation value should be managed by the state (28TT02). Furthermore, due to poor education, the long-lasting tradition of shifting cultivation (also known as swidden agriculture) and lack of skills in forestry work,[6] many indigenous forest households are unable to benefit from their forests, or are uninterested in doing so. Therefore, forest allocation is not yet effective (3EP02, 17EP05, 29IP04, Government Inspectorate 2014). In these cases, it is recommended that the indigenous people should still lawfully benefit from the forests by keeping them involved in the work of forest protection and paying these in situ protectors decent salaries.

A police officer stresses:

Much of the budgets are now being used to maintain the apparatus of forest protection forces. This money should be given to the local people to directly protect the forests. At the same time, income from their work should be reasonably paid, enough for them to have an acceptable life. If so, they will no longer illegally log. (03EP02)

Similarly a NGO executive believes:

Sometimes, some forests are so exhausted, so degraded; the only solution is to close them off for 10–15 years. Then the government has to pay the villagers to ensure their goats and cows are not coming to the forests. You need to completely protect the forest to allow it to re-generate naturally. You need to literally defend it. But then you can pay for people to defend it. You need proper benefit sharing, so people living nearby can get the most the benefits. Right now they don't. (19NG03)

The third way of prioritising forest communities is to legalise SSITH.[7] As examined in Chap. 4, SSITH is economically and culturally important for the forest communities as it helps them meet the basic needs of subsistence, such as building a traditional house and having access to farmland for food cultivation. The research evidence suggests that while all the local authorities are struggling to stop SSITH, the ban on SSITH does more harm than good. Evaluating the logging ban, a NGO member states:

I think the logging ban is probably a really bad idea when the government can't enforce it. So what it means often is that those who want to log legally require long complex documentation. It's too expensive, it's too time-consuming…It makes no sense to follow the law. So you got a kind of absurd situation where the logging ban is probably going to do more harm than good. (19NG03)

Another NGO interviewee contends:

Whatever policies that ban logging but fail to offer alternative livelihoods to ensure the lives of the local people, are problematic. I think it is necessary to have an alternative viewpoint of forest governance in Vietnam. It is needed to get them [the local people] involved in the process [of forest governance]. Their subsistence-led logging causes little environmental impact but creates great benefits, encouraging them to protect the forests. (7NG01)

There is, therefore, strong evidence to recommend that SSITH, as defined in this research, should be legalised. However, it is also emphasised that the legalisation of SSITH may create an opportunity for LSITH, similar to the way in which housing policy for poor forest ethnic minorities is illegally abused to obtain large amounts of timber (Chap. 4). It is, therefore, vital to ensure that this legalisation allows only non-commercial timber harvesting to be carried out to meet basic subsistence needs. This is doable because while the need is neither frequent nor intensive (e.g. on average one household needs 4–5 m^3 of timber to rebuild their house every ten years), the village and commune authorities can easily discern the real need of each household. At the same time, all forms of trading and exchanging the timber houses are strictly forbidden.

Revision of the Policy of Forestland Conversion

The policy of forestland conversion has been abused by timber corporations worldwide that have obtained large amounts of timber in conjunction with the support of corrupt authorities (Boekhout van Solinge 2010; EIA 2012; Hewitt 2013). In an recent effort to evaluate the (il)legality of forestland conversion projects, Hewitt (2013) concludes that "there is a high likelihood that forest conversion taking place at present includes significant elements of illegal practice, particularly around permit allocation processes". In Brazil, for example, research by Boekhout van Solinge (2010) shows that while massive projects of forestland conversion into cattle ranches are responsible for 70% of the total deforestation in the Brazilian Amazon, they often illegally clear an extra area of at least 20%, in some cases 60–80%, of the allowed forest areas. Within the Asia-Pacific region, plantations of oil palm and rubber, and large projects such as the construction of hydropower dams, while generating sizable volumes of timber, "can act as a front for illegal logging" (EIA 2012: 9). Added to the rampant exploitation of timber, many forestland conversion projects fail to provide indigenous communities with promised benefits, creating "a legacy of conflicts, both within affected communities and with the plantation companies" (EIA and Telapak 2009: 24).

A similar scenario is believed to be found in Vietnam. Indeed, in many provinces where the policy of forestland conversion has been implemented, it has been abused for large-scale illegal logging particularly in the form of rubber plantations and hydropower construction (Chap. 4). Although it is not possible to calculate exactly how much timber has been illegally harvested in 263,500 hectares of forests used in 2400 projects of forestland conversions in Vietnam from 2006 to 2013, it is believed that using a wide array of highly sophisticated techniques, as a whole, the volume of illegal timber is truly significant (Chap. 4). In addition to serving as a pretext for illegal logging, environmental and social consequences due to the concession are also vivid and harsh, particularly for forest-dwelling communities (Chap. 5). One of these consequences is that the forest inhabitants, whilst losing their livelihoods based on the converted forests, find little alternative employment, but take part in illegal logging and forest grabbing in further isolated areas. Additionally, the implementation of forestland conversion projects generates abundant opportunities for smaller-scale illegal logging (Chap. 4).

The above analysis argues that the policy of forestland conversion has created security problems and serious loopholes enabling illegal logging, especially LSITH. It is thus necessary to revise the policy to halt LSITH as well as to ensure economic, environmental and community security for the forest people. Although some Vietnamese authorities have started to revise the policy,[8] with the focus of eliminating LSITH and minimising its impacts on human security, there are three specific recommendations that could be implemented.

To begin with, as revealed in Chap. 4, falsifying the forest status (e.g. timber value and timber species) in the course of obtaining logging permits is possibly the most sophisticated technique used to carry out LSITH. Thus, while tremendously challenging, it is vital to tackle this technique (3EP02, 18EP06, 20EP07). To do so, organisations that are independent of harvesting companies, and tasked with assessing forest status, should be established (3EP02, 35FO07). These organisations should be employed and paid for by state authorities rather than the harvesting firms as is currently the case. Such independent bodies can make sure that their evaluation is objective and not influenced by the harvesting companies (3EP02, 35FO07).

Second, there are a large number of commercial timber companies that are permitted to execute forestland conversion policy, but they are virtually incapable of delivering the projects because they are not specialists in rubber plantation or hydropower construction (Chap. 4). It is, therefore, important to only allow the companies specialised in these fields to engage in policy implementation. In other words, companies whose key motivation for engaging in such projects is to obtain timber should be excluded from the policy.

Finally, in advance of granting logging permits, potential harvesting companies should be required to carry out some of their important statutory commitments. The facts have shown that after receiving forests and harvesting timber, many companies at best exercise a small part of, and at worst completely ignore, their commitments to reforest and bring benefits to the local communities that are adversely affected by their projects. To avoid this practice and minimise the risks towards the indigenous people, these companies need to actually complete some of their commitments before being given logging permits. The most significant commitments are reforestation of an area equivalent to the forest area that is going to be converted in their project and particularly the delivery of real benefits for the affected communities via programmes of adequate compensation and suitable resettlement.

Those firms that have already harvested timber, but failed to adequately meet these commitments, thereby adversely affecting human security in the affected communities, should be appropriately punished. Using the basis of seriousness of consequences caused by these firms' operations on the human security of the forest-based communities, significant administrative fines and criminal penalties should be imposed on these firms. In other word, when human security is threatened by timber traffickers, the traffickers should be firmly punished with penalties commensurate with the level of security threats they cause. Furthermore, the potential threats to human security of forest-based residents should be taken into account by forestry state agencies when granting logging permits.

More Appropriate Allocation of Manpower, Salary and Equipment

It is widely observed that in many timber-producing countries, law enforcement agencies face a severe lack of financial and personnel investment, which results in low wages, little training and insufficient equipment—far lower than the actual requirements to successfully tackle timber trafficking. The lack of resources may generate stressful working environments, low staff morale and provide the conditions for corruption. As Rosander (2008) rightly argues, if law enforcement agencies are weak, there is a lower chance of detection and so a greater tendency to participate in illegal activities. Thus, Stewart (2014: 243) recommends that the resource investment to build up law enforcement capacity to tackle forest crime "must be among the governments' highest priorities as part of any national strategy to protect forests".

In the context of Vietnam, research commissioned by IUCN indicates that in the early 2000s, the Vietnamese government spent \$3–3.5 million per annum on the conservation sector. This amount of funding "has remained relatively stable in recent years", comprising 0.13% of the country's GDP, 0.5% of the total allocation of the public budget, or \$1200 per km^2 in protected areas per year (Emerton et al. 2006: 16). Compared to other countries, the state budget investment in Vietnam in protected areas is "significantly greater than the expenditures in other parts of Asia and are comparable to (or in some cases even greater than) public budget allocations for protected area management in Europe and North America" (Emerton et al. 2006: 16). More recently, in an attempt to evaluate the funding for biodiversity and forest conservation, USAid (2013: 36) concludes that "although relatively low, funding for protected area management in Vietnam is on par with that of countries that are able to manage their protected areas adequately".

The research data shows that there are some areas where the investment in staff and equipment for Kiem Lam, as the most specialised force mandated in controlling timber trafficking, appear to be relatively sufficient. A police officer in Thanh Hoa province states that in his province "what the police have is also what Kiem Lam have" (3EP03). A Kiem Lam officer in the province also stresses "Kiem Lams are not in shortage of weapons, pepper spray, electric batons or anything else" (6FO02). An example is that with over 40 officers, the number of officers of the Kiem Lam Unit in Ba Be National Park is more than enough compared to the statutory quota. This is because the current regulation prescribes that each officer is in charge of 500 hectares of special-use or protected forest and 1000 hectares of production forest, whereas each Kiem Lam officer in the Ba Be National Park agency is only in charge of about 300 hectares (15FO04).

It can generally be assessed that the Vietnamese policy makers have offered an encouraging investment in green issues in general and forest preservation in particular. It is necessary for this investment to continue to be further prioritised in the national policy agenda. However, a more urgent solution at the moment, as has emerged from the research data and supported by other relevant research (Do 2010; Emerton et al. 2006; USAid 2013; World Bank 2005), is that it is imperative to use the current investment more effectively and appropriately.

The current mechanism used to allocate both funding and personnel is ineffective and sometimes inappropriate. An overwhelming part of the funding is spent on infrastructure, administrative and other works, whereas little is allocated for routine operations and personnel capacity building, which results in exceedingly low staff salaries and insufficient professional training and equipment (Emerton et al. 2006). In a nationwide study by Do (2010: 158), when officers were asked about the present allocation of personnel for Kiem Lam forces in Vietnam, 33% and 55.56% of the officers responded that the existing allocation is "insufficient" and "very insufficient", respectively.

Except for some cases such as Ba Be National Park and Thanh Hoa province, by and large, Kiem Lam forces in Vietnam have a severe shortage of manpower, skills and equipment. There are currently about 12,000 Kiem Lam staff in Vietnam who are in charge of around 14 million hectares of forest; thus it is easy to calculate that each staff member has to be in charge of over 1100 hectares of forest, equivalent to a total area of more than 1600 international-standard football fields. In Dak Lak province—a notorious timber trafficking hotspot—there are a total of 280 Kiem Lam staff with 35% serving as property guards, catering staff, clerical assistants, cashiers and accountants (Ha Binh—Trung Tan 2014), whereas with over 600,000 hectares of forests (MARD 2012), each frontline Kiem Lam officer is in charge of over 3500 hectares of forest. With this huge area of forests that often have rugged terrain and are located in isolated regions, it is truly difficult for Kiem Lam forces to effectively deliver their mission.

The shortage of manpower is exacerbated by low level of professionalism and especially an unclear distribution of forest protection staff. The professional skills such as questioning arrestees, taking testimony and managing records are often limited (29IP04). At the same time, it is a concern that rather than working in the protected forests, the majority of Kiem Lam work outside the forests, mostly in checkpoints along main roads. Consequently the richest forests, often located in the most isolated areas, receive the least manpower. An interesting piece of evidence is revealed by a research team that conducted a 1700-hour survey of Banteng in Vietnam spanning a three-year period. During that time, they never met any Kiem Lam patrols in the protected areas they surveyed (Brunner 2012). Returning to one of the most serious cases of illegal logging in Vietnam in 2014, which took place in a remote pristine forest in Da Nang province, there were only two Kiem Lam officers assigned to over 4700 hectares of these remote forests. This insufficient and irrational distribution of staff for the extremely valuable forests is thought to be the most substantial facilitator for illegal logging (Nguyen Cau 2014).

Additionally, the existing financial renumeration for forest protection officers does not appear to be commensurate with the risk and hardship facing these officers. There is a strong consensus among many respondents in this research that, in Vietnam, the salary of public sector staff in general, and Kiem Lam officials in particular, is extremely low. In the early 2000s, the average official monthly salary of one Kiem Lam staff member was about £18 (Nguyen 2003), recently it was raised to around £60 (Do 2010). Senior officers who have a university degree with at least 10 years' work experience may get £100 a month plus a monthly allowance of £7 (Nong nghiep Vietnam 2012). At the same time, some other professions have much better salaries. Employees in the electricity companies complain they cannot live now on their monthly salary of £230 (Nguyen Thuong 2011). The question is that if electricity staff struggle to live on £230 a month, how can Kiem Lam staff live on only £60 (Nguyen Thuong 2011).

The lack and inappropriate allocation of manpower, equipment and salaries not only undermines the strength of law enforcement officers and emboldens offenders to commit crime, but it also creates low morale, the possibility of corruption among staff, and puts personal security of officers at risk when facing violent traffickers. Thus, it is recommended that while continuing to prioritise state funding of forest protection, the funding should be used more effectively. This can be done by allocating a much greater portion to personnel capacity improvement by substantially increasing salaries, attractive incentives, professional training and equipment for frontline staff especially those working in isolated but rich forests. Concurrently, it is also important to reduce spending on infrastructure, secondary staff and administrative work.

In addition to proposing the principal solutions above, the research interviewees also suggest areas in which the current policy framework could be improved. It is, for instance, recommended that the environmental police should be provided with the powers to both issue administrative fines and to conduct formal criminal investigations, thereby strengthening the force's ability to conduct more thorough criminal investigations (8EP03, 20EP07, 40EP11). The proponents of this initiative explain that under the current legal framework, the environmental police are entrusted only with collecting information on timber trafficking. If they detect a violation that appears to be an administrative offence, the case must be transferred to the Kiem Lam force for further examination; and Kiem Lam is the only force that has power to impose administrative penalties. Meanwhile, if the incident appears to amount to a criminal case, it must be transferred to the investigative police to conducting a formal criminal investigation. These transfers require not only extra cumbersome paperwork, but also prevent environmental police from investigating the case more thoroughly (8EP03, 20EP07, 40EP11).

Additionally, under the current Penal Code, timber trafficking is considered as an economic offence, as prescribed in Article 175 of Chapter XVI on Economic Offences. It is recommended that in order to make it easier to prosecute timber traffickers, the offence should be moved to Chapter XVII on Environmental Offences (3EP02, 32IP05). This is because under timber trafficking's current classification as an economic crime, the traffickers are accordingly prosecuted primarily based on the economic consequences caused by their actions. These consequences mainly comprise the timber volumes illegally harvested, smuggled, traded and processed by the offenders.

Meanwhile, in parallel with the economic impacts, as revealed in Chap. 5, these offenders cause other significant environmental consequences such as forest loss, the ruin of forest vegetation, the degradation of biodiversity and pollution. If timber trafficking was defined as an environmental offence, all of these environmental consequences would be seriously considered in the course of prosecuting offenders. This means that illegal harvesters would be responsible not only for how many trees they have illegally cut down, but also for how much forest vegetation is ruined and how much pollution is generated by their logging. This recommendation would be particularly useful when the offenders employ the technique of dividing the volume of illegal timber into smaller portions that are less than the threshold for criminal prosecution. In such cases, environmental damage would provide major evidence to aid a successful prosecution of these offenders (3EP02, 32IP05).

Law Enforcement

Davies et al. (2009: 42) emphasise that "behaviour is not self-defining, nor are rules self-enforcing. Laws do not have an impact unless they are enforced, or unless there is an anticipation of enforcement". That said, a concrete law would yield few desired outcomes without an effective enforcement of this law. This is certainly the case in the field of animal harm where, as argued by Nurse (2013: 254), while in many cases legal provisions on animal protection are already provided, "it is in the enforcement of legislation and the nature of the criminal justice or law enforcement response that problems occur". In the literature on timber trafficking, together with problematic legislation, the lack of effective law enforcement is cited as the primary condition for the existence of timber trafficking (Tacconi 2007). In the field of environmental protection in Vietnam, the director of IUCN Vietnam states that "generally speaking, Vietnam has a good body of environmental laws and regulations but implementation is very weak" (Brunner 2012: 2). Examining the process of law enforcement of timber trafficking in the country, two solutions are suggested, which would help improve the effectiveness of the policing of this crime.

Crackdown on Corruption

Current research on environmental crime jointly commissioned by UNEP and Interpol points out that "corruption is a deeply embedded feature of environmental crime, facilitating crime across all levels of the supply chains. Comprehensive anti-corruption measures must be a key feature of the overall effort" (Nellemann et al. 2014: 14). As emphasised in Chap. 1, it is apparent worldwide that corruption plays an important role in supporting timber trafficking. Assessing the scale of the problem in Indonesia, Schloenhardt (2008: 53) affirms that "corruption and bribery are perhaps the greatest facilitators of illegal logging".

Based on information obtained from interviewees in triangulation with official documents and press releases, it is clear that the involvement of corruption in timber trafficking in Vietnam is severe and plays a significant role in the entire illicit business. This means that curbing corruption may well be a robust solution to curtailing the crime. Indeed, timber trafficking in Vietnam would not be as severe as it is now without extensive support from corrupt officers who serve as either masterminds, patrons, abettors or even leaders of traffickers (Pham 2008). It is officially confirmed by the Department of Legal Affairs—MARD (2012) that in some places, officers overlook and abet timber traffickers, and by the Environmental Police Department (2012: 9) that:

> A proportion of the contingent of officials and Party members who have authority in the work of forest management and protection either lack resistance [to bribery], have a lapse of moral virtues, support and shield [Lam Tacs], and share the illicit earnings with Lam Tacs and other perpetrators, or the officials are manipulated by these perpetrators.

In the Kiem Lam force only, in the past three years, 716 officers have received sanctions with 19 staff demoted, 19 fired and 16 criminally charged (Nam Phuong 2014). It is plausible that not all but a proportion of these punished staff members are corrupt.[9] In the proceedings of the national conference organised by the Department of Environmental Police in 2012, 10 out of the 19 papers, which mention contributors to the problem of forest crimes in Vietnam, argue that the support of law enforcement forces for Lam Tacs is one of the main factors allowing such crimes to flourish.[10]

In relation to the responses of the research participants, in th main there is confirmation that corruption in the field of timber trafficking in Vietnam exists. However, it seems there is a large gap of confidence in the degree of the problem. Some responses show more sceptical and cautious views. An environmental police officer, for instance, suspects that:

> Given the allegation that Lam Tacs are being "given the green light" by the law enforcement officers, there has not yet been a piece of legal evidence to confirm this. This is quite difficult. In fact, we have not discovered any cases. However, there is a suspicion about this. (1EP01)

In a similar cautious, albeit somewhat franker manner, another environmental police officer says:

> We know clearly but cannot talk about this because we don't have legal evidence. If we are not careful, we can get in trouble…Timber traders often have relationships with some leaders at least at district or province levels. In the public forums, these leaders urge arrest…but who knows, behind them might be something shady. How can some timber vans be smuggled uninterruptedly day and night? They must have documents. To do so, some leaders must be involved. (12EP04)

However, some are very confident about the pervasiveness and degree of the problem such as "of course, this is needless to say" (17EP05), "that Lam Tacs buy Kiem Lams off is true" (23FO05) or "basically, there will be no timber trafficking without a nod from law enforcement forces" (3EP02).

Thus, it can be seen that all data sources in this research indicate the involvement of corruption in timber trafficking in Vietnam. To control this involvement, it is thought that the first task is to gain an in-depth understanding of who are the corrupt officers, what steps of the trafficking chain are more susceptible to corruption, and what tactics are used by the involved parties to facilitate the criminal operation. Overall, it is suggested that corruption can occur in all steps of timber trafficking and within all types of anti-trafficking forces.

To begin with the step of harvesting, a typical tactic is that bribes are often offered in advance, mostly to forest rangers or forest protection staff, and then the corrupt staff members overlook the logging incidents or inform the loggers about their patrol plans[11] (5LR01, 26EP09). Thanks to the leaked information, the loggers know where and when their logging will be safest. This explains the fact that:

> In many areas, while Kiem Lam forces are still working every day, at the same time the forests they protect are continuously bleeding. A commonplace excuse of the Kiem Lam forces is that their manpower is insufficient to check thousands of hectares of forest all at once. They say that when Kiem Lams check somewhere, Lam Tacs log somewhere else. But sometimes this is not really true because Lam Tacs are already informed by the Kiem Lam. (5LR01)

The disclosure of information is also an explanation for the "symbolic and ineffective" collaboration among anti-trafficking agencies in many localities, as assessed by the Department of Legal Affairs—MARD (2012: 26). Indeed, if a provincial police agency, for example, is planning to raid a logging site in a district, then the plan should not be shared with district agencies to avoid leaking the operational deployment (26EP09, 29EP04). A provincial police officer complains:

> Sometimes, we find it difficult to collaborate with the district agencies. In these situations, we secretly carry out the arrest and write the record of the arrest first, and then inform the district agencies later. Informing them in advance would mean informing Lam Tacs as well. (26EP09)

In the step of transportation of large amounts of timber, corruption is arguably more visible because it is not possible to carry heavy and cumbersome commodities outside forests without collusion with officers (Chap. 4). An environmental police officer observes that:

> In many forested areas, the only way to bring timber out of the forest is via a unique route. Along that route, many checkpoints of different forces such

> as forest rangers, Kiem Lam, inspectors of commune people's committees and interagency forces are in operation. If the stations work actively, no stick can get out, let alone large timber trucks. (20EP07)[12]

Unlike the harvesting step where bribes are often paid in advance, timber transporters may give a payoff on the basis of single incidents because the transporters would encounter many different officers working in various stations and mobile inspectors along the route. To have officers turn a blind eye, multiple bribes are normally required in different geographical areas because each corrupt officer can only guarantee the smuggling route within his/her vested zone. Going beyond this zone, the carriers have to make further inducements to other officers (3EP02, 5LR01). An interviewee reveals an exception:

> If the first corrupt officer is a leader of the first checkpoint and has a close relationship with the leaders in the following stations, he may frankly ask the carrier for more money, then he can contact the leaders in the next stations; hence the transportation will be cleared through the next ones as well. (5LR01)

In addition to turning a blind eye, some Kiem Lam officers wearing uniforms either directly drive the illegal timber trucks through the checkpoints, or lead them through on their motorbikes (Cong Bac 2014; Le Minh 2014; Nguyen Nam 2015), or help traffickers fabricate documents to facilitate their transportation (Tuoi Tre 2011).

In parallel with the corruption at various steps of timber trafficking, the research evidence shows that corrupt officers may come from a variety of forces such as Kiem Lam, the police, the border army, and administrative authorities who work at various governmental levels whether that be at village, commune, district or provincial. An interviewee who used to partake in timber trafficking stresses that all types of staff who are in charge of timber trafficking can be corrupted: "to ăn to nhỏ ăn nhỏ" (translated verbatim as "the small eats small and the big eats big"). This means that the more powerful an officer is, the greater amount of money he receives in bribes from traffickers (5LR01). He describes an event when a village head, the lowest level of administrative leadership, is bribed by his team in a deal to buy several high-value trees that are the collective property of the village.

> We asked the leader to look for anything which is currently most desired by the village such as a kindergarten, a spiritual house, a pumping station or a canal but which they haven't got money to build. We asked him to persuade the villagers to sell the trees for money to build. He was promised a Chinese motorbike worth about six million VND [£200] if he succeeded. We told him that during his two-year term, there would not be many similar opportunities. And the head tried his best to convince the villagers and finally the deal got done. (5LR01)

Corruption is arguably most common in forest rangers and Kiem Lam forces specialised in protecting forests, but also in other institutions whose duty is in part to combat timber trafficking such as the police, local authorities and even the border army. There are cases where corrupt police officers help timber traffickers by reducing their legal liability during criminal investigation (Phap luat Viet Nam 2013), or help them to illegally appropriate the timber (Nguyen Xuan 2013). In addition, border army officers may not be immune from corruption (26EP09, 28TT02). An interviewee suspects that by taking advantage of special regulations on border management, some border army officers support timber trafficking:

> In order to prevent Kiem Lam or police forces from approaching the areas of border forests where illegal logging is suspected, the border army agencies may present very convincing reasons. For example, at that time, they are deploying a special investigation related to national security issues. Or they are waiting in ambush for a massive drug smuggling gang. So that the appearance of any other forces will adversely affect these high-profile missions. In these situations, the police and Kiem Lam forces are only granted permission to come to the forest areas after a couple of days when the Lam Tacs have enough time to escape from the logging scene. (28TT02)

The interviewee's assumption is strengthened by an occasion in 2012 when four captains and political chiefs in two border army stations located alongside Yok Don National Park were disciplined by the provincial Commanders of Dak Lak Border Army for their misconduct associated with the severe problem of timber smuggling in the park (Thai and Trung 2012). Official statistics also show that from 1997 to 2006 across Vietnam, 1845 criminal defendants were charged with timber trafficking and its associated crimes, and 86 of these were governmental staff (4.7%) who worked in SFEs, Kiem Lam, the police, army, local authorities and other state agencies (Pham 2008).

It is thus clearly evident that by and large corruption is involved in all steps of timber trafficking and no law enforcement forces are immune. The corrupt staff members offer various kinds of strong support to the criminal operation, which creates a substantial, and in many cases, invincible obstacle to the suppression of crime. To put it differently, in the current situation, timber trafficking in Vietnam will not be controlled without a crackdown on corruption.

In order to curtail corruption, many solutions should be synchronously undertaken, but two main tasks are emphasised in this research. Before anything else, it is imperative to clarify the responsibility and intensify the accountability of individual law enforcers. As concluded by Transparency International (2012) the exploitation of natural resources is one of the major fields most affected by corruption in Vietnam. This is explained by a clear dearth of transparency and accountability related to information on profit and expenditure in this field (Vietnam Chamber of Commerce and Industry 2011). Confusingly, this field is considered by the governmental stakeholders as a "secretive" field, so little information is publicly available (Global Compact Network Vietnam 2010).

As introduced earlier, there are seven main forces that participate in the fight against timber trafficking in Vietnam. Interestingly, while at the national level these agencies belong to five different ministries, at local levels they are under the administration of various local people's committees, which generates "confusing, conflicting and overlapping" institutional mandates (USAid 2013: 53). This leads to a common situation in which a specific incident of timber trafficking has occurred, but it is not possible to hold any specific officers accountable.

Thus, there should be a very clear boundary that clarifies who is in charge of a specific area, and they have to be fully accountable for illicit operations in this area. In conjunction with the conflicting responsibility problem is another common problematic situation: if any sanctions are imposed on the guilty officers, they are not severe enough to deter corrupt activities. As specified earlier, while every year in Vietnam there are tens of thousands of timber trafficking cases detected, which results in the loss of thousands of hectares of forest, in most cases, the guilty officers do not usually receive harsh punishments. Some are given warnings, some are transferred to different working units, a handful are sacked, but only a few are brought to criminal trial. Interestingly, no one has to compensate for the economic and environmental loss of the forests. As rightly described by a senior NGO officer:

> In many, many cases, Kiem Lam officers are literally inside the trucks, providing protection. There are many clear examples of collusion. But even when that happens, even when that is exposed on TV, few people lose their job, so this continues…Everyone has "legitimate" excuses. You have a system in which there is no accountability. In the Vietnamese [language] you don't even have a word for accountability. No one loses their job because of large-scale logging in a protected area…They are responsible but they are not accountable. Lots of people are responsible for lots of things but no one is accountable. (19NG03)

This excessive lenience, compounded with the opaqueness of responsibility and high-value bribes offered, may well embolden a large number of staff to engage in corrupt relationships with rich timber traffickers. These staff members may believe that even if they are sanctioned, the benefits from corruption overwhelm the significance of the penalty (5LR01). Thus it is important to impose more meaningful punishments in order to hold corrupt staff fully accountable and to have in place an effective deterrent to collusion with timber traffickers. This approach is consistent with the suggestions from other research on the involvement of corruption in timber trafficking, which considers the lack of accountability and transparency as a key explanation for corruption (Contreras-Hermosilla 2002; Kishor and Oksanen 2006; Rosander 2008).[13] Nellemann (2012: 37) recommends that the abuse of enforcement power to obtain personal benefit will not stop "unless national law-makers find ways to strengthen their accountability to their constituents, as well as to national authorities".

The second solution to reduce corruption is better financial compensation for forest protection staff. As shown earlier in this chapter, the existing salaries are neither commensurate with the risk and hardship encountered by Kiem Lam officials nor sufficient to meet basic living needs for many officers and their families. It is observed that in response to low salaries, some Kiem Lam officers find alternative legitimate ways to get extra income such as working as consultants to forest owners about forestry management or renting forestlands for plantations (3EP02); some do the job because of their "real love [of] and responsibility for forests" (Nong nghiep Vietnam 2012); many have to quit their jobs (Nguyen Thuong 2011),[14] but others struggle to provide sustenance for their whole family. For the struggling ones, obviously, the quickest and most lucrative way to "have an ordinary life and support for their family" is to "handshake" with Lam Tac (Do 2010: 161). An officer believes that "simply overlooking only one truck [of illegal timber] can yield another month's salary" (26EP09).

There is consensus among the research participants that at the outset all Kiem Lam officers are entirely honest and committed to the respected mission of forest protection; they actually want to combat timber traffickers with a zero-tolerance approach. Yet, in order to keep their jobs, and for the sake of "miếng cớm, manh áo [bowls of rice and pieces of cloths]" (26EP09), coupled with the repeated experience of the "glamorous lure" offered by timber traffickers (32IP04), some officers reluctantly give up their initial determination. An interviewee sadly believes that in order to cope with their "starvation", Kiem Lam officers have to actively "invite" Lam Tacs to log in the forests of which they are in charge (28TT02). The discussion points to the possibility that as long as the economic security of law enforcers is not guaranteed, it is difficult for them to resist an illegitimate but lucrative relationship with timber traffickers. As such, the traffickers still enjoy great favours offered by the bribed staff so they can maintain and extend their illicit businesses.

This scenario is consistent with the principal recommendation from research by VanRijckeghem and Weder (2001: 324), which is considered to be the first empirical assessment of the effect of payment in the civil service on corruption, that "an active wage policy can help in fighting corruption". It is also in agreement with the literature on timber trafficking that because of receiving low salaries, while being assigned to manage forests with highly valuable timbers, the probability of malfeasance and the motivation to accept bribes among the officers is clearly high (Kishor and Oksanen 2006; Seneca Creek Associates 2004).

However, it has also been found that increasing salaries for law enforcers should not be seen as the definitive solution to stop corruption. There has been research indicating that an increase in salary does not necessarily lead to a decrease in corruption. Via the examination of a large number of case studies on corruption in developing countries, Olken and Pande (2012), for instance, show that in some cases increasing the wages of governmental officials results in a short-term decline in corruption levels, but the effect lessens in the long-term; whereas in other cases, the effect of a salary increase is insignificant.

Furthermore, since corruption is a widespread phenomenon in Vietnam,[15] this approach should be taken together with the efforts to curb corruption taking place outside the field of timber trafficking as well. Research by Wyatt and Cao (2015) suggests that tackling corruption in wildlife trafficking in Vietnam is highly complex, and should be incorporated into the elimination of various economic, legal, political, social and cultural conditions on which corruption in Vietnam flourishes.

Reduction in the Consumption of Endangered Timber

Demand that surpasses the legal supply is often a major driver of trafficking in environmentally sensitive goods such as wild animals, timber, fish and hazardous waste; thus demand reduction is an increasingly important approach to police the trafficking (OECD 2012). Nellemann et al. (2014: 11) point out that since consumption remains "the most important driver" of illegal wildlife and timber trade, it is important to "identify end-user markets and systematically design, support and implement where appropriate consumer awareness campaigns focusing on high consumer end-markets".

In Vietnam, from the discussions in previous chapters, it can be seen that there is an increasingly high demand for timber.[16] An NGO executive firmly stresses that with "huge domestic markets, Vietnam has an irreducible minimum demand for timber" (19NG03). Estimates by To et al. (2014) indicate that the annual domestic demand for timber in Vietnam is about 19 million m^3, and it continues to display an upward trend. The demand comes from various sectors including household furniture, house-building, infrastructure construction and the shipbuilding industry (To et al. 2014). In any of these sectors, there is constantly an opportunity for unlawful timber to be used, thereby creating conditions conducive to a thriving illicit trade in timber.

While the demand for all timber species is high, this research is particularly concerned with the demand for endangered timber that originates from the traditional consumption of this timber[17] (Chap. 3). It is observed by some interviewees that over the past two decades, while plain timber species in Vietnam have not varied much in price, the values of endangered timber have been increasing significantly (1EP01, 8EP03, 19NG03). Due to a limited legal supply, illegal trade in the endangered timber becomes very lucrative. Accordingly, the high demand for endangered timber is a considerable contributor to the problem of timber trafficking in Vietnam. This suggests that reducing the consumption of endangered timber is an important means of tackling timber trafficking in the country (1EP01, 8EP03, 19NG03).

This demand-reduction solution typically employs methods of awareness raising and marketing campaigns (Drury 2009). It is suggested that, in the context of Vietnam, it is important to carry out awareness-raising campaigns about endangered timber amongst Vietnamese consumers, which could contribute to addressing the ethical and legal concerns associated with their consumption. These campaigns would be particularly effective for "accidental" consumers who, as identified by Wyatt (2013),[18] buy and use products made from endangered timber, but are neither aware of nor able to discern the illicit origin of the timber used to make the purchased products and the environmentally destructive implications of their consumption.

As UNODC (2013) notices, significantly different from other forms of trade in illegal goods, in timber trafficking, once the products are finished, it is extremely difficult for buyers to distinguish whether or not these products are made from illegally cut timber. Meanwhile, like buyers of timber products in other countries (Schloenhardt 2008), when deciding to purchase timber products, many consumers pay heed to aspects such as what timber species the product was made from, how much it is worth and how beautiful its colours and annual rings are, rather than the legitimacy of the source or the environmental implications of their consumption (4TT01, VTV2 2013).

During the research fieldwork, it was noticed that products such as *Lộc Bình* (Bottle for Fortune), *Thần Tài* (Statue for Intelligence), *Tượng Phật* (Buddha statue) and *Phúc Lộc Thọ* (Happiness-Fortune-Longevity statue), mostly made from endangered timber, are mainly displayed in middle- and upper-class households, large restaurants, coffee shops, hotels and souvenir shops (see Figs. 6.1 and 6.2). Even inside some high-quality cars, drivers hang small decorative items made from endangered timber as psychological therapy to bring fortune to their driving and as a comfort. Some consumers are unaware of the endangered status and illegality of the timber. Awareness-raising campaigns, therefore, would help these accidental consumers to be more mindful and responsible about their consumption.

Fig. 6.1 A Happiness-Fortune-Longevity statue (*in front*) and a Fortune bottle (*behind*) are placed in the living room of a Central Highlands family, Source: own photo taken in September 2013

Fig. 6.2 A small Buddha statue was on sale at £170 in Pleiku Airport, Gai Lai province, Source: own photo taken in September 2013

Nonetheless, there is another type of buyer of endangered timber who should be treated more seriously than simply fed with the platitudes of environmental protection. They are "committed" buyers who "ideologically believe in" and "are committed to the consumption of wildlife regardless of the illegality and the negative consequences involved" (Wyatt 2013: 97). As mentioned in Chap. 3, there are a number of rich Vietnamese who are passionate about products made from endangered timber and are willing to pay enormous amounts of money to purchase these products. These consumers are clearly aware that the timber may be unlawfully traded and that the timber species is on the brink of extinction.

In an investigation into the traditional consumption of endangered timber in Vietnam, Vietnamese Television indicated that many consumers are clearly aware of the endangered status of the timber species they buy. They believe that because the timber is on the brink of extinction, it is now economically prudent to buy products made from such species; otherwise there will be no chance in the future to possess such rare and precious commodities (VTV2 2013). A police officer describes the situation:

> Many users [of endangered timber] believe that the bigger and taller are the items made from more endangered timbers, the classier the purchaser can be and the more magical effects these items can have. Clearly such endangered timber species are incredibly expensive. A window frame or sleeping mat made from Sua timber is worth 4–5 hundreds million VND [£14,000–16,000]. It is expensive but they still buy because they enjoy the rare and exclusive status of the timber. (3EP02)[19]

Another interviewee gives a further insight into the committed buyers in Vietnam:

> I know of many cases when the products made from extremely endangered timber are used as bribes for the rich leaders. These leaders are often less keen on nice houses, cars, lands or money for their children. But they prefer something like these timber products that are precious and unique for the purpose of showing off their class. In these cases, both the bribers and the leaders know that the rarer the timber is, the more meaningful the bribes are. There is even a "race" among these leaders to collect the endangered timber. You and I are both district chairmen, so in my living room I don't feel able to show the [timber] items less rare than yours. I cannot be less posh than you. (5LR01)

For these committed consumers, the effectiveness of the awareness-raising campaigns on the consumption of endangered timber may be limited, particularly in the short-term. Indeed, some studies have argued that the effectiveness of awareness-raising campaigns regarding the consumption of wildlife, including endangered species, may be limited in Vietnam (Brunner 2012; Drury 2009; Milliken and Shaw 2012; TRAFFIC 2008; USAid 2013). Brunner (2012: 1), for example, observes that there have already been "numerous" awareness campaigns, but they "have had no discernible impact".

This research argues that such campaigns can have considerable impact, but it is a relatively sluggish process. This is first because the traditional consumption of endangered wildlife including timber in Vietnam is "deep-rooted, sanctioned at the highest levels of society" (Brunner 2012: 1), and has been already "absorbed into the blood" of the buyers (3EP02, 8EP03, 20EP07, Ba Thang 2014). For these consumers, despite their awareness of the environmentally harmful implications and unclear origin of the timber, they still enjoy the consumption. An environmental police officer frankly sats that:

> Many influential people repeatedly mouth platitudes about forest protection, but they buy and really enjoy wood products made from highly endangered timber. Their act is opposite to their words and clearly facilitates illegal logging. (26EP09)

Furthermore, in traditional Vietnamese society, individualism is dwarfed by the need to conform to communal norms. This means that even though some individuals are aware of the endangered status of timber species, their individual behaviour is nevertheless influenced by group behaviours that still follow traditional modes and methods of consumption (Drury 2009). That said, until the majority of Vietnamese are made aware of the environmentally harmful consequences of their consumption, their behaviour will not change. Thus it will take a prolonged period of time for the majority of Vietnamese consumers to voluntarily give up their environmentally unfriendly consumption.

Accordingly, merely raising awareness about the protection of endangered timber species may be insufficient at present, particularly in the context that the majority of endangered timber species in Vietnam are in old-growth forests that now account for only 8% of the country's forests, but are the most popular locations for timber trafficking (MNRE 2010). Right now, it is, therefore, imperative to have a parallel solution to reduce demand for endangered timber. To this end, it is recommended that the consumption of endangered timber should be explicitly forbidden, and that the law enforcement agencies need to seriously target this unlawful consumption.

Currently, there tends to be inconsistent and ambiguous legislation pertaining to the consumption of timber including endangered species. While the Forest Protection and Development Act 2004 generally stipulates that the illegal use of timber is strictly prohibited,[20] this "illegal use" is not included in either the Penal Code (specified in Joint Circular 19/2007/TTLT/BNN&PTNT-BTP-BCA-VKSNDTC-TANDTC) for criminal charges or Decree 157/2013/ND-CP for administrative sanctions.[21] Likewise, Point 3, Article 5 of Decree 32/2006 generally stipulates that the use of endangered timber that is "in contravention of the provisions of this Decree and current legal provisions" is strictly prohibited. However, subsequently the Decree only specifies the acts of harvesting, smuggling, processing and trading in endangered timber. Thus, all the relevant legislation fails to clarify the consumption of timber in general and endangered timber in particular. In other words, what constitutes the illegal consumption of endangered timber is ambiguous.

With respect to the enforcement of law against the consumption of endangered timber, it is frequently disregarded, if not completely ignored, by the responsible authorities. As shared by the research interviewees, most law enforcers show little determination in actually decreasing the consumption of endangered timber. All the statistics published by the authorities exclude illegal timber consumption violations, but rather only include harvest, transport, processing and trade data.

Altogether, the traditional use of endangered timber, in conjunction with the lack of alternative products,[22] the unclear legal provisions and the low priority for enforcement, are believed to be an important factor that leads to a very high demand for endangered timber, thereby intensifying the problem of timber trafficking in Vietnam. Clarifying, therefore, the current legislation on illegal consumption of timber by explicitly banning the buying and use of endangered timber, especially critically endangered species listed in Group IA of Decree 32/2006/ND-CP, and then concurrently tightening enforcement on the consumption of endangered timber would be effective solutions to tackle timber trafficking in Vietnam. That said, in order to better tackle timber trafficking, it is now time for Vietnamese policy makers and law enforcers to start targeting the demand side more seriously, rather than merely attempting to control the supply side, as is currently the case.

In Brief: How to Better Control Timber Trafficking in Vietnam

In the field of wildlife crime, there are a number of theories employed by different scholars to explain the crime, which includes differential association, techniques of neutralisation, rational choice, routine activities and conflict perspectives (Crow et al. 2014). Rather than using these theories as such, most studies on timber trafficking attempt to explore different underlying drivers of the crime, and based on this exploration, corresponding solutions are suggested. Seneca Creek Associates (2004: 6), for instance, contends that the crime is "a symptom of corruption, graft, lax enforcement, and poor social conditions". Meanwhile, using a case in Cameroon, Alemagi and Kozak (2010) identify six drivers of illegal logging: systemic corruption; conflicts; poverty; licensing schemes; inadequate institutional support; and usurpation of property rights. Contreras-Hermosilla (2001), however, emphasises that identifying general causes of illegal logging is an arduous job because the causes are complicated and vary greatly across the cultures and governance systems of different countries. The country's socio-legal strategy, traditions and the degree of democracy are all factors that lead to this transnational discrepancy (Contreras-Hermosilla 2001). In line with Contreras-Hermosilla, when looking for solutions to forest crimes occurring worldwide, UNODC (2012: 1, 2) concludes:

> Wildlife and forest offences are a complex phenomenon with many layers and dimensions. Wildlife and forest offences often result from the interplay of a multitude of factors—cultural, economic, social and environmental… There is no "one size fits all" solution to this issue. In formulating effective countermeasures, it is important that local patterns of wildlife and forest offences and the concerns of local communities be recognised and integrated into policy and legislation.

Taking into account these discussions, this chapter looks into five main drivers of the current problem of timber trafficking in Vietnam, which are grouped into two dimensions: the policy framework and law enforcement. The insights into these drivers lead to recommending five corresponding solutions to help curb timber trafficking in Vietnam. Among these solutions, this research highlights that in the context of Vietnam, an appropriate overarching policy approach should be a prerequisite. Despite some success in launching a decentralised policy of forestland ownership, there is still a fundamental drawback to the policy approach of forest governance, which creates good conditions for illegal timber harvesting.

The current policy basically marginalises forest dwellers, leading to the vast majority of high-quality forests in Vietnam not being owned by, or benefiting, local households. It is highly probable that this marginalisation not only erodes these locals' attempts to protect the forests, but also contributes to them participating in unlawful logging with a view to improving their scant living conditions. There is strong evidence that illegal logging tends to take place most rampantly in rich forests that are owned by state actors such as SFEs, forest management boards and commune people's committees. Therefore, a policy framework of forest governance that essentially considers forest-based inhabitants as being the most important owners, protectors and beneficiaries of forests should be instantly and thoroughly adopted.

This solution is strongly consistent with the suggestion of previous policy studies on forestry and conservation in many other parts of the world. Indeed, the existing literature recommends that in order to successfully control illegal logging, it is fundamental to reform the forestry policy framework that marginalises forest-based residents. This framework is considered as socially unacceptable because under it forests are owned by state agencies, meaning that a great number of poor local households who rely on forest products are not allowed to access forest resources, which leads them into illegal activities to meet their basic needs (Kaimowits 2007; Wells et al. 2007).[23] In an in-depth, year-long research project that scrutinises illegal logging in the USA, Pendleton (2007: 18) critically argues:

> Much of the literature on illegal logging starts from an assumption that it is universally believed that illegal logging is harmful, wrong and otherwise without merit...Yet, this view fails to explain why illegal logging persists at such an astounding level. How could something that is so wrong flourish?

Pendleton (2007: 19) posits that irrespective of labelling logging as "illegal" or "deviant", illegal logging still persists because it is possibly "a stabilising influence on the social order of the forest community", and hence criminalising it would "disrupt the shared identity of this community". A key lesson from Pendleton's research, therefore, is that "the forest community matters", and that timber trafficking persists not merely owing to financial incentives, but it also undertakes "critical social functions essential to the preservation of the forest community" (Pendleton 2007: 40). Likewise, Mexico is "a remarkable case study in what some consider as the best form of forest management" since 75% of the country's forests are controlled by, and greatly benefit, local communities (*The Economist* 2010: 1).

The solution from this research also echoes the results of previous studies on the counter-productive fallout of narrow legislative approaches, particularly the intensive criminalisation policy in Vietnam in dealing with subsistence-led logging (McElwee 2004; Sikor and To 2011; To and Canby 2011)[24] and the proven effectiveness of the meaningful involvement of forested communities in the process of forest management and conservation (Boissiere et al. 2009; de Jong et al. 2006; FSIV and FAO 2009; Nguyen et al. 2007; Pham et al. 2012; To and Tran 2014).[25]

Thus, there has already been sufficient evidence, both empirical and theoretical, to prove the appropriateness and effectiveness of the policy of forest allocation to in situ forest communities. It is now time for Vietnamese policy makers to stop their "experimentation" with a decentralised policy and implement it thoroughly and urgently for the purpose of curbing timber trafficking, protecting the remaining intact forests, and improving various aspects of human security for the country's forest dwellers.

Conclusion

This chapter has provided the answer to the third main question of how to better control timber trafficking in Vietnam by recommending five main solutions to tackle the crime. The first and possibly most fundamental solution is to adopt an overall forestry policy that genuinely prioritises forest-dwelling inhabitants. This can be achieved by (1) thoroughly shifting the ownership of forests, particularly rich ones, from state agencies to indigenous households, (2) regarding the local inhabitants as the most effective protectors of the forest and/or (3) legalising their small-scale logging for subsistence needs.

With regard to the second solution, it is observed that in the past decade, hundreds of thousands of hectares of Vietnamese forests have been converted. In many cases, the policy has been abused so that massive volumes of timber have been illegally harvested, which has had profound consequences for forest communities. It is thus urgent to revise this policy both to prevent the conditions for LSITH, and to ensure vital aspects of human security among the potentially affected villages. To do so, a threefold solution is proposed: (1) create independent organisations free from the influence of timber harvesting firms to undertake forest status assessments, (2) choose robust business organisations specialising in the relevant domains of rubber plantation or hydropower construction to deliver the policy and (3) require the harvesting companies to adequately conduct some of the important commitments on afforestation, compensation and resettlement before allowing the harvesting of timber.

Concerning the third solution, this research finds an echo in a common evaluation of some international commentators on the relatively generous funding supplied by the Vietnamese government for the forest preservation sector as compared to many other nations. Although the field should still receive higher financial priority in the state budget expenditure, it is now more important to make use of the current funding allocation in a much more effective way. The fact is that in timber trafficking hotspots, the professional skills, equipment and numbers of forest protection officers still lag far behind the scope and sophistication of the criminals. In consequence, a much greater portion of the funding should be used for personnel capacity improvement via substantially increasing salaries, financial incentives and professional training of frontline staff.

The fourth solution, which is related to interventions in the law enforcement process, is an effective crackdown on corruption in timber trafficking in Vietnam. This strategy is strongly emphasised in this research. The research evidence points to a high possibility that corruption creeps into all steps of timber trafficking with all law enforcement agencies implicated. There are plenty of ways the corrupt officers can give support to timber traffickers such as turning a blind eye, taking advantage of power to interfere with policing procedures, helping produce falsified documents and even orchestrating the criminal operation themselves. Since these corrupt practices are profound and pervasive, a crackdown on corruption is central to any efforts to stop timber trafficking in Vietnam. There is much work to be done to curtail this involvement, in which two tasks are key: (1) to specify individual officer's responsibilities in each area of forest and then to impose much stiffer punishments on corrupt officials and (2) to provide suitable financial rewards that guarantee the economic security of law enforcement staff.

Finally, while continuing to carry out campaigns to raise public awareness of the environmental value of forests, the Vietnamese law-makers and enforcers should pay serious attention to the illegal consumption of endangered timber. The criminal justice agencies need to convey a clear message that such consumption is not only outdated and environmentally harmful, but also illegal, meaning that consumers must bear the legal liability for their purchases. Only when a solution that takes into account both formal and informal enforcement is administered, will the demand for endangered timber in Vietnam decrease, thus contributing to the breakdown of timber trafficking in the country. The solutions proposed in this research to curb timber trafficking in Vietnam are demonstrated in Fig. 6.3.

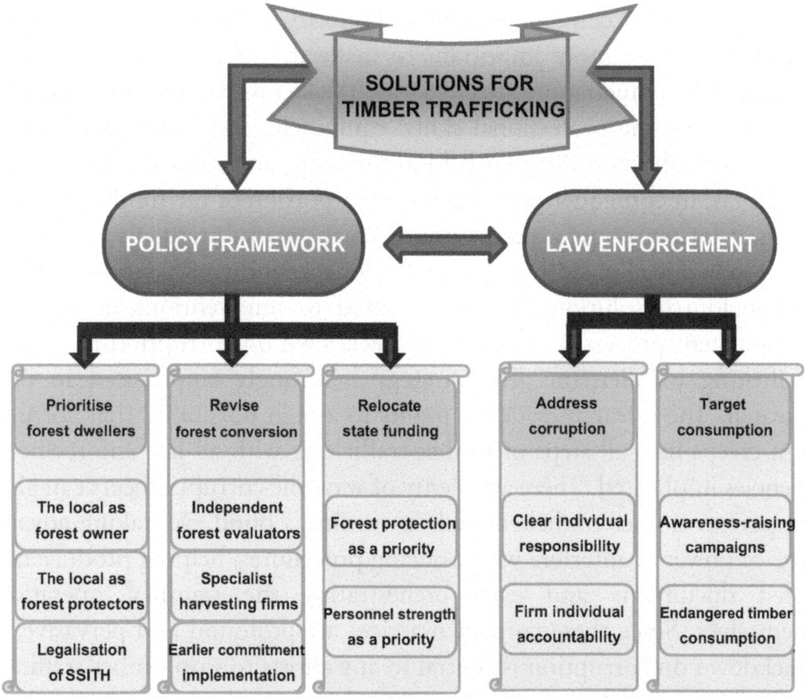

Fig. 6.3 Main solutions for curbing timber trafficking in Vietnam

Notes

1. In the implementation of Resolution 28/NQ-TW in 2003 by the Politburo of the Communist Party of Vietnam and Decree 200/2004/ND-CP in 2004 by the Vietnam Government, state forestry enterprises (SFEs) have gradually been shifting mainly to state forest companies (FCs) and forest management boards (MBs) (Chap. 3). However, because SFEs have existed for so long, most Vietnamese still call FCs (công ty lâm nghiệp), SFEs (lâm trường).
2. According to the result from a recent workshop organised by the Vietnam Administration of Forestry, there are many reasons for the excessively slow progress of allocating forests to local people. Some of the main reasons are a lack of determination from local authorities, a shortage of funding for the implementation, unexpected complains from forest recipients and the inability to handle post-allocation issues such as unlawful trade in the resources from allocated forests (VAF 2014). Another problem with forest allocation in Vietnam is that there is inconsistency in the progression of forest allocation between different regions in the country with the Central Highlands showing a greatly slower progress. Only less than 2% of the forests in four provinces in the Central Highlands have been allocated to households (World Bank 2009).
3. According to FPD official statistics, by 2013, some 1.17 million hectares of forests, accounting for only 23% of the total allocated forestland areas, have been given certificates of land-use right (FPD 2013).
4. The most notable recent cases of illegal logging in Vietnam are detected in the protected forest owned by the Management Board of Song Luy Protection Forest, Binh Thuan province (Viet Quoc 2015), in the ecological forest owned by Dak Lak Rubber Company, Dak Lak province (Le Phuoc 2014), in the special-use forest owned by the Management Board of Ba Na—Nui Chua Forest, Da Nang province (Nguyen Cau 2014) and in the watershed forest owned by Huong Son Forestry Enterprise, Ha Tinh province (Thanh Chau 2013).
5. This is a Vietnamese proverb used to describe the pointlessness of doing something. In this example, it means if the arrested toads are placed on a plate without having anything to keep them there, they will definitely jump out of the plate, meaning that the arrest is worthless. Likewise, if the dwellers are still poor and unemployed, finding no other ways to maintain their livelihoods, they will continue to log illegally in the nearby forests. Thus, without improving the livelihood of the forest-

based people, the fight over illegal logging carried out by these dwellers would be pointless, just like the arrest of toads.

6. Some interviewees believe that in order to manage and ensure that forests directly benefit people, the forest owners should have basic knowledge and skills in forestry fields such as understanding the forest terrain, characteristics of forestland soil, biological properties of the trees and the skill of forestry plantation. At the moment, many, if not most, of the forest ethnic minorities lack such understanding and skill (3EP02, 17EP05, 29IP04).
7. As introduced in Chap. 3, via Programme 134 and Programme 167, the Vietnamese government has attempted to legalise some of the logging of poor forest-based households. Yet, only poor households residing in some forested localities that have officially approved annual logging plans are allowed to harvest timber for building their houses (Chap. 3). This means that households living in other forested localities, particularly in protected forests, are excluded from this policy. Furthermore, the policy requires complex paperwork that challenges the ability of poorly educated mountain inhabitants (3EP02, 31LR05).
8. Lately, provincial authorities in the Central Highlands have revoked 76 rubber plantation projects, which covered 8000 hectares of forest, and have also suspended 38 hydropower construction projects that spanned 1300 hectares (Uyen Thu 2013). Similarly, in November 2013, the National Assembly of Vietnam issued a resolution on revamping the planning of construction of hydropower projects; 424 projects were removed from the planning process, 136 projects were suspended and 158 projects were subject to revision (VOV 2014).
9. Most well-known recent corruption cases relating to timber trafficking in Vietnam that were brought to criminal courts took place in the provinces of Ha Giang (Phap luat Viet Nam 2013), Ha Tinh (Thanh Chau 2013), Nghe An (Khanh Hoan 2013; Nguyen Duy 2013) and Khanh Hoa (Nguyen Xuan 2013; V.Tao 2013).
10. In the proceedings, nine remaining papers presented by other anti-trafficking agencies address the drivers of forest offences, but do not mention corruption. This does not necessarily mean that in realty there is no corruption involved in the offences. Perhaps, it is simply because there is no formal legal evidence to confirm the occurrence of such allegations since there has yet to be an officer in these agencies officially charged with corruption.

11. There was a case in Ha Tinh province where the head of a forest protection station and a deputy chief of a forest management board of Hong Linh Forest, in collusion with illegal loggers, overlooked 58 incidents of illegal logging, which occurred between late 2010 and early 2012 with almost 280 m^3 of timber illegally harvested (Xuan Hong 2012).
12. The route into Hong Linh Forest in Ha Tinh province, for example, only lasts about 5 km but has six checkpoints operated by different agencies, namely the Border Army Unit, Kiem Lam Station, Forest Protection Board of Hong Linh Forest, Khe Sinh Forest Protection Station, Custom Unit and Son Hong commune. Irrespective of the dense supervision, Hong Ling Forest is still being illegally harvested and illicit timber is being brought out of the forest (Duy Tuan 2012). In this case, five corrupt forest protection staff were charged with collusion with timber smugglers (Xuan Hong 2012).
13. Different researchers propose different explanations for the prevalence of corruption in timber trafficking. Rosander (2008) presumes the three major factors are lack of accountability and transparency, failing policies and institutions, and lack of strong and organised civil society groups. Meanwhile, Contreras-Hermosilla (2002) and Kishor and Oksanen (2006) share a similar argument that a mixture of discretionary powers, little oversight, low accountability, high profit from illegal operations, and remote logging locations are the essential drivers of corruption in timber trafficking.
14. In 2011, in Quang Binh alone, 129 Kiem Lam officers, whose monthly salary was only £24, collectively signed a resignation form (Nguyen Thuong 2011).
15. Corruption is recognised both nationally and internationally as a widespread social and criminal problem in Vietnam (Dao Le 2010; Nguyen 2013; Norad 2011; Phan and Pham 2010; Transparency International 2012). In the 2011 Corruption Perceptions Index, Vietnam performed below average with a score of 2.9 on a 0 (highly corrupt) to 10 (highly clean) scale. The country ranked 112 out of 182 assessed countries worldwide and 21 out of 35 countries in the Asia-Pacific region. The monetary cost of corruption in Vietnam is estimated at 2% of the country's GDP, whereas the number of corruption cases detected may account for only 5% of the total cases actually occurring (Phan and Pham 2010).
16. This book focuses on the domestic consumption of timber.
17. As noted in Chap. 3, *"endangered timber"* is the short name for "rare, precious and endangered timber species" that are listed in either Group IA or Group IIA in Decree 32/2006/ND-CP. Also, the phase *"consumption*

of endangered timber" in this research refers to the consumption of timber products such as decorative items, materials for house construction and wood furniture that are made from rare, precious and endangered timber species.

18. Wyatt (2013) identifies three types of buyers interested in wildlife products: accidental, denial and committed. The denial buyers are mostly those who work in zoos, aquariums, gardens and museums. In the context of timber trafficking in Vietnam, because timber is not commonly used for such businesses, denial consumption is not especially prevalent. However, the two other kinds of buyers indeed warrant attention.
19. A medium-sized plank of Sua timber (50 × 10 × 200 cm), which weighs about 110 kg, was sold in Phong Nha Ke Bang National Park with a price-tag of £87,000 (3EP02). It is quite shocking to know that at 1 kg of Ky Nam timber in Vietnam may be worth £300,000 although a leading scientist studying such timber warns buyers that its utility is modest, considering the timber is simply a kind of firewood (Pham 2011).
20. The current Vietnamese legislations often use the term *"forest plants"* (thực vật rừng), which is interpreted as comprising timber. Article 12 of the 2004 Forest Protection and Development Act specifies 16 groups of "prohibited acts" including illegal use and consumption of forest plants and animals.
21. Section 3 in Decree 157/2013/ND-CP specifies "violations on management of forest products" and consists of four articles (from Article 21 to Article 24) stipulating violations on illegally transporting, buying, selling, storing, processing and trading forest products. Accordingly, the illegal consumption of forest products is neither included nor clarified.
22. As largely agreed by the interviewees, the high demand for endangered timber in Vietnam is compounded by the lack of alternative materials and products especially for construction and decoration. That said, if there are sufficient amounts of plastic or metal products at a reasonable price, of a good quality and of a suitable design for decorations and furniture, there would certainly be a decrease in the demand for timber in general and endangered timber in particular.
23. In most countries worldwide, the vast majority of forest is owned by the state. In Asian countries such as Cambodia, India, Indonesia, Malaysia, Myanmar, the Philippines and Thailand, over 80% of forests are publicly owned, which essentially means that state agencies have uppermost rights over the use of the forest resources (ITTO 2011).

24. A study by Forest Trends (2013: 5) on the forestry policy in Vietnam, for example, concludes that "further criminalisation of logging would provide added impetus to illegal activities by empowering corrupt local officials, increasing the profits made by traders and wholesalers, and diminishing the benefits accruing to villagers".
25. To reduce forest degradation in Vietnam, Boissiere et al. (2009: 2743), for instance, argue that: "Communities can and should be actively involved in building a shared understanding of what the forest provides, how it can be conserved and the benefits to be obtained. Conservation in Vietnam has much to gain from local participation". In a similar vein, in an attempt to evaluate the work of forestland allocation in the current context of restructuring the forestry sector in Vietnam, To and Tran (2014: 75) propose that "the current system in which forest companies are given priority and free reign should be replaced by a system in which households and communities govern their own forests. In order to realise this vision, proactive steps must be taken to decentralise natural resource use and management towards household and community actors by affording them better access to forestland".

References

Alemagi, D., & Kozak, R. A. (2010). Illegal logging in Cameroon: Causes and the path forward. *Forest Policy and Economics, 12*(8), 554–561.

Ba Thang. (2014). Gốc cây gây sốt [Tree stumps become best sellers]. *Cong an Da Nang*, 1/11/2014. Retrieved March 25, 2015, from http://cadn.com.vn/news/65_122617_go-c-cay-gay-so-t.aspx

Boekhout van Solinge, T. (2010). Equatorial deforestation as harmful practice and criminological issue. In R. White (Ed.), *Global environmental harm*. Cullompton, UK: Willan.

Boissiere, M., Sheil, D., Basuki, I., Wan, M., & Le, H. (2009). Can engaging local people's interests reduce forest degradation in Central Vietnam? *Biodiversity and Conservation, 18*(10), 2743–2757.

Brunner, J. (2012). *Biodiversity conservation in Vietnam: A perfect storm*. IUCN Vietnam. Retrieved from http://cmsdata.iucn.org/downloads/iucn_nbsap_workshop_march_2012_paper_final.pdf

Cong Bac. (2014). 12 tháng tù giam cho cán bộ kiểm lâm tiếp tay cho lâm tặc [12 months imprisonment for Kiem Lam officer supporting Lam Tac].

VOV—Đài tiếng nói Việt Nam, 12/12/2014. Retrieved December 21, 2014, from http://vov.vn/phap-luat/12-thang-tu-giam-cho-can-bo-kiem-lam-tiep-tay-cho-lam-tac-370594.vov

Contreras-Hermosilla, A. (2001). *Law compliance in the forestry sector: An overview*. Washington, DC: World Bank.

Contreras-Hermosilla, A. (2002). Conference paper: Policy and legal options to improve law compliance in the forest sector—*Reforming Government Policies and the Fight against Forest Crime*, pp. 43–91. Rome: FAO.

Crow, M., Shelley, T. O. C., & Stretesky, P. (2014). Camouflage-collar crime: An examination of wildlife crime and characteristics of offenders in Florida. In T. Spapens, R. White, & M. Kluin (Eds.), *Environmental crime and its victims: Perspectives within green criminology* (pp. 185–204). Surrey: Ashgate.

Dao Le, T. (2010). *Bribery offences under Vietnamese criminal law in comparision with Swedish and Australian criminal law.* Ph.D. thesis, Hanoi Law University—Lund University.

Davies, M., Tyrer, J., & Croall, H. (2009). *Criminal justice*. New York: Pearson Longman.

de Jong, W., Do, D. S., & Trieu, V. H. (2006). *Forest rehabilitation in Vietnam: Histories, realities, and future*. Bogor, Indonesia: Center for International Forestry Research—CIFOR.

Department of Legal Affairs—Mard. (2012). Conference paper: The status of forest protection and solutions to enhance the work of forest protection. In *Preventing and combatting violations against legislations on forest protection* (pp. 18–29). Hanoi: The Department of Preventing and Combating Environmental Crimes (in Vietnamese).

Do, A. T. (2010). *Measuring the effectiveness of the enforcement of organised wildlife trade crime: A comparative assessment between Vietnam and England and Weles*. A thesis submitted in partial fulfilment of the requirements of the University of the West of England, Bristol for the degree of Doctor of Philosophy, University of the West of England.

Drury, R. C. (2009). *Identifying and understanding consumers of wild animal products in Hanoi, Vietnam: Implication for conservation management*. The Degree of Doctor of Philosophy, University College London.

Duy Tuan. (2012). 'Xã lâm tặc' và những khu rừng bị chảy máu ['Lam Tac commune' and the bleeding forests]. 09/03/2012. Retrieved February 6, 2015, from http://vietnamnet.vn/vn/xa-hoi/63333/-xa-lam-tac--va-nhung-khu-rung-bi-chay-mau.html

EIA. (2012). *Appetite for destruction: China's trade in illegal timber*. London: Environmental Investigation Agency.

EIA & TELAPAK. (2009). *Up for grabs: Deforestation and exploitation in Papua's plantations boom*. London and Bogor, Indonesia: Environmental Investigation Agency and Telapak. Retrieved from http://www.eia-international.org/wp-content/uploads/up-for-grabs.pdf

Emerton, L., Bishop, J., & Thomas, L. (2006). *Sustainable financing of protected areas: A global review of challenges and options*. Gland, Switzerland; Cambridge, UK: IUCN. Retrieved from http://www.europarc.org/uploaded/documents/128.pdf

Environmental Police Department. (2012). Conference paper: Opening paper—The work of preventing and combating the violations on forest protection. In *Preventing and combatting violations against legislations on forest protection* (pp. 7–10). Hanoi: The Department of Preventing and Combating Environmental Crimes (in Vietnamese).

Forest Trends. (2013). *Small-scale illegal logging in Vietnam: Implications for FLEGT and REDD+*. The Forest Trends.

FPD. (2012). *Report on the operational performance in 2011 and the working programme in 2012*. Hanoi: The Forest Protection Department (in Vietnamese).

FPD. (2013). *Forest protection force of Vietnam: Forty years of construction and development (1973–2013)*. Hanoi: The Forest Protection Department (in Vietnamese).

FSIV & FAO. (2009). *Vietnam forestry outlook study*. Bangkok: Forest Science Institute of Vietnam and FAO Regional Office for Asia and the Pacific.

Global Compact Network Vietnam. (2010). *A review of the social and environmental conditions of industries in Vietnam against the global compact principles*. Hanoi, Vietnam. Retrieved from http://www.globalcompactvietnam.org/upload/attach/GCNV%20Research%201.pdf

Government Inspectorate. (2014). *The declaration No.2490/TB-TTCP, 15/10/2014 of the inspection conclusion on the land management and utilisation with focus on land allocation and forest allocation in a number of projects in Dak Nong province in the 2004–2011 period*. Hanoi: Government Inspectorate of Vietnam (in Vietnamese).

Ha Binh—Trung Tan. (2014). Giảm 130.000 ha rừng, cỏ quan nào chịu trách nhiệm? [The redution of 130,000 ha of forests: Which authorities are accountable?]. *Tuoi Tre*, 13/12/2014. Retrieved January 31, 2015, from http://tuoitre.vn/tin/chinh-tri-xa-hoi/20141213/giam-130000-ha-rung-co-quan-nao-chiu-trach-nhiem/684381.html

Hewitt, D. (2013). *Potential legality issues from forest conversion timber*. Forest Trends. Retrieved from http://www.forest-trends.org/documents/files/doc_4137.pdf

ITTO. (2011). *Status of tropical forest management 2011*. Yokohoma: International Tropical Timber Organization.

Kaimowits, D. (2007). Forest law enforcement and rural livelihoods. In L. Tacconi (Ed.), *Illegal logging: Law enforcement, livelihoods and the timber trade* (pp. 110–138). London: Earthscan.

Khanh Hoan. (2013). Vụ lật xe chở gỗ lậu khiến 10 người chết: 8 bị cáo lãnh án tù [The case of flipped truck of illegal timber killing 10 people: 8 defendants are sentenced to prison]. *Laodong*, 24/1/2013. Retrieved June 2, 2015, from http://www.thanhnien.com.vn/pages/20130124/vu-lat-xe-cho-go-lau-khien-10-nguoi-chet-8-bi-cao-lanh-an-tu.aspx

Kishor, N., & Oksanen, T. (2006). *Combating illegal logging and corruption in the forestry sector. Strengthening forest law enforcement and governance*. Environment Matters, The World Bank Group.

Le Minh. (2014). Lâm tặc là ai? [Who are Lam Tacs?]. *Tien Phong*, 2/12/2014. Retrieved December 21, 2014, from http://www.tienphong.vn/xa-hoi-chuyen-hom-nay/lam-tac-la-ai-790928.tpo

Le Phuoc. (2014) (10/12/2014). Vụ rừng sinh thái Buôn Đôn bị phá—Trách nhiệm thuộc về ai? Tai Nguyen & Moi Truong. Retrieved January 30, 2015, from http://tainguyenmoitruong.com.vn/vu-rung-sinh-thai-buon-don-dak-lak-bi-pha-trach-nhiem-thuoc-ve-ai.html

MARD. (2012). *National forest status declaration*. Development, M. O. A. A. R. Hanoi: Ministry of Agriculture and Rural Development.

MARD. (2014). *Decision 3322/QD-BNN-TCLN on the national forest status declaration by 2013*. Hanoi: Ministry of Agriculture and Rural Development (in Vietnamese).

McElwee, P. (2004). You say illegal, I say legal: The relationship between 'illegal' logging and land tenure, poverty, and forest use rights in Vietnam. *Journal of Sustainable Forestry, 19*(1-3), 97–135.

Milliken, T., & Shaw, J. (2012). *The South Africa-Viet Nam Rhino Horn trade nexus: A deadly combination of institutional lapses, corrupt wildlife industry professionals, and Asian crime syndicates*. Johannesburg, South Africa: Traffic.

MNRE. (2010). *National Environmental Report—2010—Overview of the environmental in Vietnam*. Hanoi: Ministry of Natural Resources and Environment (in Vietnamese).

N.Hung. (2014). Về "thủ phủ vàng" nghe chuyện xử lâm tặc [Visit "the capital of gold" to listen the story of treating illegal loggers] *Gia Dinh & Xa Hoi* [*Family and Society*], 3/9/2014. Retrieved February 5, 2015, from http://giadinh.net.vn/xa-hoi/ve-thu-phu-vang-nghe-chuyen-xu-lam-tac-2014090308289738.htm

Nam Phuong. (2014). Làm trong sạch lứợng kiểm lâm [Purifying the Kiem Lam force]. *Nong Nghiep Viet Nam*, 2/12/2014. Retrieved December 21, 2014, from http://nongnghiep.vn/lam-trong-sach-luc-luong-kiem-lam-post135413.html

Nellemann, C. (2012). *Green carbon, black trade: Illegal logging, tax fraud and laundering in the worlds tropical forests. A rapid response assessment*. Birkeland Trykkeri AS, Norway: Interpol Environmental Crime Programme—United Nations.

Nellemann, C., Henriksen, R., Raxter, P., Ash, N., & Mrema, E. (2014). *The environmental crisis—Threats to sustainable development from illegal exploitation and trade in wildlife and forest resources*. United Nations Environment Programme and GRID-Arendal, Nairobi and Arendal. Retrieved from http://www.grida.no/

Ngoc Linh. (2014). Chung quanh việc giao đất, giao rừng cho đồng bào dân tộc thiểu số [Regarding the issue of allocating lands and forests for the ethnic minority residents]. *Nhan Dan*, 22/11/2014. Retrieved March 26, 2015, from http://www.nhandan.com.vn/mobile/_mobile_kinhte/_mobile_tintuc/item/24901102.html

Nguyen Cau. (2014). Rừng đặc dụng Bà Nà—Núi Chúa bị tàn phá do đâu [Why is the special use forest of Ba Na—Nui Chua destroyed?]. *Cong An Nhan Dan Online*, 31/10/2014. Retrieved January 20, 2015, from http://cand.com.vn/Xa-hoi/Rung-dac-dung-Ba-Na---Nui-Chua-bi-tan-pha-do-dau-276935/

Nguyen Duy. (2013). Vụ lật xe chở gỗ đè chết 10 người: Các bị cáo lĩnh 206 tháng tù [The case of flipped truck of illegal timber killing 10 people: The defendants get 206 month in prison]. *Dantri*, 25/1/2013. Retrieved June 2, 2015, from http://dantri.com.vn/phap-luat/vu-lat-xe-cho-go-de-chet-10-nguoi-cac-bi-cao-linh-206-thang-tu-689243.htm

Nguyen, M. H., Vu, V. D., Nguyen, V. S., Hoang, V. T., Nguyen, H. D., Pham, N. T., et al. (2007). *Report on the review of Vietnam's wildlife trade policy*. Hanoi: CRES/FPD/UNEP/CITES/IUED.

Nguyen Nam. (2015). Khởi tố, bắt giam kiểm lâm viên huyện Hàm Tân [Charging and arresting a Kiem Lam staff in Ham Tan district]. *Tuoi Tre*, 28/3/2015. Retrieved March 29, 2015, from http://tuoitre.vn/tin/chinh-tri-xa-hoi/20150328/khoi-to-bat-giam-kiem-lam-vien-huyen-ham-tan/726599.html

Nguyen Thuong. (2011). Bao giờ kiểm lâm sống đúợc bằng lúóng? [When can Kiem Lam live with their salary?]. *Con Nguoi & Thien Nhien*, 28/11/2011.

Retrieved June 2, 2015, from http://www.thiennhien.net/2011/11/28/bao-gio-kiem-lam-song-duoc-bang-luong/

Nguyen, T. N. (2013). Tham nhũng—nguyên nhân và biện pháp ngăn ngừa [Corruption—Causes and preventive measures]. *Communist Review—Tap chi Cong san*, 17/1/2013. Retrieved June 2, 2015, from http://www.tapchicongsan.org.vn/Home/PrintStory.aspx?distribution=19790&print=true

Nguyen, V. S. (2003). *Vietnam's illegal trade in wildlife: Why it flourishes*. Hanoi, Vietnam: Economy and Environment Program for Southeast Asia.

Nguyen Xuan. (2013). Hành trình sa ngã của một sĩ quan [The wrong journey of a officer]. *Khanh Hoa Online*, 23/11/2013. Retrieved June 2, 2015, from http://baokhanhhoa.com.vn/phap-luat/201311/hanh-trinh-sa-nga-cua-mot-si-quan-2277047/

Nong Nghiep Vietnam. (2012). Kiểm lâm—nghề đặc biệt [Kiem Lam—The special occupation]. *Nong nghiep Vietnam*, 22/05/2012. Retrieved June 2, 2015, from http://nongnghiep.vn/nongnghiepvn/72/1/15/94960/kiem-lamnghe-dac-biet.aspx

NORAD. (2011). *Joint evaluation of support to anti-corruption efforts—Viet Nam country report*. Oslo, Norway: Norwegian Agency for Development Cooperation.

Nurse, A. (2013). *Animal harm: Perspectives on why people harm and kill animals*. Farnham, Surrey, UK; Burlington, VT: Ashgate.

OECD. (2012). *Illegal trade in environmentally sensitive goods*. Paris: Organisation for Economic Co-operation and Development.

Olken, B. A., & Pande, R. (2012). Corruption in developing countries. *Annual Review of Economics, 4*, 479–509.

Pendleton, M. R. (2007). *The social basis of illegal logging and forestry law enforcement in North America*. In L. Tacconi (Ed.), *Illegal logging, law inforcement, livelihoods and the timber trade*. London: Earthcan.

Pham, D. X. (2008). *Investigation delivered by the investigation police on crimes of economic management and official position over the violations against regulations on forest exploitation and protection*. The thesis of PhD in Law, The People's Police Academy (in Vietnamese).

Pham, S. D. (2011). Điều gì khiến kỳ nam đắt khủng khiếp? [Why Ky Nam is extremely expensive?]. *Dan Tri*, 18/07/2011. Retrieved June 2, 2015, from http://dantri.com.vn/su-kien/dieu-gi-khien-ky-nam-dat-khung-khiep-499653.htm

Pham, T. T., Moeliono, M., Nguyen, T. H., Nguyen, H. T., & Vu, T. H. (2012). *The context of REDD+ in Vietnam—Drivers, agents and institutions*. Occasional

Paper 75, CIFOR, Bogor, Indonesia. Retrieved from http://www.cifor.org/publications/pdf_files/OccPapers/OP-75.pdf

Phan, X. S., & Pham, T. L. (2010). *Identifying corruption and solutions to prevent and combat corruption in Vietnam nowadays*. Hanoi: Chinh tri quoc gia (in Vietnamese).

Phap Luat Viet Nam. (2013). Nguyên Phó Văn phòng Cỏ quan CSĐT tù 8 năm vì nhận hối lộ [The former chief of investigative police office sentenced 8 years in prison for receiving bribe]. *Dantri*, 24/4/2013. Retrieved June 2, 2015, from http://dantri.com.vn/phap-luat/nguyen-pho-van-phong-co-quan-csdt-tu-8-nam-vi-nhan-hoi-lo-723047.htm

Rosander, M. N. (2008). *Illegal logging: Current issues and opportunities for Sida/SENSA engagement in Southeast Asia*. Bangkok, Thailand: RECOFTC & Sida.

Schloenhardt, A. (2008). *The illegal trade in timber and timber products in the Asia-Pacific region—Research and Public Policy Series No. 89*. Canberra: Australian Institute of Criminology.

Seneca Creek Associates, LLC. (2004). *"Illegal" logging and global wood markets: The competitive impacts on the US wood products industry*. American Forest & Paper Association.

Sikor, T., & To, P. X. (2011). Illegal logging in Vietnam: Lam Tac (forest hijackers) in practice and talk. *Society & Natural Resources, 24*(7), 688–701.

Stewart, D. (2014). Project leaf, and interpol's work on illegal logging and forest crime. In T. Spapens, R. White, & M. Kluin (Eds.), *Environmental crime and its victims: Perspectives within green criminology* (pp. 236–247). Surrey: Ashgate.

Tacconi, L. (2007). The problem of illegal logging. In L. Tacconi (Ed.), *Illegal logging: Law enforcement, livelihoods and the timber trade* (pp. 1–16). London: Earthscan.

Thai, B. D., & Trung, T. (2012). Khởi tố vụ án phá rừng nghiêm trọng: Xử lý trách nhiệm nhiều cá nhân để xảy ra phá rừng [Prosecution of the serious case of illegal logging: Handling many individuals responsible for illegal logging]. *Tuoi Tre*, 27/08/2012. Retrieved June 2, 2015, from http://tuoitre.vn/Chinh-tri-xa-hoi/Phap-luat/509006/xu-ly-trach-nhiem-nhieu-ca-nhan-de-xay-ra-pha-rung.html

Thanh Chau. (2013). TAND tỉnh Hà Tĩnh xét xử vụ phá rừng đầu nguồn Sỏn Hồng : Tuyên phạt 15 bị cáo 36 năm 6 tháng tù [Ha Tinh Province People's Court adjudicates the illegal logging case in the Son Hong watershed forest: 15 accused sentenced to 36 years and 6 months imprisonment. *Nhan Dan*, 31/01/2013. Retrieved June 2, 2015, from http://www.nhandan.com.vn/phapluat/thoi-su-phap-luat/item/225502-.html

The Economist. (2010). Keeping it in the community—Well-organised locals often make the best forest managers, but they need help. *The Economist*, 23/9/2010. Retrieved February 4, 2015, from http://www.economist.com/node/17062703

To, P. X., Tran, L. H., & Nguyen, T. Q. (2014). *The overview of the Vietnamese timber sector: Current status and trends*. Forest Trends/VIFORES/FPA Binh Dinh.

To, X. P., & Canby, K. (2011). *Vietnam: Overview of forest governance and trade*. Kuala Lumpur: European Forest Institute—FLEGT Asia Regional Office.

To, X. P., & Tran, H. N. (2014). *Forest land allocation in the context of forestry sector restructuring: Opportunities for forestry development and upland livelihood improvement*. Tropenbos International Viet Nam, Hue, Vietnam.

TRAFFIC. (2008). *What's driving the wildlife trade? A review of expert opinion on economic and social drivers of the wildlife trade and trade control efforts in Cambodia, Indonesia, Lao PDR and Vietnam*. Washington, DC: TRAFFIC.

Transparency International. (2012). *Overview of corruption and anti-corruption in Vietnam No.315*. U4 & The Transparency International. Retrieved from http://www.u4.no/publications/overview-of-corruption-and-anti-corruption-in-vietnam/

Tuoi Tre. (2011). Kiểm lâm "làm luật"—Kỳ 2: Làm sai lệch hồ số [Kiem Lam "making laws"—Vol 2: Falsifying documentations]. *Tuoi Tre*, 4/10/2011. Retrieved June 3, 2015, from http://tuoitre.vn/Tuoi-tre-cuoi-tuan/Van-de-Su-kien/458808/Kiem-lam-%E2%80%9Clam-luat%E2%80%9D---Ky-2-Lam-sai-lech-ho-so.html

UNODC. (2012). *Wildlife and forest crime analytic toolkit—Revised edition*. New York: United Nations Office for Drugs and Crime. Retrieved from http://www.unodc.org/documents/southeastasiaandpacific//Publications/wildlife/CJS_Response_-_VIETNAM_01_13_Dec_201.pdf

UNODC. (2013). *Transnational organized crime in East Asia and the Pacific*. Bangkok, Thailand: United Nations Office on Drugs and Crime—Regional Office for Southeast Asia and the Pacific.

USAID. (2013). *Vietnam tropical forest and biodiversity assessment*. Retrieved from http://www.usaid.gov/sites/default/files/documents/1861/Vietnam118_119FinalDraft%2015Oct2013%20(Public%20Version).pdf

Uyen Thu. (2013). *Vì sao hơn 130,000 ha rừng Tây Nguyên bị bức tử: Kỳ: Kỳ 1 "Cơn lốc" cao su "tàn sát" rừng [Why more than 130,000 ha of forests in Central Highlands are forced to suicide: Vol 1: The rubber "cyclones" massacre the forests]*. Gia Lai, p. 3, released 16/9/2013: Gia Lai Newspaper.

V.Tao. (2013). "Trảm" 2 sĩ quan huyện đội "ăn chặn kỳ nam" [Handling 2 officers of district army appropriating agarwood]. *Nguoi Lao Dong*,

19/06/2013. Retrieved June 3, 2015, from http://nld.com.vn/phap-luat/tram-2-si-quan-huyen-doi-an-chan-ky-nam-20130619080619l.htm
VAF. (2014). Workshop outcomes: "The shortcoming of the policy and practice in the work of forest and land allocation and post allocation"—Kết quả Tọa đàm: "Những bất cập, tồn tại về chính sách và thực tiễn trong công tác giao đất, giao rừng và sau giao đất, giao rừng", 8/5/2014. Hanoi: Vietnam Administration of Forestry.
Vanrijckeghem, C., & Weder, B. (2001). Bureaucratic corruption and the rate of temptation: Do wages in the civil service affect corruption, and by how much? *Journal of development economics, 65*(2), 307–331.
Viet Quoc. (2015). Bình Thuận khó giữ đúợc rừng giáp ranh [It is hard for Binh Thuan to protect the adjacent forests]. *VOV*, 7/1/2015. Retrieved January 30, 2015, from http://vov.vn/phap-luat/binh-thuan-kho-giu-duoc-rung-giap-ranh-376264.vov
Vietnam Chamber of Commerce and Industry. (2011). *Extractive Industries Transparency Initiative (EITI)—Solution to sustainable development*. Retrieved April 04, 2013, from http://vccinews.com/news_detail.asp?news_id=23299
VOV. (2014). Cảnh báo tình trạng phá đá "săn" Hoàng Đàn [Warning the situations of quarrying to "hunt" Hoang Dan timber]. *Lang Son Television and Radio*, 31/10/2014. Retrieved January 19, 2015, from http://langsontv.vn/node/46079
VTV2. (2013). *Timber villages in Vietnam in the context of international integration. TV Programme of Green Vietnam*. Education Science Board, Vietnam Television Broadcast.
Wells, A., Gatto, F. D., Richards, M., Pommier, D., & Contreras-Hermosilla, A. (2007). Rural livelihoods, forest law and the illegal timber trade in Honduras and Nicaragua. In L. Tacconi (Ed.), *Illegal logging: Law enforcement, livelihoods and the timber trade* (pp. 139–167). London: Earthscan.
White, R. (2011). *Transnational environmental crime: Toward an eco-global criminology*. London: Routledge.
White, R., & Heckenberg, D. (2014). *Green criminology: An introduction to the study of environmental harm*. Oxon, UK; New York: Routledge.
World Bank. (2005). *Vietnam environment monitor 2005—Biodiversity*. Hanoi: The Vietnamese Ministry of Environment and Natural Resources, the World Bank, and the Swedish International Development Cooperation Agency. Retrieved from http://siteresources.worldbank.org/INTEAPREGTOPENVIRONMENT/Resources/VN_Env_Monitor_05.pdf
World Bank. (2009). *Country social analysis: Ethnicity and development in Vietnam—summary report*. Washington, DC: World Bank. Retrieved from

http://siteresources.worldbank.org/INTEAPREGTOPSOCDEV/Resources/499760ESW0Whit1C10VietnamSummary1LR.pdf

Wyatt, T. (2013). *Wildlife trafficking—A deconstruction of the crime, the victims and the offenders*. Basingstoke, UK: Palgrave Macmillan.

Wyatt, T., & Cao, N. A. (2015). *Corruption and wildlife trafficking* (pp. 1–54). Bergen, Norway: U4 Anti-Corruption Resource Centre.

Xuan Hong. (2012). Khi chủ rừng và kiểm lâm cùng nhau... phá rừng [When forest owner and Kiem Lam together... deforest]. *Nguoi Dua Tin*, 27/12/2012. Retrieved May 29, 2015, from http://www.nguoiduatin.vn/khi-chu-rung-va-kiem-lam-cung-nhau-pha-rung-a42177.html

Concluding Remark

While it can be seen that "never in the recent history of humankind have issues of ecological and social justice been so intertwined or so urgent" (White 2013c: 8), it is predicted that the importance of green crime, and wider green harms, "looks set to increase in the near future" (Hall and Farrall 2013: 131). The increasing imperativeness of green crime, in conjunction with the long-lasting lack of interest by traditional criminology in the crime, suggests that much remains to be done to attain an adequate understanding of various forms of green crime particularly in the context of Eastern developing societies that "have often been side-lined by Western criminology" (Hall 2013a: 150). Without such understanding, it is not possible to effectively tackle green crime. In the case of timber trafficking in the Asia-Pacific region, for example, Schloenhardt (2008: 146) observes:

> One of the most immediate observations about the illegal trade in timber and timber products in the Asia–Pacific region is the absence of systematic scholarly research on the topic. It is difficult, if not impossible, to analyse the patterns and magnitude of the illicit trade without further study of the core issues and the surrounding circumstances.

N.A. Cao, *Timber Trafficking in Vietnam*, Palgrave Studies in Green Criminology,
DOI 10.1007/978-3-319-64280-2

To reiterate, the over-arching aim of this book is to obtain a detailed and systematic understanding of timber trafficking in the current context of Vietnam. Based on this understanding, another aim is to make empirical and theoretical contributions to the relatively new body of knowledge of green criminology. These aims can be accomplished by answering and discussing in depth three research questions:

1. How is timber trafficking currently occurring in Vietnam? In other words, what is the current typology of timber trafficking in Vietnam?
2. How is timber trafficking affecting Vietnamese society or what is the victimisation from timber trafficking in Vietnam?
3. What can be done to curb timber trafficking in Vietnam or what are the solutions to better control timber trafficking in Vietnam?

It is expected that the information provided throughout the previous six chapters of this book will have answered in detail these three questions and made some contributions to knowledge, particularly in the discipline of green criminology that, despite existing as a "niche area" in criminology as commented by Floyd (2015: 278), has now become "an impressive body of scholarly work" (Shearing 2015: 259).

It is also expected that this book has substantially filled in the four gaps in knowledge about timber trafficking identified at the beginning of Chap. 1. Indeed, this book has: (1) paid profound attention to revealing various impacts of timber trafficking on the well-being of underprivileged groups; (2) employed green criminological perspectives to investigate the green crime of timber trafficking; (3) used a variety of techniques to collect both primary and secondary data including in-depth interviews with those who directly fight against, and those who are victimised by, timber trafficking; and (4) examined the crime in Vietnam—an important Eastern location of timber trafficking, but overlooked by international criminologists.

Nevertheless, this book is only one exploratory investigation of timber trafficking with its focus on illegal activities that take place inside the boundary of Vietnam. It is undoubtedly worth undertaking further empirical research in Vietnam revolving around either the entirely legal but ecologically harmful practices of logging or cross-border timber trafficking. It is also important to carry out research focusing on non-human species and the environment as the primary sources of victimisation. One possibility would be examining the extensive ecological degradation and pollution produced by the excessive and irresponsible operations in the harvesting, transportation and processing of timber.

Further studies would be particularly meaningful and fruitful if they employed critical conceptual frameworks in green criminology to define and conceptualise the nature and the root of the studied phenomena. Likewise, as recommended by South (2015: 270) "a criminological imagination for the twenty-first century should find it stimulating and relevant to think about employing the concept of 'security' in a broader sense than usual", so future studies could integrate a specific perspective in security studies, either human security or environmental security, for the purpose of examining the crime victimisation, which is "an important direction for green criminology" (Stretesky et al. 2014: 153).

Finally, since time immemorial, the Vietnamese have firmly believed that forests were one of the most valuable assets of nature, saying "rừng vàng, biển bạc" (forests are gold and seas are silver). Forests, for generations, have brought great benefits to a substantial proportion of the Vietnamese population as well as the country's biodiversity. It is important to remember that currently the remaining old-growth forests only account for 1% of Vietnamese territory, but they contain 85% of the globally, 16th-ranked, biodiversity of the country (USAid 2013). This means that even a small forest loss would have significantly harmful and regrettable impacts on both human and non-human species. Thus, forest protection is now a crucial mission for Vietnam. This mission cannot be accomplished without considerable effort from all related actors: the Vietnamese, national politicians, legislators, enforcers, international supporters and researchers.

Appendix A: Interview List

Code	Date of interview	Interviewee cohort
1EP01	2 August 2013	Environmental police
2FO01	3 August 2013	Forest protection
3EP02	7 August 2013	Environmental police
4TT01	8 August 2013	Timber trader
5LR01	10 August 2013	Local resident
6FO02	14 August 2013	Forest protection
7NG01	19 August 2013	NGO staff
8EP03	25 August 2013	Environmental police
9FO03	26 August 2013	Forest protection
10IP01	27 August 2013	Investigative police
11IP02	27 August 2013	Investigative police
12EP04	29 August 2013	Environmental police
13LA01	30 August 2013	Local authority
14LR02	30 August 2013	Local resident
15FO04	31 August 2013	Forest protection
16NG02	20 August 2013	NGO staff
17EP05	22 August 2013	Environmental police
18EP06	22 August 2013	Environmental police
19NG03	23 August 2013	NGO staff
20EP07	6 September 2013	Environmental police
21IP03	7 September 2013	Investigative police
22IP03	7 September 2013	Investigative police

(continued)

N.A. Cao, *Timber Trafficking in Vietnam*, Palgrave Studies in Green Criminology,
DOI 10.1007/978-3-319-64280-2

(continued)

Code	Date of interview	Interviewee cohort
23FO05	9 September 2013	Forest protection
24LA02	10 September 2013	Local authority
25LR03	13 September 2013	Local resident
26EP09	13 September 2013	Environmental police
27FO06	14 September 2013	Forest protection
28TT02	15 September 2013	Timber trader
29IP04	17 September 2013	Investigative police
30LR04	18 September 2013	Local resident
31LR05	18 September 2013	Local resident
32IP05	19 September 2013	Investigative police
33LA03	19 September 2013	Local authority
34NG04	26 September 2013	NGO staff
35FO07	26 September 2013	Forest protection
36LR06	15 September 2013	Local resident
37TT03	27 September 2013	Timber trader
38EP10	9 August 2013	Environmental police
39 NG 05	23 September 2013	NGO staff
40EP11	17 September 2013	Environmental police
41IP06	10 September 2013	Investigative police

Appendix B: Main Policies and Legislation on Environmental Protection and Forest Governance in Vietnam

Number and Time	Names	Summaries
1. *International conventions*		
Joined in 1994	Convention on International Trade in Endangered Species of Wild Fauna and Flora (CITES)	Established in 1973, CITES is an international agreement between governments, aiming to ensure sustainable international trade in wildlife. After joining CITES in 1994, Vietnam has enacted a number of obligatory laws, and established CITES Authority of Vietnam as the administrative agency, with the Institute of Ecology and Biological Resources and the Centre for Natural Resources and Environmental Studies as the two scientific authorities

(*continued*)

N.A. Cao, *Timber Trafficking in Vietnam*, Palgrave Studies in Green Criminology,
DOI 10.1007/978-3-319-64280-2

(continued)

Number and Time	Names	Summaries
1994	The Convention on Biological Diversity Vienna Convention for the Protection of the Ozone Layer and the Montreal Protocol The United Nations Framework Convention on Climate Change	After joining each agreement, the corresponding administrative agencies are established, and laws and regulations are issued to put the agreement commitments into practice
1995	Basel Convention on the Control of Transboundary Movements of Hazardous Wastes and Their Disposal	
2002	Kyoto Protocol as a Non-Annex 1 Party	
2006	Stockholm Convention on Persistent Organic Pollutants	
2. *Resolutions by Central Committee and Politburo of Communist Party of Vietnam*		
Resolution 41/NQ/TW Politburo (VIII), declared on 15/11/2004	"The Environmental Protection in the Era of Strengthening the Country's Industrialisation and Modernisation"	Emphasises that environmental protection is a matter of survival for humanity, one that ensures people's health and welfare, contributes to socio-economic development, political stability, national security and boosts economic global integration of our country. Given forest protection, the Resolution demands the work of strengthening forest protection and development and promoting land and forest allocation and appropriate contractual forms for individuals, families and collectives

Number and Time	Names	Summaries
Resolution 24-NQ/ TW—7th Plenary, (XI) Central Committee, declared on 3/6/2013	"The Proactive Response to Climate Change and the Enhancement of the Governance of Natural Resources and Environmental Protection"	Aims that by 2020, Vietnam is basically ready to respond to climate change, prevent natural disasters, reduce greenhouse gas emissions, and improve the utilisation of natural resources in a reasonable, effective and sustainable manner. It also aims to restrain the increasing levels of pollution and biodiversity degradation, ensure residential environment and ecological harmony, and build a green, environmentally friendly economy
3. *Acts by the National Assembly of Vietnam*		
No.15/1999/ QH10 on 21/12/1999, partly amended by No.37/2009/ QH12 passed on 19/6/2009	Penal Code	Offers the entire Chapter XVII with 11 articles defining 11 environmental offences, including the offence of "Destroying Forests" (Article 189) as the conducts of burning and destroying forests. The most relevant article regarding timber trafficking is Article 175 namely "Breaching Regulations on Forest Exploitation and Protection" that is placed in Chapter XVII about "Crimes Against Economic Management Order". This article defines two groups of criminal conduct: (1) Illegally exploiting forest trees or committing other acts of violating the state's regulations on forest exploitation and protection, if not falling under the cases specified in Article 189 of this Code; (2) Illegally transporting and/or trading in timber, if not falling into the cases specified in Article 153 and Article 154 of this Code

(*continued*)

(continued)

Number and Time	Names	Summaries
No.29/2004/ QH11, passed on 3/12/2004	Forest Protection and Development Act	Classifies the forests in Vietnam, clarifies the governance, development and exploitation of the forests; and the rights and obligations of the forest owners. It defines 16 groups of prohibited acts, including 1. Illegally logging or exploiting forests. ... 9. Illegally transporting, processing, advertising, trading in, using, consuming, storing, exporting or importing forest plants and animals. ... 16. Other acts of harming forest resources and ecosystems
No.52/2005/ QH11, passed on 29/11/2005	Environmental Protection Act	Regulates the measures and resources for environmental protection, stipulates the rights and obligations of different actors comprising state agencies, organisations, family households and individuals in regard to the mission of environmental protection
No.20/2008/ QH12 passed on 13/11/2008	Biodiversity Act	Regulates the work of conservation and sustainable development of biodiversity and the roles of different actors involved in the work
4. *Decrees by the government and decisions by the Prime Minister*		
Decision 134/2004/ QD-TTg, dated 20/7/2004	A Number of Policies to Support the Households of Poor Ethnic Minorities with Cultivable, Residential Lands, Residential Houses and Clean Water	Offers various measures to support production lands, residential lands, residential houses and clean water for poor ethnic minorities to improve their living conditions and livelihoods, to overcome poverty sooner. One of the measures is to allow local authorities to harvest timber for building houses for ethnic minority residents

Number and Time	Names	Summaries
Decree 23/2006/ ND-CP, dated 30/3/2006	The Implementation of the Forest Protection and Development Act	Specifies the key issues in forest governance such forest planning, protection, development, exploration, allocation, revocation, conversion and ownership
Decree 32/2006/ ND-CP, dated 30/3/2006	The Management of Forest Endangered, Rare and Precious Fauna and Flora	Provides the list of endangered, rare and precious (called endangered for short) forest animals and plants in Vietnamese forests and the management of these species. Endangered forest plants and animals are divided into two groups: – Group I: strictly banned from exploitation and use for commercial purposes. Forest plants and animals of Group I are further divided into: Group IA, which consists of 15 species of forest plants (e.g Sua, Thuy Tung) and Group IB, which consists of 62 species of forest animals. – Group II: restricted from exploitation or use for commercial purposes. Forest plants and animals of Group II are further divided into: Group IIA, which consists of 37 species of forest plants (e.g. Lim, Trac, Nghien, Cam Lai) and Group IIB, which consists of 89 species of forest animals
Decision 186/2006/ QD-TTg, dated 14/8/2006	The Enactment of the Regulations on Forest Management	Regulates the management, protection, development and exploitation of special-use forests, protective forests and production forests

(continued)

(continued)

Number and Time	Names	Summaries
Decree 157/2013/NĐ-CP, dated 11/11/2013	Punishing Administrative Violations in the Field of Forest Management, Protection and Forest Product Management	Stipulates in detail all administrative violations in the field of forest governance, the forms and levels of sanction, the powers and procedures of handling the violations. Article 12, for example, define the conduct of "illegal harvesting of forest" for three different types of forest (special-use, protective and production forests) with three different type of timber (plain, Group IIA and Group IA), and the punishments for each conduct. Under this article, the minimum monetary fine is £18 for illegal harvesting of less than 0.5 m^3 of plain timber in productive forests (Point a, Clause 2). Meanwhile, the maximum monetary fine is £6700 for illegal harvesting of 7–12.5 m^3 of Group IIA timber or 1.5–2 m^3 of Group IA timber in production forests, or 5–10 m^3 of Group IIA timber or 1–1.5 m^3 of Group IA timber in production forests, or 2.5–5 m^3 Group IIA timber or 0.7–1 m^3 of Group IA timber in special-use forests. Illegal harvesting of more than these maximum administrative volumes will be the subject of criminal law procedures

Number and Time	Names	Summaries
5. *Circulars and Joint Circulars by ministries*		
Joint Circular 19/2007/TTLT/ BNN&PTNT-BTP-BCA-VKSNDTC-TANDTC, dated 8/3/2007	Guiding the Implementation of Articles in the Penal Code on Offences in the Field of Forest Management, Forest Protection and Forest Product Management	Jointly issued by MARD, Ministry of Justice, Ministry of Public Security, Supreme People's Procuracy and Supreme People's Court. It explains and specifies some terms, articles and criminal thresholds for the offences in the field of forest and forest products. For example, it clarifies that *"causing serious consequences"* in illegal timber harvesting will be the subject of criminal law procedures, and means the illegal timber amounts will be between more than the maximum administrative volumes and less than two times these volumes. Illegal timber amounts of more than two times and less than four times the maximum administrative volumes are considered as *"causing very serious consequences"*
Circular 35/2011/ TT-BNNPTNT, dated 20/5/2011	Guiding the Harvest of Timber and NTFPs	Issued by MARD, the circular guides the conditions and procedures for planned logging, salvage logging and harvest of NTFPs
Circular 01/2012/ TT-BNNPTNT, dated 4/1/2012	Regulations on the Inspection of Legality and Originality of Forest Products	Issued by MARD, the circular regulates what is considered to be legal documentation for a given timber volume and how to inspect the timber's legality and originality
6. *National programmes and strategies*		
The 327 Programme issued by Prime Minister Decision No.327, dated 15/9/1992	Greening the Bare Land and the Denuded Hills	Receiving $213 million in funding from the Vietnamese government, it aims at establishing new protection and special-use forests. As a result of this programme, 466,000 households were contracted to protect 1.6 million hectares of forests: 299,000 hectares of forest were naturally regenerated; and 397,000 hectares of forest were newly planted forests

(*continued*)

(continued)

Number and Time	Names	Summaries
The 661 Programme issued by Prime Minister Decision No.661 dated 29/07/1998	Five Million Hectare Reforestation Programme	Receiving $250–300 million in funding from the Vietnamese government and $450 million from foreign direct investments, it aimed at obtaining a national forest area of 14.3 million hectares in 2010 (the same level of forest cover in 1943). The outcome was that the forest area in 2010 in Vietnam was about 13.5 million hectares, 0.8 million less than the expected aim
Forestry Development Strategy, approved by Prime Minister's Decision 18/2007/ QD-TTg dated 5/2/2007	Vietnam Forestry Development Strategy 2006–2020	Recognises that the forestry sector plays an important role in protecting the environment, conserving biodiversity and reducing poverty, especially for mountainous inhabitants, contributing to social stability, security and defence. It aims to increasing the forest cover rate to 42–43% by 2010 and 47% by 2020. The production value of the forestry sector will account for 2–3% of national GDP by 2020

The Vietnamese government introduced several initiatives targeting effective forest management and sustainable forest harvesting such as the Programme of Exploitation, Processing and Trading of Forest Products, the Programme of Renovation of Forestry State Enterprises, and the Forest Sector Support Partnership. Additionally, the Prime Minister of Vietnam has issued four different directives for the same purpose of implementing urgent measures to prevent the cutting, burning and illegal exploitation of forests, including Directive No.286/1997, No.12/2003, No.08/2006 and No.1685/2011. For more discussion on forestry policies please see:FSIV and FAO 2009; MARD 2001, 2012b; Morris et al. 2004; Pham 2010; To and Canby 2011; USAid, 2013.

Appendix C: Criminal Case Records Consulted in this Book

Type of record	Judicial agencies	Description
Investigation Conclusion of Criminal Case No.07/BC-PC15 dated 7/9/2009	Investigation Police Agency, Public Security Department of Kon Tum province	The two accused who were H'Lăng ethnic minority illegally harvested three Giổi trees (Group III) and one Du Sam tree (Group IIA) in Zone 101, Dak Glei Protective Forest with a total of 13.4 m^3 of timber for the purpose of building a house. This can be seen as a typical SSITH case
Investigation Conclusion of Criminal Case No.02/KLDT-PC46 dated 30/9/2011	Investigation Police Agency, Public Security Department of Kon Tum province	A group of 16 accused illegally harvested 7.8 m^3 of Trac timber (Group IIA) in Dak Uy Special-used Forest. The group discussed and assigned the tasks of preparing the tools, cutting trees and porting the timber for each member. The illegal timber was sold to a timber trader in Bac Ninh province. This can be seen as a typical MSITH case

(continued)

N.A. Cao, *Timber Trafficking in Vietnam*, Palgrave Studies in Green Criminology,
DOI 10.1007/978-3-319-64280-2

(continued)

Type of record	Judicial agencies	Description
Investigation Conclusion of Criminal Case No.25/KLDT-PC46 dated 12/1/2012	Investigation Police Agency, Public Security Department of Quang Binh province.	A director of a carpentry company in Nam Dinh province bought 400 planks of illegal Gu Lau timber (Group IA) in Quang Binh with a price of £500/m³. This case underpins the argument that processing factories in lowland localities may use illegal timber
Court Judgments of Criminal Case No.13/2013/HSST, dated 28/1/2013	People's Court of Bo Trach district, Quang Binh province	Forest rangers of Bo Trach Forest Enterprise detected four people transporting illegal timber by buffalos. The transporters then came to the enterprise premises to reclaim the seized timber. Prevented by the rangers, they used knives, stones and sticks to beat the rangers. This case demonstrates the violence against on-duty staff
Investigation Conclusion of Criminal Case No.30/ANDT, dated 29/1/2013	Security Investigation Agency—The Ministry of Public Security	Nine commercial companies in three provinces, Dak Nong, Binh Phuoc and Tien Giang, illegally traded a total of almost 2500 m³ of timber. This is a notable case indicating the techniques used in the illegal trade and processing of timber: illegal timber is bought from local people, and then laundered in nearby processing factories before being sold to further markets
Investigation Conclusion of Criminal Case No.18/KL-DT dated 12/3/2013	Investigation Police Agency of the Ba Be District Public Security, Bac Kan Province	Six separate groups of local loggers illegally cut down a total 15 Nghien trees (Group IIA) in 83 Zone, 50 Point, Ba Be National Park. The illegal timber planks were brought and concealed in the loggers' village, then sold to local timber traders. This is an instance of MSITH. It also indicates that timber traders play an important role in the criminal operation, but it is difficult to hold them accountable

Type of record	Judicial agencies	Description
Investigation Conclusion of Criminal Case No.24/KLĐT dated 10/5/2013	Investigation Police Agency of the Ba Be District Public Security, Bac Kan Province	In order to have timber to build a house, a villager asked four other villagers coming to 51 Zone, 104 Point, Ba Be National Park to illegally cut down two Nghien trees (17.5 m^3). This is a typical case of SSITH
Investigation Conclusion of Criminal Case No.13/KLDT-PC46 dated 15/1/2013	Investigation Police Agency, Public Security Department of Quang Binh province	12 defendants illegally harvested three Sua trees (Group IA) in Phong Nha Ke Bang National Park. Sua is a critically endangered and exceptionally valuable timber species, and the three logged trees were worth VND71 billion (£2.4 million). The illegal logging case triggered other serious illegal activities taking place inside the park and nearby villages such as timber robbery, assault, vandalism and fraud. This is an ample example of the severe multifaceted impacts of timber trafficking in Vietnam
Court Judgments of Criminal Case No.41/2013/HSST, dated 5/6/2013	People's Court of Bo Trach district, Quang Binh province	

References

Abbott, C. (2008). *An uncertain future: Law enforcement, national security and climate change*. London: Oxford Research Group.

ADB. (2005). *Vietnam: Country environmental analysis*. Hanoi, Vietnam: Asia Development Bank.

Agnew, R. (2012a). Dire forecast: A theoretical model of the impact of climate change on crime. *Theoretical Criminology, 16*(1), 21–42.

Agnew, R. (2012b). It's the end of the world as we know it: The advance of climate change from a criminological perspective. In R. D. White (Ed.), *Climate change from a criminological perspective* (pp. 13–25). New York: Springer.

Agnew, R. (2013). The ordinary acts that contribute to ecocide—A criminological analysis. In N. South & A. Brisman (Eds.), *The Routledge international handbook of green criminology* (pp. 58–72). London: Routledge.

Agrawal, A. (2007). Forests, governance, and sustainability: Common property theory and its contributions. *International Journal of the Commons, 1*(1), 111–136.

Akella, S., & Cannon, J. (2004). *Strengthening the weakest link—Strategies for improving the enforcement of environmental laws globally*. Arlington, VA: Center for Conservation and Government at Conservation International.

Alemagi, D., & Kozak, R. A. (2010). Illegal logging in Cameroon: Causes and the path forward. *Forest Policy and Economics, 12*(8), 554–561.

N.A. Cao, *Timber Trafficking in Vietnam*, Palgrave Studies in Green Criminology,
DOI 10.1007/978-3-319-64280-2

Alkire, S. (2003). *A conceptual framework for human security*. Oxford: Centre for Research on Inequality, Human Security and Ethnicit—University of Oxford.

Allenby, B. R. (2000). Environmental security: Concept and implementation. *International Political Science Review, 21*(1), 5–21.

Anh The. (2013). Công ty lâm nghiệp mở đường xuyên rừng để... phá rừng [The forestry company opens the road through forests to ... deforest]. *Dan Tri*, 17/10/2013. Retrieved January 4, 2015, from http://dantri.com.vn/kinh-doanh/cong-ty-lam-nghiep-mo-duong-xuyen-rung-de-pha-rung-790968.htm

ANTV. (2014). Tận thu gỗ lòng hồ thủy điện hay phá rừng? [The salvage of timber in hydropower reservoirs or deforestation?]. *ANTV*, 4/9/2014. Retrieved January 2, 2015, from http://antv.gov.vn/xahoi/quang-ngai-nguy-trang-tinh-vi-cho-xe-go-lau/63201.html

Ba Be Investigation Police Agency. (2013). *Investigative conclusion of criminal case No.24/KLĐT dated 10/5/2013 by the Investigation Police Agency of the Ba Be District Public Security,on the criminal case of violating the regulations on forest exploitation and protection taking place on 25/11/2012 in 51 Zone, 104 Point, the Ba Be National Park*. Ba Be, Bac Kan (in Vietnamese).

Ba Thang. (2014). Gốc cây gây sốt [Tree stumps become best sellers]. *Cong an Da Nang*, 1/11/2014. Retrieved March 25, 2015, from http://cadn.com.vn/news/65_122617_go-c-cay-gay-so-t.aspx

Bac Ninh Environmental Police. (2012). Conference paper: The work of protecting, fighting and handling illicit activities in trading and transporting forest products in Bac Ninh province. In *Preventing and combatting violations against legislations on forest protection* (pp. 30–34). Hanoi: The Department of Preventing and Combating Environmental Crimes (in Vietnamese).

Bailey, C. A. (2007). *A guide to qualitative field research*. Thousand Oaks, CA: Pine Forge.

Bajpai, K. P. (2000). *Human security: Concept and measurement*. Notre Dame, IN: University of Notre Dame Joan B. Kroc Institute for International Peace Studies.

Baldwin, D. A. (1997). The concept of security. *Review of International Studies, 23*(1), 5–26.

Banks, D., & Rice, M. (2008). *Environmental crime: A threat to our future*. London: Environmental Investigation Agency.

Barnett, J. (2007). Environmental security. In A. Collins (Ed.), *Contemporary security studies* (pp. 182–203). New York: Oxford University Press.

Barnett, J., & Adger, W. N. (2010). Environmental change, human security, and violent conflict. In R. A. Matthew, J. Barnett, B. McDonald, & K. L. O'Brien (Eds.), *Global environmental change and human security* (pp. 119–136). Cambridge, MA: MIT Press.

Barnett, J., Matthew, R. A., & O'Brien, K. L. (2010). Global environmental change and human security: An Introduction. In R. A. Matthew, J. Barnett, B. McDonald, & K. L. O'Brien (Eds.), *Global environmental change and human security*. MIT Press.

Baulch, B., Chuyen, T. T. K., Haughton, D., & Haughton, J. (2007). Ethnic minority development in Vietnam. *The Journal of Development Studies, 43*(7), 1151–1176.

Baulch, B., Pham, H. T., & Phuong, N. T. T. (2008). *The Economic development of ethnic minorities in Vietnam*. Institute of Development Studies and the Department of Economics—University of Sussex and Centre for Analysis and Forecasting of Vietnam Academy of Social Sciences. Retrieved from http://www.ids.ac.uk/files/policybrief-final-Eng.pdf

BBC News. (2004). Global warming—Biggest threat. *BBC News*. Retrieved June 2, 2015, from http://news.bbc.co.uk/1/hi/sci/tech/3381425.stm

Beirne, P. (2007). Animal rights, animal abuse and green criminology. In P. Beirne & N. South (Eds.), *Issues in green criminology: Confronting harms against environments, humanity and other animals*. Cullompton, UK: Willan.

Beirne, P. (2009). *Confronting animal abuse: Law, criminology, and human-animal relationships*. Lanham, MD: Rowman & Littlefield Publishers.

Beirne, P., & South, N. (2006). *Green criminology*. Aldershot: Ashgate.

Beirne, P., & South, N. (2007). Introduction: Approaching to green criminology. In P. Beirne & N. South (Eds.), *Issues in green criminology: Confronting harms against environments, humanity and other animals*. Cullompton, UK: Willan.

Bell, S., & McGillivray, D. (2008). *Environmental law*. Oxford and New York: Oxford University Press.

Bennett, O., & Ares, E. (2011). *Illegally logged timber: EU and UK legislation, Standard Note: SN/SC/5858*. Retrieved from file:///C:/Users/w10036832/Downloads/sn05858.pdf

Berge, E., & van Laerhoven, F. (2011). Editorial: Governing the commons for two decades—A complex story. *International Journal of The Commons, 5*(2), 160–187.

Bernard, H. R. (2013). *Social research methods: Qualitative and quantitative approaches*. Los Angeles, CA: Sage.

Bhattacherjee, A. (2012). *Social Science research: Principles, methods, and practices*. Tampa, FL: University of South Florida. Open Access Textbooks.

Bich Ngoc. (2013a). GS.TSKH Nguyễn Ngọc Lung: Cao su và thủy điện tận lực phá rừng! [Prof. Dr. Nguyen Ngoc Lung: Rubber and hydropower deforestate by their best!]. *Đất Việt*, 02/10/2013. Retrieved March 25, 2015, from http://www.baodatviet.vn/chinh-tri-xa-hoi/binh-luan/gstskh-nguyen-ngoclungcao-su-va-thuy-dien-tan-luc-pha-rung-2356287/

Bich Ngoc. (2013b). Lộ rõ lợi ích nhóm 'phù phép' phá rừng trồng cao su [Revealing the group interests in mythologising to deforest for rubber plantation]. *Đất Việt*, 01/10/2013. Retrieved June 2, 2015, from http://www.baodatviet.vn/kinh-te/doanh-nghiep/lo-ro-loi-ich-nhom-phu-phep-pha-rung-trong-cao-su-2356205/

Binh An. (2014). Thi hành Bộ luật Hình sự: Bất lực với tội phạm môi trường? [Enforcing the Penal Code: Being powerless over green crime]. *Ministry of Justice*. Retrieved June 2, 2015, from http://www.moj.gov.vn/ct/tintuc/Pages/thong-tin-khac.aspx?ItemID=5291

Binh Duong Company. (2010). *Environmental impact evaluation report of the project in rubber development in 935, 936 Zones Ia Puch commune, Chu Prong district, Gia Lai province*. Gia Lai: Binh Duong Company—Binh Doan 15 (in Vietnamese).

Bisschop, L. (2012). Out of the woods: The illegal trade in tropical timber and a European trade hub. *Global Crime, 13*(3), 191–212.

Bisschop, L., & Walle, G. V. (2013). Environmental victimisation and conflict resolution: A case study of e-Waste. In R. Walters, D. Westerhuis, & T. Wyatt (Eds.), *Emerging issues in green criminology: Exploring power, justice and harm* (pp. 34–54). Houndmills, Basingstoke, UK: Palgrave Macmillan.

Biswas, N. R. (2011). Is the environment a security threat? Environmental security beyond securitization. *International Affairs Review, XX*(1).

Biswas, N. R. (2011a). Is the environment a security threat? Environmental security beyond securitization. *International Affairs Review, 20*(1).

Blevins, K. R., & Edwards, T. D. (2009). Wildlife crime. In J. M. Miller (Ed.), *21st century criminology: A reference handbook* (pp. 557–563). Thousand Oaks, CA: Sage.

Boekhout van Solinge, T. (2008). Eco-crime: The tropical timber trade. In D. Siegel & H. Nelen (Eds.), *Organized crime: Culture, markets and policies* (pp. 97–111). New York: Springer.

Boekhout van Solinge, T. (2010a). Deforestation crimes and conflicts in the Amazon. *Critical Criminology, 18*(4), 263–277.

Boekhout van Solinge, T. (2010b). Equatorial deforestation as harmful practice and criminological issue. In R. White (Ed.), *Global environmental harm*. Cullompton, UK: Willan.

Boekhout van Solinge, T., & Kuijpers, K. (2013). The Amazon rainforest—A green criminological perspective. In N. South & A. Brisman (Eds.), *The Routledge international handbook of green criminology*. London: Routledge.

Bogataj, N., & Krč, J. (2014). A forest commons revival in Slovenia. *Society & Natural Resources, 27*(8), 867–881.

Boissiere, M., Sheil, D., Basuki, I., Wan, M., & Le, H. (2009). Can engaging local people's interests reduce forest degradation in Central Vietnam? *Biodiversity and Conservation, 18*(10), 2743–2757.

Brack, D. (2003). Illegal logging and the illegal trade in forest and timber products. *International Forestry Review, 5*(3), 195–198.

Brack, D. (2004). The growth and control of international environmental crime. *Environmental Health Perspectives, 112*(2), A80–A81.

Brack, D. (2005). Controlling illegal logging and the trade in illegally harvested timber: The EU's Forest Law Enforcement, Governance and Trade initiative. *Review of European Community & International Environmental Law, 14*(1), 28–38.

Brack, D. (2007). *Illegal logging*. London: Energy, Environment and Development Programme—Chatham House.

Brack, D., & Hayman, G. (2001). *Intergovernmental actions on illegal logging: Options for intergovernmental action to help combat illegal logging and illegal trade in timber and forest products*. London: Chatham House.

Brack, D., Marijnissen, C., & Ozinga, S. (2002). *Controlling imports of illegal timber: Options for Europe*. London: The Royal Institute of International Affairs and FERN.

Brisman, A. (2012). Crime-environment relationships and environmental justice. *Seattle Journal for Social Justice, 6*(2), 6.

Brisman, A. (2014). Of theory and meaning in green criminology. *International Journal for Crime, Justice and Social Democracy, 3*(2), 22–35.

Brisman, A., & South, N. (2014). *Green cultural criminology: Constructions of environmental harm, consumerism, and resistance to ecocide*. London: Routledge.

Brook, D. (1998). Environmental genocide: Native Americans and toxic waste. *American Journal of Economics and Sociology, 57*(1), 105–113.

Brown, L. R. (1986). Redefining national security. *Challenge, 29*(3), 25–32.

Brunner, J. (2012). *Biodiversity conservation in Vietnam: A perfect storm*. IUCN Vietnam. Retrieved from http://cmsdata.iucn.org/downloads/iucn_nbsap_workshop_march_2012_paper_final.pdf

Bryant, B. I. (1995). *Environmental justice: Issues, policies, and solutions*. Washington, DC: Island Press.

Bryman, A. (2012). *Social research methods*. Oxford: Oxford University Press.

Bullard, R. D. (1996). Environmental justice for all. In R. D. Bullard (Ed.), *Environmental justice and communities of color* (pp. 3–22). San Francisco: Sierra Club Books.

Buttoud, G. (2000). How can policy take into consideration the "full value" of forests? *Land Use Policy, 17*(3), 169–175.

Buzan, B. (1991). *People, states and fear: An agenda for international security studies in the post-Cold War era*. Colchester: ECPR Press.

Cao, N. A., & Wyatt, T. (2013). A green criminological exploration of illegal wildlife trade in Vietnam. *Asian Journal of Criminology, 8*(2), 129–142.

Carew-Reid, J., Kempinski, J., & Clausen, A. (2010). *Biodiversity and development of the hydropower sector: Lessons from the Vietnamese experience—Volume I—Review of the effects of hydropower development on biodiversity in Vietnam*. Hanoi: ICEM—International Centre for Environmental Management.

Carrabine, E., Cox, P., Lee, M., Plummer, K., & South, N. (2009). *Criminology: A sociological introduction*. New York, NY: Routledge.

Carter, S. M., & Little, M. (2007). Justifying knowledge, justifying method, taking action: Epistemologies, methodologies, and methods in qualitative research. *Qualitative Health Research, 17*(10), 1316–1328.

Casson, A., & Obidzinski, K. (2007). From new order to regional autonomy: Shifting dynamics of illegal logging in Kalimantan, Indonesia. In L. Tacconi (Ed.), *Illegal logging, law inforcement, livelihoods and the timber trade*. London: Earthcan.

Castle, A. (1997). *Transnational organized crime and international security*. Vancouver: Institute of International Relations, University of British Columbia.

Chan, A. (2010). *Illegal logging in Indonesia: The environmental, economic and social costs*. Minneapolis: BlueGreen Alliance.

Chavis, B. F. (1994). Preface. In R. Bullard (Ed.), *Unequal protection: Environmental justice and communities of color* (pp. xi–xii). San Francisco, CA: Sierra Club Books.

China National People's Congress. (1997). *Criminal Law of the People's Republic of China*. Adopted at the Second Session of the Fifth National People's

Congress on July 1, 1979. Revised at the Fifth Session of the Eighth National People's Congress on March 14, 1997.
Cho, J., & Trent, A. (2006). Validity in qualitative research revisited. *Qualitative Research, 6*(3), 319–340.
CIP & EIA. (2005). *The illegal logging crisis in Honduras: How U.S. and E.U. imports of illegal Honduran wood increase poverty, fuel corruption and devastate forests and communities*. Washington, DC: Center for International Policy & Environmental Investigation Agency.
Clifford, M., & Edwards, T. D. (2012). *Environmental crime*. Burlington, MA: Jones & Bartlett Publishers.
Coleman, E. A., & Liebertz, S. S. (2014). Property rights and forest commons. *Journal of Policy Analysis and Management, 33*(3), 649–668.
Collins, A. (2007). *Contemporary security studies*. Oxford: Oxford University Press.
Commission on Human Security. (2003). *Human security now*. New York: United Nations.
Cong Bac. (2014). 12 tháng tù giam cho cán bộ kiểm lâm tiếp tay cho lâm tặc [12 months imprisonment for Kiem Lam officer supporting Lam Tac]. *VOV—Đài tiếng nói Việt Nam*, 12/12/2014. Retrieved December 21, 2014, from http://vov.vn/phap-luat/12-thang-tu-giam-cho-can-bo-kiem-lam-tiep-tay-cho-lam-tac-370594.vov
Contreras-Hermosilla, A. (2001). *Law compliance in the forestry sector: An overview*. Washington, DC: World Bank.
Contreras-Hermosilla, A. (2002). Conference paper: Policy and legal options to improve law compliance in the forest sector—*Reforming Government Policies and the Fight against Forest Crime*, pp. 43–91. Rome: FAO.
Crow, I., & Semmens, N. (2008). *Researching criminology*. Maidenhead: Open University Press.
Crow, M., Shelley, T. O. C., & Stretesky, P. (2014). Camouflage-collar crime: An examination of wildlife crime and characteristics of offenders in Florida. In T. Spapens, R. White, & M. Kluin (Eds.), *Environmental crime and its victims: Perspectives within green criminology* (pp. 185–204). Surrey: Ashgate.
Curran, E. (2015). Asia's about to spawn a new tiger economy: Good morning, Vietnam. *Bloomberg*, 22/3/2015. Retrieved May 1, 2015, from http://www.bloomberg.com/news/articles/2015-03-22/asia-s-about-to-spawn-a-new-tiger-economy-good-morning-vietnam
Dak Nong Environmental Police. (2012). Conference paper: The work of preventing and combatting violations of laws on forest protection in Dak Nong

province. In *Preventing and combatting violations against legislations on forest protection* (pp. 158–166). Hanoi: The Department of Preventing and Combating Environmental Crimes (in Vietnamese).

Dang, T. K. (2013a). Doanh nghiệp buôn gỗ lậu bằng hồ só giả [Firms smuggling timber by fabricated documentations]. *Lao Dong*, 4/10/2013. Retrieved October 25, 2014, from http://laodong.com.vn/phap-luat/doanh-nghiep-buon-go-lau-bang-ho-so-gia-141181.bld

Dang, T. K. (2013b). Kiểm lâm "biến" gỗ lậu thành... gỗ hợp pháp [Kiem lam "mythecising" illegal timber into legal timber]. *Lao Dong*, 22/6/2013. Retrieved October 25, 2014, from http://laodong.com.vn/phap-luat/kiem-lam-bien-go-lau-thanh-go-hop-phap-122941.bld

Dang, T. K. (2014). Nhiều doanh nghiệp phá rừng, bán đất dự án [Many companies destroy forests, sell the project land]. *Lao Dong*, 11/12/2114. Retrieved January 1, 2015, from http://laodong.com.vn/xa-hoi/nhieu-doanh-nghiep-pha-rung-ban-dat-du-an-277157.bld

Daniel, T. C., Muhar, A., Arnberger, A., Aznar, O., Boyd, J. W., Chan, K. M., et al. (2012). Contributions of cultural services to the ecosystem services agenda. *Proceedings of the National Academy of Sciences, 109*(23), 8812–8819.

Dao Le, T. (2010). *Bribery offences under Vietnamese criminal law in comparision with Swedish and Australian criminal law.* Ph.D. thesis, Hanoi Law University—Lund University.

Dasgupta, S., Laplante, B., Meisner, C., Wheeler, D., & Jianping Yan, D. (2007). *The impact of sea level rise on developing countries: A comparative analysis.* World Bank Policy Research Working Paper, 4136.

Dauvergne, P., & Lister, J. (2011). *Timber.* Cambridge: Polity.

Davies, M., Tyrer, J., & Croall, H. (2009). *Criminal justice.* New York: Pearson Longman.

Davies, P., & Francis, P. (2011). Doing criminological research. In P. Davies, V. Jupp, & P. Francis (Eds.), *Doing criminological research* (2nd ed., pp. 99–107). Los Angeles, CA; London: Sage.

Davies, P. A. (2014). Green crime and victimization: Tensions between social and environmental justice. *Theoretical Criminology, 18*(3), 300–316.

Dawson, C. (2009). *Introduction to research methods: A practical guide for anyone undertaking a research project.* Oxford: How to Books Limited.

Dawson, C. (2013). *Advanced research methods: A practical guide for anyone undertaking research projects.* London: How to Books.

de Jong, W., Do, D. S., & Trieu, V. H. (2006). *Forest rehabilitation in Vietnam: Histories, realities, and future*. Bogor, Indonesia: Center for International Forestry Research—CIFOR.

Denzin, N. K., & Lincoln, Y. S. (1994). Entering the field of qualitative research. In N. K. Denzin & Y. S. Lincoln (Eds.), *Handbook of qualitative research* (pp. 1–22). London: Sage.

Department of Legal Affairs—Mard. (2012). Conference paper: The status of forest protection and solutions to enhance the work of forest protection. In *Preventing and combatting violations against legislations on forest protection* (pp. 18–29). Hanoi: The Department of Preventing and Combating Environmental Crimes (in Vietnamese).

Detraz, N. (2009). Environmental security and gender: Necessary shifts in an evolving debate. *Security Studies, 18*(2), 345–369.

DFID. (2002). *Poverty and environment*. London: Department for International Development.

DFID. (2007). *Crime and persuasion: Tackling illegal logging, improving governance*. London: Department for International Development.

Dinh Thieu. (2014). Quảng Trị: Lợi dụng chính sách giao rừng để...phá rừng [Quang Tri: Abusing forest allocation policy to deforesting]. *VOV—Radio Voice of Vietnam*, 26/02/2014. Retrieved June 2, 2015, from http://vov.vn/xa-hoi/moi-truong/loi-dung-chinh-sach-giao-rung-depha-rung-312450.vov

Do, A. T. (2010). *Measuring the effectiveness of the enforcement of organised wildlife trade crime: A comparative assessment between Vietnam and England and Weles*. A thesis submitted in partial fulfilment of the requirements of the University of the West of England, Bristol for the degree of Doctor of Philosophy, University of the West of England.

Do, D. H. (2011). Yok Đôn, hổ thẹn với rừng! [Yok Don, what a shame with forests]. *Laodong*, 30/07/2011. Retrieved June 2, 2015, from http://laodong.com.vn/lao-dong-cuoi-tuan/yok-don-ho-then-voi-rung-44208.bld

Do, V. B., Le, T. H., & Hoang, T. S. (2007). *Studying scientific foundations for timber classification [Nghiên cứu cỏ sở khoa học phân loại gỗ]*. Ha Noi: The Institute of Forestry Science of Vietnam.

Dodds, F., & Pippard, T. (2012). *Human and environmental security: An agenda for change*. Routledge.

Dollar, D., & Litvack, J. (1998). Macroeconomic reform and poverty reduction in Vietnam. In D. Dollar, P. Glewwe, & J. Litvack (Eds.), *Household welfare and Vietnam's transition* (pp. 1–28). Washington, DC: The World Bank.

Drury, R. C. (2009). *Identifying and understanding consumers of wild animal products in Hanoi, Vietnam: Implication for conservation management*. The Degree of Doctor of Philosophy, University College London.

Du Rees, H. (2001). Can criminal law protect the environment? *Journal of Scandinavian Studies in Criminology and Crime Prevention, 2*(2), 109–126.

Duc Trung & Tan Phat. (2014). Những công ty "đầu độc" môi trường [The companies that "poison" the environment]. *Sai Gon Giai Phong*, 9/11/2014. Retrieved January 8, 2015, from http://www.sggp.org.vn/moitruongdothi/2014/11/366461/

Duong, D. V. (2012a). Conference paper: Fighting against on-duty officers in the work of forest protection in Quang Binh—Current status and solutions. In *Preventing and combatting violations against legislations on forest protection* (pp. 118–122). Hanoi: The Department of Preventing and Combating Environmental Crimes (in Vietnamese).

Duong, V. L. (2012b). Conference paper: Current status of violations of laws on forest protection in the Central Highland and solutions to prevent and combat. In *Preventing and combatting violations against legislations on forest protection* (pp. 11–17). Hanoi: The Department of Preventing and Combating Environmental Crimes.

Duy Hau & Quang Tao. (2011). Lâm tặc "mở hội" ở Vườn quốc gia Yok Đôn [Lam tac "opening festival" inside the Yok Don National Park]. *DanViet*, 12/5/2011. Retrieved February 6, 2015, from http://danviet.vn/thoi-su/lam-tac-mo-hoi-o-vuon-quoc-gia-yok-don/42754p1c24.htm

Duy Tuan. (2012). 'Xã lâm tặc' và những khu rừng bị chảy máu ['Lam Tac commune' and the bleeding forests]. 09/03/2012. Retrieved February 6, 2015, from http://vietnamnet.vn/vn/xa-hoi/63333/-xa-lam-tac--va-nhung-khu-rung-bi-chay-mau.html

Dybing, S. S. (2012). Environmental harm: Social causes and shilting dynamics. In R. Ellefsen, G. Larsen, & R. Sollund (Eds.), *Eco-global crimes: Contemporary problems and future challenges* (pp. 273–294). Farnham: Ashgate Publishing Ltd.

Earl, C. (2013). Saigon style: Middle-class culture and transformations of urban lifestyling in post-reform Vietnamese media. *Media International Australia, 147*, 85–97.

EIA. (2007a). *Attention Wal-Mart shoppers: How Wal-Mart's sourcing practices encourage illegal logging and threaten endangered species*. McKinleyville, CA: Environmental Investigation Agency.

EIA. (2007b). *No question asked—The impacts of US market demand for illegal timber and the potential for change*. Washington, DC: Environmental Investigation Agency. Retrieved from http://www.illegal-logging.info/sites/default/files/uploads/eia_no_questions_asked.pdf

EIA. (2011). *Crossroads: The illicit timber trade between laos and Vietnam*. London: Environmental Invetsigation Agency.

EIA. (2012). *Appetite for destruction: China's trade in illegal timber*. London: Environmental Investigation Agency.

EIA. (2013). *First class connecions—Log smuggling, illegal logging, and corruption in Mozambique*. London: Environmental Investigation Agency.

EIA & TELAPAK. (2001). *Timber trafficking: Illegal logging in Indonesia, South East Asia and international consumption of illegally sourced timber*. London and Bogor, Indonesia: Environmental Investigation Agency and Telapak.

EIA & TELAPAK. (2009). *Up for grabs: Deforestation and exploitation in Papua's plantations boom*. London and Bogor, Indonesia: Environmental Investigation Agency and Telapak. Retrieved from http://www.eia-international.org/wp-content/uploads/up-for-grabs.pdf

Ellefsen, R., Larsen, G., & Sollund, R. (2012). *Eco-global crimes: Contemporary problems and future challenges*. Farnham: Ashgate Publishing Ltd.

Elliott, L. (2007). Transnational environmental crime in the Asia Pacific: An un(der)securitized' security problem? *Pacific Review, 20*(4), 499–522.

Eman, K., Meško, G., & Fields, C. B. (2009). Crimes against the environment: Green criminology and research challenges in Slovenia. *Journal of Criminal Justice and Security, 11*(4), 574–592.

Emerton, L., Bishop, J., & Thomas, L. (2006). *Sustainable financing of protected areas: A global review of challenges and options*. Gland, Switzerland; Cambridge, UK: IUCN. Retrieved from http://www.europarc.org/uploaded/documents/128.pdf

Engler, M. (2008). The value of international wildlife trade. *Traffic Bulletin, 22*(1), 4–5.

Environmental Audit Committee. (2006). *Government response to the Committee's Second Report of Session 2005–06 on Sustainable Timber—Third Special Report of Session 2005–06*. Environmental Audit Committee—House of Commons. London: The Stationery Office.

Environmental Audit Committee. (2009). *Reducing greenhouse gas emissions from deforestation: No hope without forests, fifth report of session 2008–09, report, together formal minutes, oral and written evidence*. London: The Stationery Office.

Environmental Performance Index. (2014). *A global ranking for the environment: 2014 EPI results*. Retrieved June 18, 2015, from http://epi.yale.edu/

Environmental Police Department. (2010). *Ministry of Public Security issued Decision No.449 regulating the functions, duties, powers and organizational structure of the Department of Environmental Crime Prevention and Combat Police*. Department of Environmental Crime Prevention and Combat Police. Retrieved June 2, 2015, from http://www.canhsatmoitruong.gov.vn/default.aspx?tabid=441&ID=3029&CateID=479

Environmental Police Department. (2012). Conference paper: Opening paper—The work of preventing and combating the violations on forest protection. In *Preventing and combatting violations against legislations on forest protection* (pp. 7–10). Hanoi: The Department of Preventing and Combating Environmental Crimes (in Vietnamese).

European Commission. (2003). *Communication from the Commission to the Council and the European Parliament: Forest Law Enforcement, Governance and Trade (FLEGT): Proposal for an EU action plan*. Brussels: Commission of the European Communities.

FAO. (2001). *Illegal activities and corruption in the forest sector*. The Food and Agriculture Organization of the United Nations. Retrieved from ftp://ftp.fao.org/docrep/fao/003/y0900e/

FAO. (2007). *State of the world's forests 2007*. Rome: Food and Agriculture Organization of the United Nations.

FERN. (2002). *Illegal logging, and the global trade in illegally sourced timber; a crime against forests and peoples*. Forests and the European Union Resource Network. Retrieved February 15, 2013, from http://www.sinkswatch.org/sites/fern.org/files/pubs/ngostats/logging.pdf

Florini, A., & Simmons, P. (1998). *The new security thinking: A review of the North American literature*. New York: Rockefeller Bros. Fund.

Floyd, R. (2015). Environmental security and the case against rethinking criminology as 'security-ology'. *Criminology and Criminal Justice, 15*(3), 277–282.

Flyvbjerg, B. (2006). Five misunderstandings about case-study research. *Qualitative Inquiry, 12*(2), 219–245.

Flyvbjerg, B. (2011). Case study. In Y. S. Lincoln & N. K. Denzin (Eds.), *The Sage handbook of qualitative research* (4th ed., pp. 301–329). Thousand Oaks, CA: Sage.

Forest Trends. (2010). *Timber markets and trade between Laos and Vietnam: A commodity chain analysis of Vietnamese-driven timber flows*. Forest Trends & DFID. ISBN: 1-932928-41-3.

Forest Trends. (2013). *Small-scale illegal logging in Vietnam: Implications for FLEGT and REDD+*. The Forest Trends.

FPD. (2009). *Report on the operational performance in 2008 and the working programme in 2009*. Hanoi: The Forest Protection Department (in Vietnamese).

FPD. (2010). *Report on the operational performance in 2009 and the working programme in 2010*. Hanoi: The Forest Protection Department (in Vietnamese).

FPD. (2011). *Report on the operational performance in 2010 and the working programme in 2011*. Hanoi: The Forest Protection Department (in Vietnamese).

FPD. (2012a). *Functions and duties of the Forest Protection Department [Chức năng, nhiệm vụ của Cục Kiểm lâm]*. The Forest Protection Department. Retrieved April 11, 2013, from http://www.kiemlam.org.vn/Desktop.aspx/List/Gioithieu/Chuc_nang_nhiem_vu_cua_Cuc_Kiem_lam/

FPD. (2012b). *Report on the operational performance in 2011 and the working programme in 2012*. Hanoi: The Forest Protection Department (in Vietnamese).

FPD. (2013a). *Forest protection force of Vietnam: Forty years of construction and development (1973–2013)*. Hanoi: The Forest Protection Department (in Vietnamese).

FPD. (2013b). *Report on the operational performance in 2012 and the working programme in 2013*. Hanoi: The Forest Protection Department (in Vietnamese).

FPD. (2014). *Report on the operational performance in 2014 and the working programme in 2015*. Hanoi: The Forest Protection Department (in Vietnamese).

FPD Bac Kan. (2009). *Report on the operational performance in 2009 and the working programme in 2010*. Bac Kan: The Forest Protection Sub-Department (in Vietnamese).

FPD Bac Kan. (2011). *Report on the operational performance in 2011 and the working programme in 2012*. Bac Kan: The Forest Protection Sub-Department (in Vietnamese).

FSIV & FAO. (2009). *Vietnam forestry outlook study*. Bangkok: Forest Science Institute of Vietnam and FAO Regional Office for Asia and the Pacific.

Fukuda-Parr, S., & Messineo, C. (2012). *Human security: A critical review of the literature*. Working Paper 11, Centre for Research on Peace and Development (CRPD).

Gaarder, E. (2013). Evading responsibility for green harm: State-coeporate exploitation of race, class, and gender inequiality. In N. South & A. Brisman (Eds.), *The Routledge international handbook of green criminology*. London: Routledge.

Gasper, D., & Truong, T.-D. (2005). Deepening development ethics: From economism to human development to human security. *The European Journal of Development Research, 17*(3), 372–384.

Gerring, J. (2007). *Case study research: Principles and practices*. Cambridge: Cambridge University Press.

Gibbs, C., Gore, M. L., McGarrell, E. F., & Rivers, L. (2010). Introducing conservation criminology towards interdisciplinary scholarship on environmental crimes and risks. *British Journal of Criminology, 50*(1), 124–144.

Glasbeek, H. (2002). *Wealth by stealth: Corporate crime, corporate law, and the perversion of democracy*. Between the Lines.

Global Compact Network Vietnam. (2010). *A review of the social and environmental conditions of industries in Vietnam against the global compact principles*. Hanoi, Vietnam. Retrieved from http://www.globalcompactvietnam.org/upload/attach/GCNV%20Research%201.pdf

Global Witness. (2001). *The logs of war, the timber trade and armed conflict*. London: Global Witness.

Global Witness. (2003). *The usual suspects: Liberia's weapons and mercenaries in Côte d'Ivoire and Sierra Leone—Why it's still possible, How it works and How to break the trend*. London: Global Witness.

Global Witness & EIA. (2009). *Investigation into the illegal felling, transport and export of precious wood in SAVA Region Madagascar*. London: Global Witness and Environmental Investigation Agency.

Gomm, R., Hammersley, M., & Foster, P. (2000). *Case study method: Key issues, key texts*. London: Sage.

Goncalves, M. P., Panjer, M., Greenberg, T. S., & Magrath, W. B. (2012). *Justice for forest: Improving criminal justice efforts to combat illegal logging*. Washington, DC: World Bank.

Gorard, S. (2013). *Research design: Creating robust approaches for the social sciences*. London: Sage.

Gorte, R. W., & Sheikh, P. A. (2010). *Deforestation and climate change*. Congressional Research Service.

Government Inspectorate. (2014a). *The declaration No.2490/TB-TTCP, 15/10/2014 of the inspection conclusion on the land management and utilisation with focus on land allocation and forest allocation in a number of projects in Dak*

Nong province in the 2004–2011 period. Hanoi: Government Inspectorate of Vietnam (in Vietnamese).

Government Inspectorate. (2014b). *The declaration No.2834/TB-TTCP, 19/11/2014 of the inspection conclusion on the land management and utilisation and construction investment management in a number of projects in Gia Lai province*. Hanoi: Government Inspectorate of Vietnam (in Vietnamese).

Government Office. (2013). *Announcement No.159/TB-VPCP, dated 11/4/2013—The conclusion of the Deputy Prime Minister Hoang Trung Hai in the meeting on forest protection and development of the Central Highland provinces*. Hanoi: Officer of the Government (in Vietnamese).

Graife, M. B., & Stretesky, P. B. (2013). Crude laws: Treadmill of production and state violations in civil and criminal liability for oil discharges in navigable water. In N. South & A. Brisman (Eds.), *The Routledge international handbook of green criminology* (pp. 150–166). London: Routledge.

Grainger, A. (1993). *Controlling tropical deforestation*. London: Earthscan Publications.

Grant, C. (2013). *Written journalism in Vietnam*. Independent Study Project (ISP) Collection, Paper 1723. Retrieved from http://digitalcollections.sit.edu/isp_collection/1723

Graycar, A., & Felson, M. (2010). Situational prevention of organized timber theft and related corruption. In R. V. Clarke & N. Tilley (Eds.), *Situational prevention of organised crimes* (1st ed., pp. 81–92). Hoboken: Taylor and Francis.

Green, P., Ward, T., & McConnachie, K. (2007). Logging and legality: Environmental crime, civil society, and the state. *Social Justice, 34*(2), 94–110.

Greenpeace. (2007). *Carving up the Congo—Destruction by numbers logging roads to climate ruin*. Amsterdam.

Groombridge, N. (1998). Masculinities and crimes against the environment. *Theoretical Criminology, 2*(2), 249–267.

GSO. (2009). *Migration and urbanization in Vietnam: Situation, trends and differences*. Hanoi: The General Statistics Office.

Gubler, D. J., Reiter, P., Ebi, K. L., Yap, W., Nasci, R., & Patz, J. A. (2001). Climate variability and change in the United States: Potential impacts on vector- and rodent-borne diseases. *Environmental Health Perspectives, 109*(2), 223–233.

Guertin, C. É. (2003). *Illegal logging and illegal activities in the forestry sector: Overview and possible issues for the UNECE Timber Committee and FAO European Forestry Commission*. Quebec, Canada: Quebec Wood Export Bureau.

Gutierrez-Velez, V. H., & MacDicken, K. (2008). Quantifying the direct social and governmental costs of illegal logging in the Bolivian, Brazilian, and Peruvian Amazon. *Forest Policy and Economics, 10*(4), 248–256.

Guymon, C. L. D. (2000). International legal mechanisms for combating transnational organized crime: The need for a multilateral convention. *Berkeley Journal of International Law, 18*, 53.

H.Nhut. (2008). Tuổi Trẻ và những người bạn đồng hành [Tuoi Tre and the fellow]. *Tuoitre*, 19/6/2008. Retrieved June 2, 2015, from http://tuoitre.vn/Chinh-tri-Xa-hoi/264325/Tuoi-Tre-va-nhung-nguoi-ban-dong-hanh.html#ad-image-0

Ha An. (2012). Hoàn thiện các quy định pháp luật hình sự về tội phạm môi trường [Improving criminal legislations on environmental crimes] *Dai Bieu Nhan Dan*, 19/9/2012. Retrieved May 27, 2015, from http://daibieunhandan.vn/default.aspx?tabid=81&NewsId=258465

Ha Binh—Trung Tan. (2014). Giảm 130.000 ha rừng, cơ quan nào chịu trách nhiệm? [The redution of 130,000 ha of forests: Which authorities are accountable?]. *Tuoi Tre*, 13/12/2014. Retrieved January 31, 2015, from http://tuoitre.vn/tin/chinh-tri-xa-hoi/20141213/giam-130000-ha-rung-co-quan-nao-chiu-trach-nhiem/684381.html

Ha, D. B., & Nguyen, C. T. (2014) Conference paper: Analysis about the influence of Vietnamese cultures and Vietnam's general educational system on Cdio implementation. In *The 10th international CDIO conference*. Universitat Politècnica de Catalunya, Barcelona.

Ha Thanh. (2014). Giữ rừng ở Bạch Thông, thực trạng và giải pháp [Keeping the Bach Thong forest. Current situation and solutions]. *Bac Kan Online*, 22/9/2014. Retrieved June 2, 2015, from http://www.baobackan.org.vn/channel/2262/201409/giu-rung-o-bach-thong-thuc-trang-va-giai-phap-2339944/

Ha Tinh People's Council. (2012). *The Program: Sustainable protection and development of forests—The period 2012–2015 with orientation forwards 2020 (Issued together with Resolution No. 32/2012/NQ-HĐND dated 19.07.2012 of the Province People's Council)*. XVIth Ha Tinh People's Council—3rd Section, Ha Tinh (in Vietnamese).

Ha Vy. (2011). Một giám đốc "ăn" hơn 220m3 gỗ [A director "eats" over 220m3 of timber]. *Cong An Nhan Dan*, 1/10/2011. Retrieved May 29, 2015, from http://cand.com.vn/ANTT/Mot-giam-doc-an-hon-220m3-go-186617/

Hall, M. (2011). Environmental victims: Challenges for criminology and victimology in the 21st century. *Varstvoslovje: Journal of Criminal Justice & Security, 13*(4), 371–391.

Hall, M. (2013a). *Victims of environmental harm: Rights, recognition and redress under national and international law*. London: Routledge.

Hall, M. (2013b). Victims of environmental harms and their role in national and international justice. In R. Walters, D. Westerhuis, & T. Wyatt (Eds.), *Emerging issues in green criminology: Exploring power, justice and harm* (pp. 218–241). Houndmills, Basingstoke, UK: Palgrave Macmillan.

Hall, M. (2014a). Environmental harm and environmental victims: Scoping out a 'green victimology'. *International Review of Victimology, 20*(1), 129–143.

Hall, M. (2014b). Victims of environmental crime: Routes for recognition, restitution and redress. In T. Spapens, R. White, & M. Kluin (Eds.), *Environmental crime and its victims: Perspectives within green criminology* (pp. 104–118). Surrey: Ashgate.

Hall, M., & Farrall, S. (2013). The criminogenic consequences of climate change—Blurring the boundaries between offenders and victims. In N. South & A. Brisman (Eds.), *The Routledge international handbook of green criminology* (pp. 120–133). London: Routledge.

Halsey, M. (2004). Against 'green' criminology. *British Journal of Criminology, 44*(6), 833–853.

Halsey, M. (2013). Conservation criminology and the "General Accident" of climate change. In N. South & A. Brisman (Eds.), *The Routledge international handbook of green criminology*. London: Routledge.

Halsey, M., & White, R. (1998). Crime, ecophilosophy and environmental harm. *Theoretical Criminology, 2*(3), 345–371.

Hameiri, S., & Jones, L. E. E. (2013). The politics and governance of non-traditional security. *International Studies Quarterly, 57*(3), 462–473.

Hammersley, M., & Gomm, R. (2000). Introduction. In M. Hammersley, R. Gomm, & P. Foster (Eds.), *Case study method: Key issues, key texts* (pp. 1–16). London: Sage.

Hampson, F. O. (2008). Human security. In P. Williams (Ed.), *Security studies: An introduction*. London: Routledge.

Hampson, F. O., Daudelin, J., Hay, J., Reid, H., & Marting, T. (2002). *Madness in the multitude: Human security and world disorder*. Don Mills, ON: Oxford University Press.

Hanh Nguyen. (2014). Đại gia dưỡng thân: Giường gỗ trăm năm, đá ngàn tuổi [The super rich take care themselves: Hundred-year-old timber or thousand-year-old stone beds]. *VietnamNet*, 4/10/2014. Retrieved December 8, 2014, from http://vietnamnet.vn/vn/kinh-te/200435/dai-gia-duong-than--giuong-go-tram-nam--da-ngan-tuoi.html

Hardin, G. (1968). The tragedy of the commons. *Science, 162*(3859), 1243–1248.

Harding, J. (2013). *Qualitative data analysis from start to finish*. London: Sage.

Hartley, R. D. (2008). *Corporate crime: A reference handbook*. Santa Barbara, CA: ABC-CLIO.

Hauck, M. (2007). Non-compliance in small-scale fisheries: A threat to security. In P. Beirne & N. South (Eds.), *Issues in green criminology: Confronting harms against environments, humanity and other animals* (pp. 270–289). Devon: Willan.

Hawksworth, J., & Tiwari, A. (2011). *The World in 2050: The accelerating shift of global economic power—Challenges and opportunities*. London: PricewaterhouseCoopers. Retrieved from http://www.pwc.com/en_GX/gx/psrc/pdf/world_in_2050_jan2011.pdf

Hayton, B. (2010). *Vietnam: Rising dragon*. Yale University Press.

Heckenberg, D., & White, R. (2013). Innovative approaches to researching environmental crime. In N. South & A. Brisman (Eds.), *The Routledge international handbook of green criminology* (pp. 85–103). London: Routledge.

Henn, M., Weinstein, M., & Foard, N. (2009). *A critical introduction to social research*. London: Sage.

Herbig, F., & Joubert, S. (2006). Criminological semantics: Conservation criminology—Vision or vagary? *Acta Criminologica, 19*(3), 88–103.

Hewitt, D. (2013). *Potential legality issues from forest conversion timber*. Forest Trends. Retrieved from http://www.forest-trends.org/documents/files/doc_4137.pdf

Hirschberger, P. (2008). *Illegal wood for the European market: An analysis of the EU import and export of illegal wood and related products*. WWF-Germany.

Hoai Nam. (2014). Ngang nhiên vận chuyển gỗ lậu bằng xe máy ở Quảng Nam [Blatanly transporting illegal timber by motorbikes in Quang Nam]. *VOV*, 21/9/2014. Retrieved June 2, 2015, from http://vov.vn/doi-song/ngang-nhien-van-chuyen-go-lau-bang-xe-may-o-quang-nam-353231.vov

Hoai Nhon. (2012). "Xẻ thịt" rừng làm rẫy ["Butchering" forests for cultivation]. *Nguoi Lao Dong*, 22/08/2012. Retrieved June 2, 2015, from http://nld.com.vn/ban-doc/xe-thit-rung-lam-ray-20120822112765911.htm

Hoang, C. N. (2012a). Kon Tum: Lâm tặc táo tợn bắt và tra tấn chủ rừng [Kon Tum, Lam tac audaciously arrest and torture forest owners]. *Vietnam+*, 24/2/2012. Retrieved March 2, 2014, from http://www.vietnamplus.vn/kon-tum-lam-tac-tao-ton-bat-va-tra-tan-chu-rung/130869.vnp

Hoang Nam. (2011). Khó ngăn phá rừng [Difficult to prevent timber trafficking]. *Tienphong*, 10/7/2011. Retrieved June 2, 2015, from http://www.tienphong.vn/phap-luat/kho-ngan-pha-rung-544555.tpo

Hoang Nam. (2014). Lộ đường dây vận chuyển gỗ lậu quý hiếm bằng đường sắt [Revealing the network of transporting illegal timber by railway]. *Tien Phong*, 29/5/2014. Retrieved December 22, 2014, from http://www.tienphong.vn/xa-hoi/lo-duong-day-van-chuyen-go-lau-quy-hiem-bang-duong-sat-710266.tpo

Hoang Quan. (2012). Cán bộ chỉ đạo con cháu... phá rừng [Officials direct their descedants ... to deforest]. *Cong an Thanh Pho Ho Chi Minh*, 6/7/2012. Retrieved January 2, 2015, from http://www.congan.com.vn/?mod=detnews&catid=703&p=&id=474017

Hoang, T. D. (2012b). Conference paper: The work of preventing and combatting violations of laws on forest protection in Thua Thien Hue province. In *Preventing and combatting violations against legislations on forest protection* (pp. 123–127). Hanoi: The Department of Preventing and Combating Environmental Crimes (in Vietnamese).

Hoang, T. T., & Nguyen, T. D. (2012). Conference paper: Violations against legislations on forest protection and arising issues over the work of education and training. In *Preventing and combatting violations against legislations on forest protection* (pp. 112–117). Hanoi: The Department of Preventing and Combating Environmental Crimes (in Vietnamese).

Holdaway, S., & Rock, P. (1998). *Thinking about criminology*. London: UCL Press.

Holifield, R. (2001). Defining environmental justice and environmental racism. *Urban Geography, 22*(1), 78–90.

Holmgren, E., Keskitalo, E. C. H., Lidestav, G., Kulturgeografiska, I., Samhällsvetenskapliga, F., & Umeå, U. (2010). Swedish forest commons—A matter of governance? *Forest Policy and Economics, 12*(6), 423–431.

Human Rights Watch. (2009). *"Wild Money": The human rights consequences of illegal logging and corruption in Indonesia's forestry sector*. New York: Human Rights Watch. Retrieved from http://www.hrw.org/sites/default/files/reports/indonesia1209web_0.pdf

Human Security Centre. (2005). *Human security report 2005: War and peace in the 21st century*. New York: Oxford University Press.

Huong Giang. (2014). Bắt giữ nhiều vụ dùng thuốc nổ khai thác gỗ sưa [Arresting many cases using explosives to harvest Sua timber]. *Nhan Dan*

Online, 5/5/2014. Retrieved January 19, 2015, from http://www.nhandan.com.vn/mobile/_mobile_phapluat/_mobile_thoisu/item/23104402.html

Huong Nguyen. (2010). Xử lý tội phạm môi trường: "Bó" bởi nhiều cái khó [Handling environmental crime: "Obstructed" by many difficulties]. *Nhan Dan*, 30/11/2010. Retrieved June 2, 2015, from http://www.nhandan.com.vn/phapluat/item/12390302-.html

Huu Thanh. (2014). Vụ thảm sát 5 phu trầm: Các bị can bị truy tố 4 tội danh [The massacre of five agarwood hunters: The accused are prosecuted with four offences]. *Tienphong*, 12/02/2014. Retrieved June 2, 2015, from http://www.tienphong.vn/Phap-Luat/vu-tham-sat-5-phu-tram-cac-bi-can-bi-truy-to-4-toi-danh-677867.tpo

Hydropower No.7 Management Board. (2006). *The Project of An Khe—Ka Nak Hydropower: Environmental impact assessment report*. Phu Yen: Vietnam Electricity Corporation (in Vietnamese).

ICEM. (2003). *Vietnam national report on protected areas and development—Review of protected areas and development in the lower Mekong River region*. Queensland, Australia: International Center for Environmental Management. Retrieved from http://cmsdata.iucn.org/downloads/vietnam_pad.pdf

International Monetary Fund. (2013). *World economic outlook database—Vietnam*. International Monetary Fund.

Interpol. (2012). *Environmental crime, It's global theft*. Interpol—Environmental Crime Programme. Retrieved from http://www.interpol.int/Crime-areas/Environmental-crime/Environmental-crime

Interpol & World Bank. (2009). *Chainsaw project: An Interpol perspective on law enforcement in illegal logging*. Rural Development and Natural Resources—World Bank & Interpol.

Investigation Security Agency. (2013). *The investigative conclusion on the case of violation of regulation on forest exploitation and protection, taking place in Dak Nong, Binh Phuoc and Tien Giang in 2012*. Security Investigation Agency—The Ministry of Public Security, Ho Chi Minh City (in Vietnamese).

ITTO. (2011). *Status of tropical forest management 2011*. Yokohoma: International Tropical Timber Organization.

ITTO. (2013). *What are illegal logging and illegal trade?*. Bangkok, Thailand: International Tropical Timber Organisation. Retrieved April 24, 2013, from http://www.itto.int/feature06_01/

Jagger, P., Shively, G., & Arinaitwe, A. (2012). Circular migration, small-scale logging, and household livelihoods in Uganda. *Population and Environment, 34*(2), 235–256.

Joint Circular. (2009). *Joint Circular No.08/2009/TTLT-BXD-BTC-BKHĐT-BNNPTNT-NHNN on guiding the implementation of the Decision No.167/2008/QĐ-TTg*. Ministry of Construction, Ministry of Finance, Ministry of Planning and Investment, State Bank, Ha Noi.

Josephi, B. (2013). How much democracy does journalism need? *Journalism, 14*(4), 474–489.

Julian, P. (2004). Inequality, social differences and environmental resources. In R. D. White (Ed.), *Controversies in environmental sociology*. Cambridge: Cambridge University Press.

Kaimowits, D. (2007). Forest law enforcement and rural livelihoods. In L. Tacconi (Ed.), *Illegal logging: Law enforcement, livelihoods and the timber trade* (pp. 110–138). London: Earthscan.

Kaimowitz, D. (2003). Forest law enforcement and rural livelihoods. *International Forestry Review, 5*(3), 199–210.

Kaldor, M. (2007). *Human security: Reflections on globalization and intervention*. Cambridge: Polity.

Karesh, W. B., Cook, R. A., Gilbert, M., & Newcomb, J. (2007). Implications of wildlife trade on the movement of avian influenza and other infectious diseases. *Journal of Wildlife Diseases, 43*(3), S55–S59.

Kendall, R. E. (1998). Interpol alert. *Our Planet*. Retrieved June 2, 2015, from http://www.ourplanet.com/imgversn/105/kendall.html

Kerr, P. (2007). Human security. In A. Collins (Ed.), *Contemporary security studies* (pp. 91–108). Oxford: Oxford University Press.

Khanh Hoan. (2013). Vụ lật xe chở gỗ lậu khiến 10 người chết: 8 bị cáo lãnh án tù [The case of flipped truck of illegal timber killing 10 people: 8 defendants are sentenced to prison]. *Laodong*, 24/1/2013. Retrieved June 2, 2015, from http://www.thanhnien.com.vn/pages/20130124/vu-lat-xe-cho-go-lau-khien-10-nguoi-chet-8-bi-cao-lanh-an-tu.aspx

Khatchadourian, R. (2008). The stolen forests—Inside the covert war on illegal logging. *The New Yorker*. http://www.newyorker.com/magazine/2008/10/06/the-stolen-forests

Kieu Thang. (2013). Thực hiện tốt các chương trình nhà ở cho hộ nghèo. *Nhan Dan*, 11/04/2013. Retrieved June 2, 2015, from http://www.nhandan.com.vn/chinhtri/cung-suy-ngam/item/20084202-.html

King, G., & Murray, C. J. L. (2001). Rethinking human security. *Political Science Quarterly, 116*(4), 585–610.

Kishor, N., & Oksanen, T. (2006). *Combating illegal logging and corruption in the forestry sector. Strengthening forest law enforcement and governance*. Environment Matters, The World Bank Group.

Kohm, S., & Greenhill, P. (2013). 'This is the North, where we do what we want': Popular green criminology and 'Little Red Riding Hood' films. In N. South & A. Brisman (Eds.), *The Routledge international handbook of green criminology* (pp. 365–378). London: Routledge.

Kokko, A. (2004). Growth and reform since the Eighth Party Congress. In D. McCargo (Ed.), *Rethinking Vietnam*. New York and London: RoutledgeCurzon.

Kon Tum Investigation Police Agency. (2009). *The investigative conclusion No.07/BC-PC15 dated 7/9/2009 on the case: "Violation of regulations on forest exploitation and protection", take place in Zone 101, protection forest, Dak Glei district, Kon Tum province*. Kon Tum: Investigating Police Agency, Kon Tum Public Security Department (in Vietnamese).

Kon Tum Investigation Police Agency. (2011). *The investigative conclusion No.02/KLDT-PC46 dated 30/9/2011 on the case: Violation of regulations on forest exploitation and protection, take place in Dak Uy special-used forest, Dak Ha district, Kon Tum province*. Kon Tum: Investigating Police Agency, Kon Tum Public Security Department (in Vietnamese).

Korsell, L. E. (2001). Big sick, litle sick: Strategies for combaing environmental crime. *Journal of Scandinavian Studies in Criminology and Crime, 2*, 127.

Kramer, R. (2014). Climate change: A state-corporate crime perspective. In T. Spapens, R. White, & M. Kluin (Eds.), *Environmental crime and its victims: Perspectives within green criminology* (pp. 23–10). Surrey: Ashgate.

Kramer, R. C., Michalowski, R. J., & Kauzlarich, D. (2002). The origins and development of the concept and theory of state-corporate crime. *Crime & Delinquency, 48*(2), 263–282.

Kummer, D. M. (1992). *Deforestation in the postwar Philippines*. Chicago and London: University of Chicago Press.

Kvale, S. (1996). *InterViews: An introduction to qualitative research interviewing*. Thousand Oaks, CA: Sage.

Kwong, W. (2013). *'Persons of versatility': Private security officers and private policing in residential estates in Hong Kong*. Doctoral dissertation, University of Portsmouth.

Kwong, W. Y. H. (2013). *'Persons of versatility': Private security officers and private policing in residential estates in Hong Kong*, Doctoral dissertation, University of Portsmouth.

Lam Dong Environmental Police. (2012). Conference paper: Shortcomings and violations in implementing the forestland conversion policies in Lam Dong province. In *Preventing and combatting violations against legislations on forest protection* (pp. 137–140). Hanoi: The Department of Preventing and Combating Environmental Crimes (in Vietnamese).

Lam Dong Province Inspection. (2013). *Conclusion No.42 /KL-TTr, 11/3/2013 of the inspection on the projects of rubber plantation of the non-state companies in the districts of Da Teh, Da Huoai and Bao Lam*. Department of Inspection—Lam Dong Province People's Committee, Lam Dong (in Vietnamese).

LandNet. (2013). Không để Rừng Tây Nguyên trở thành hoài niệm! [Don't let the Central Highland forests became nostalgia]. *LandNet*, 25/9/2013. Retrieved January 21, 2015, from http://land.net.vn/info/Tin-tuc/Khong-de-Rung-Tay-Nguyen-tro-thanh-hoai-niem!-61-317.html

Lawson, S., & MacFaul, L. (2010). *Illegal logging and related trade indicators of the global response*. The Royal Institute of International Affairs.

Le Chi. (2013). Timber export exceeds $4 million for the first time [Xuất khẩu gỗ lần đầu tiên vượt mốc 4 tỷ USD]. *VN-Express*. Retrieved June 2, 2015, from http://vnexpress.net/gl/kinh-doanh/2013/01/xuat-khau-go-lan-dau-tien-vuot-moc-4-ty-usd/

Le, D. Q. (1998). *The fight against illegal timber harvest and transport in the Central Highlands provinces*. The Dissertation of Master in Law, The People's Police Academy (in Vietnamese).

Le Minh. (2014). Lâm tặc là ai? [Who are Lam Tacs?]. *Tien Phong*, 2/12/2014. Retrieved December 21, 2014, from http://www.tienphong.vn/xa-hoi-chuyen-hom-nay/lam-tac-la-ai-790928.tpo

Le Trang. (2011). Legends of timber as precious as ... pearl [Huyền thoại về thứ gỗ quý… nhúngọc]. *Zing News*, 26/08/2011. Retrieved June 2, 2015, from http://news.zing.vn/xa-hoi/huyen-thoai-ve-thu-go-quy-nhu-ngoc/a125063.html

Le, T. T. 2009. Tăng cường hợp tác quốc tế đấu tranh phòng, chống tội phạm xuyên quốc gia ở Việt Nam trong thời kỳ hội nhập quốc tế [Strengthening international cooperation in the fight against transnational crimes in Vietnam during the international integration]. *Tạp chí Cộng sản*, 26/8/2009. Retrieved June 2, 2015, from http://www.tapchicongsan.org.vn/Home/Tieu-diem/2009/2498/Tang-cuong-hop-tac-quoc-te-dau-tranh-phong-chong-toi-pham.aspx

Lee, M. R., & Blanchard, T. C. (2010). *Health impacts of deepwater horizon oil disaster on coastal Louisiana residents*. Unpublished manuscript, Department of

Sociology, Louisiana State University, Baton Rouge, Louisiana. Retrieved from http://www.lsu.edu/pa/mediacenter/tipsheets/spill/publichealthreport_2.pdf

Lescuyer, G. (2007). Livelihood and the adaptive application of the law in the forests in Cameroon. In L. Tacconi (Ed.), *Illegal logging: Law enforcement, livelihoods and the timber trade* (pp. 167–190). London: Earthscan.

Levitsky, M. (2003). Transnational criminal networks and international security. *Syracuse Journal of International Law and Commerce, 30*(2), 227.

Liddick, D. (2011). *Crimes against nature: Illegal industries and the global environment.* Santa Barbara, CA: Praeger.

Lin, J. (2005). Tackling Southeast Asia's illegal wildlife trade. *Sybil, 9*, 191.

Little, L., & Cocklin, C. (2007). Environment and human security. In J. Pretty, A. S. Ball, T. Benton, J. S. Guivant, D. R. Lee, D. Orr, M. J. Pfeffer, & H. Ward (Eds.), *The Sage handbook of environment and society*. London: Sage.

Lodgaard, S. (2000). *Human security: Concept and operationalization*. Norwegian Institute of International Affairs.

Lonergan, S. (1999). *Global Environmental Change and Human Security (GECHS): Science plan.* International Human Dimensions Programme on Global Environmental Change.

Long, M., Stretesky, P., & Lynch, M. (2014). The treadmill of production, planetary boundaries and green criminology. In T. Spapens, R. White, & M. Kluin (Eds.), *Environmental crime and its victims: Perspectives within green criminology* (pp. 263–276). Surrey: Ashgate.

Long, M. A., Stretesky, P. B., Lynch, M. J., & Fenwick, E. (2012). Crime in the coal industry: Implications for green criminology and treadmill of production. *Organization & Environment, 25*(3), 328–346.

Luttrell, C., Obidzinski, K., Brockhaus, M., Muharrom, E., Petkova, E., Wardell, A., et al. (2011). *Lessons for REDD+ from measures to control illegal logging in Indonesia.* CIFOR.

Lynch, M. (1990). The greening of criminology: A perspective on the 1990s. *The Critical Criminologist, 2*(3), 1–4.

Lynch, M. J. (2000). The power of oppression: Understanding the history of criminology as a science of oppression. *Critical Criminology, 9*(1-2), 144–152.

Lynch, M. J. (2013). Reflections of green criminology and its boundaries: Comparing environmental victimization and considering crime from an eco-city perspective. In N. South & A. Brisman (Eds.), *The Routledge international handbook of green criminology* (pp. 43–57). London: Routledge.

Lynch, M. J., Long, M. A., Barrett, K. L., & Stretesky, P. B. (2013). Is it a crime to produce ecological disorganization? Why green criminology and political economy matter in the analysis of global ecological harms. *British Journal of Criminology, 53*(6), 997–1016.

Lynch, M. J., & Stretesky, P. (2001). Toxic crimes: Examining corporate victimization of the general public employing medical and epidemiological evidence. *Critical Criminology, 10*(3), 153–172.

Lynch, M. J., & Stretesky, P. (2007). Green criminology in the United States. In P. Beirne & N. South (Eds.), *Issues in green criminology: Confronting harms against environments, humanity and other animals* (pp. 248–269). Cullompton, UK: Willan.

Lynch, M. J., & Stretesky, P. (2014). *Exploring green criminology: Toward a green criminological revolution.* Farnham: Ashgate Publishing Ltd.

Lynch, M. J., & Stretesky, P. B. (2003). The meaning of green: Contrasting criminological perspectives. *Theoretical Criminology, 7*(2), 217–238.

MacFarlane, S. N., & Khong, Y. F. (2006). *Human security and the United Nations.* Indiana University Press.

MARD. (2001). *Five million hectare reforestation program partnership—Synthetic report.* Hanoi: Ministry of Agriculture and Rural Development (in Vietnamese).

MARD. (2009). *Circular No.34/2009/TT-BNNPTNT dated 10 June 2009: Defining the criteria for identifying and classifying forests.* Hanoi: Minister of Agriculture and Rural Development (in Vietnamese).

MARD. (2011). *Circular No.35/2011/TT-BNNPTNT dated 20 May 2011: Guiding the implementation of harvesting and salvage of timber and NTFPs.* Hanoi: Minister of Agriculture and Rural Development (in Vietnamese).

MARD. (2012a). *National forest status declaration.* Development, M. O. A. A. R. Hanoi: Ministry of Agriculture and Rural Development.

MARD. (2012b). *Overall review on policy arrangement regarding the implemention of plan in forest protection and devellopment in the period 2012–2020.* State Steering Comittee Office of Planning in Forest Protection and Development in the Period 2011–2020, MARD (in Vietnamese).

MARD. (2014a). *Decision 3322/QD-BNN-TCLN on the national forest status declaration by 2013.* Hanoi: Ministry of Agriculture and Rural Development (in Vietnamese).

MARD. (2014b). *Official letter 1895/BNN-TCLN on strengthening the management of timber processing facilities.* Hanoi: Ministry of Agriculture and Rural Development (in Vietnamese).

MARD & FSSP. (2014). *Report on the development of the forestry sector in 2013.* Hanoi: Ministry of Agriculture and Rural Development & Forest Sector Support Program and Partnership. Retrieved from http://www.vietnamforestry.

org.vn/mediastore/fsspco/2014/04/01/Bao_cao_PT_nganh_2013-ENGLISH.pdf.

Mares, D. (2010). Criminalizing ecological harm: Crimes against carrying capacity and the criminalization of eco-sinners. *Critical Criminology, 18*(4), 279–293.

Martini, M. (2012). *Overview of corruption and anti-corruption in Vietnam*. Anti-Corruption Resource Centre (U4), Transparency International & Chr. Michelsen Institute.

Mathews, J. T. (1989). Redefining security. *Foreign Affairs, 68*(2), 162–177.

Matthew, R. A., Barnett, J., McDonald, B., & O'Brien, K. L. (2010). *Global environmental change and human security*. MIT Press.

Mautner, G. (2008). Analysing newspapers, magazines and other print media. In R. Wodak & M. Krzyżanowski (Eds.), *Qualitative discourse analysis in the social sciences* (pp. 30–53). Basingstoke, UK: Palgrave Macmillan.

Mawby, R. C. (2011). Using media to understand crime and criminal justice. In P. Davies, V. Jupp, & P. Francis (Eds.), *Doing criminological research* (2nd ed., pp. 223–244). Los Angeles, CA; London: Sage.

McElwee, P. (2004). You say illegal, I say legal: The relationship between 'illegal' logging and land tenure, poverty, and forest use rights in Vietnam. *Journal of Sustainable Forestry, 19*(1-3), 97–135.

Mehta, T. P. (2002). Community forestry in India and Nepal. *PERC Reports, 20*, 3–5.

Meyfroidt, P., & Lambin, E. F. (2009). Forest transition in Vietnam and displacement of deforestation abroad. *Proceedings of the National Academy of Sciences, 106*(38), 16139–16144.

Michalowski, R. J., & Kramer, R. C. (2006). The critique of power. In R. J. Michalowski & R. C. Kramer (Eds.), *State-corporate crime: Wrongdoing at the intersection of business and government* (pp. 1–17). Piscataway, NJ: Rutgers University Press.

Milliken, T., & Shaw, J. (2012). *The South Africa-Viet Nam Rhino Horn trade nexus: A deadly combination of institutional lapses, corrupt wildlife industry professionals, and Asian crime syndicates*. Johannesburg, South Africa: Traffic.

Minh Dat. (2014). Tăng cường công tác quản lý và bảo vệ rừng. *Cong An Kon Tum*, 6/11/2014. Retrieved January 2, 2015, from http://congankontum.gov.vn/hdccat/bao-dam-trat-tu-xh/52885-tang-cuong-cong-tac-quan-ly-va-bao-ve-rung.html

Minh Thi—Minh Trieu. (2014). Chuyện những người giữ Giáng hương [The stories on protectors of Giang Huong forest]. *Gia Lai Online*, 2/10/2014.

Retrieved June 2, 2015, from http://baogialai.com.vn/channel/1622/201410/chuyen-nhung-nguoi-giu-giang-huong-2342405/

Minh Thuan & Que Ha. (2007). Kết thúc phiên só thẩm xét xử băng nhóm "Đồi Hoa Mai": Hai Chi bị tuyên phạt 19 năm tù [The end of the trial of "Doi Hoa Mai" gang: Hai Chi was sentenced to 19 years in prison]. *Thanhnien Online*, 4/10/2007. Retrieved February 6, 2015, from http://www.thanhnien.com.vn/news1/pages/200740/211212.aspx

Minh Tuan—Huy Toan. (2014). Bài 2: Tàn phá hàng trăm hecta rừng [Article 2: Destroying hundreds of ha of forests]. *Kinh Te Nong Thon*, 25/11/2014. Retrieved January 18, 2015, from http://www.kinhtenongthon.com.vn/Bai-2-Tan-pha-hang-tram-hecta-rung-140-48347.html

Ministry of Planning and Investment. (2008). *Vietnam continues to achieve the millennium development goals*. Hanoi: Ministry of Planning and Investigation

MNRE. (2010). *National Environmental Report—2010—Overview of the environmental in Vietnam*. Hanoi: Ministry of Natural Resources and Environment (in Vietnamese).

Morgan, P. (2007). Security in international politics: Traditional approaches. In A. Collins (Ed.), *Contemporary security studies* (pp. 13–34). Oxford: Oxford University Press.

Morris, J., Phi, L. T., Ingles, A., Raintree, A., & Doung, N. V. (2004). *Linking poverty reduction with forest conservation: Case studies from Vietnam*. Bangkok, Thailand: International Union for Conservation of Nature and Natural Resources.

My Ngoc & Nguyen Tran. (2004). Xét xử vụ án phá rừng lớn nhất Tây Nguyên: Giám đốc lâm trường "bắt tay" doanh nghiệp để phá rừng [Adjudicating the biggest illegal logging case in the Central Highland: Forestry enterprise directors "handshake" firms to destroy forests]. *Thanh Nien*, 29/11/2013. Retrieved February 6, 2015, from http://www.thanhnien.com.vn/news1/pages/200447/102302.aspx

N.Hung. (2014). Về "thủ phủ vàng" nghe chuyện xử lâm tặc [Visit "the capital of gold" to listen the story of treating illegal loggers] *Gia Dinh & Xa Hoi* [*Family and Society*], 3/9/2014. Retrieved February 5, 2015, from http://giadinh.net.vn/xa-hoi/ve-thu-phu-vang-nghe-chuyen-xu-lam-tac-20140903082897 38.htm

Nam Phuong. (2014). Làm trong sạch lực lượng kiểm lâm [Purifying the Kiem Lam force]. *Nong Nghiep Viet Nam*, 2/12/2014. Retrieved December 21, 2014, from http://nongnghiep.vn/lam-trong-sach-luc-luong-kiem-lam-post135413.html

Natali, L. (2013). The contemporary horizon of green criminology. In N. South & A. Brisman (Eds.), *The Routledge international handbook of green criminology* (pp. 73–84). London: Routledge.

NAV. (1999). *The penal code*. Hanoi: The National Assembly of Vietnam.

NAV. (2003). *The criminal procedure code of Vietnam*. Hanoi: The National Assembly of Vietnam.

NAV. (2005). *People's Public Security Act*. Hanoi: The National Assembly of Vietnam.

Nef, J. (1999). *Human security and mutual vulnerability—An exploration into the global political economy of development and underdevelopment*. Ottawa: International Development Research Centre.

Nellemann, C. (2012). *Green carbon, black trade: Illegal logging, tax fraud and laundering in the worlds tropical forests. A rapid response assessment*. Birkeland Trykkeri AS, Norway: Interpol Environmental Crime Programme—United Nations.

Nellemann, C., Henriksen, R., Raxter, P., Ash, N., & Mrema, E. (2014). *The environmental crisis—Threats to sustainable development from illegal exploitation and trade in wildlife and forest resources*. United Nations Environment Programme and GRID-Arendal, Nairobi and Arendal. Retrieved from http://www.grida.no/

Newman, E. (2010). Critical human security studies. *Review of International Studies, 36*(1), 77–94.

Ngo, H. N. (2008). Giao khoán bảo vệ rừng, thực hiện chính sách dân tộc và tác động đối với bảo vệ rừng ở Bình Định [The allocation of forest protection, the implementation of ethnic policies and the impacts on the work of forest protection in Binh Dinh]. *Kiem Lam Newsletter*. Retrieved June 2, 2015, from http://kiemlam.org.vn/Desktop.aspx/News/So_1_2_nam_2008/Giao_khoan_bao_ve_rung_thuc_hien_chinh_sach_dan_toc_va_tac_dong_doi_voi_bao_ve_rung_o_Binh_Dinh/

Ngoc Linh. (2014). Chung quanh việc giao đất, giao rừng cho đồng bào dân tộc thiểu số [Regarding the issue of allocating lands and forests for the ethnic minority residents]. *Nhan Dan*, 22/11/2014. Retrieved March 26, 2015, from http://www.nhandan.com.vn/mobile/_mobile_kinhte/_mobile_tintuc/item/24901102.html

Nguoi Lao Dong. (2007). Trùm xã hội đen Hai Chi hầu tòa [Gangster boss Hai Chi in court]. *Nguoi Lao Dong*, 03/10/2007. Retrieved June 2, 2015, from http://nld.com.vn/phap-luat/trum-xa-hoi-den-hai-chi-hau-toa-203966.htm

Nguyen Cau. (2014). Rừng đặc dụng Bà Nà—Núi Chúa bị tàn phá do đâu [Why is the special use forest of Ba Na—Nui Chua destroyed?]. *Cong An Nhan Dan Online*, 31/10/2014. Retrieved January 20, 2015, from http://cand.com.vn/Xa-hoi/Rung-dac-dung-Ba-Na---Nui-Chua-bi-tan-pha-do-dau-276935/

Nguyen, C. C. (2010). *The theoretical and practical issues on environmental crimes in the Vietnamese criminal law*. Master degree (in Vietnamese).

Nguyen, D. N. V., & Nguyen, T. (2008). *An overview of the use of plants and animals in traditional medicine systems in Vietnam*. TRAFFIC Southeast Asia, Greater Mekong Programme, Ha Noi, Viet Nam.

Nguyen, D. H., & Nguyen, N. S. (2010). *Đảm bảo an ninh môi trường cho phát triển bền vững [Ensuring environmental security for sustainable development]*. Hanoi: Nha Xuat Ban Khoa Hoc Va Ky Thuat.

Nguyen, D. Q. (2014). Bảo tồn văn hóa Tây Nguyên, phải giữ rừng [Preserving the culture of Central Highland, must maintain forests]. *Tien Phong*, 25/5/2014. Retrieved January 21, 2015, from http://www.tienphong.vn/xa-hoi/bao-ton-van-hoa-tay-nguyen-phai-giu-rung-708775.tpo

Nguyen Dung. (2013). Khánh Hòa: Cuộc truy đuổi kinh hoàng xe lâm tặc [Khanh Hoa: The terrifying chase of a Lam Tac van]. *Dan Tri*, 14/07/2013. Retrieved April 23, 2014, from http://dantri.com.vn/xa-hoi/cuoc-truy-duoi-kinh-hoang-xe-lam-tac-1374228200.htm

Nguyen Duy. (2013). Vụ lật xe chở gỗ đè chết 10 người: Các bị cáo lĩnh 206 tháng tù [The case of flipped truck of illegal timber killing 10 people: The defendants get 206 month in prison]. *Dantri*, 25/1/2013. Retrieved June 2, 2015, from http://dantri.com.vn/phap-luat/vu-lat-xe-cho-go-de-chet-10-nguoi-cac-bi-cao-linh-206-thang-tu-689243.htm

Nguyen Le. (2014). Tình hình tội phạm tiếp tục phức tạp [Criminal situation continues to be complex]. *VnEconomy*, 24/10/2014. Retrieved June 2, 2015, from http://vneconomy.vn/thoi-su/tinh-hinh-toi-pham-tiep-tuc-phuc-tap-20141024100925546.htm

Nguyen Luan. (2014). Phan Dũng: "Nóng" chuyện lâm tặc nơi khác đến phá rừng [Phan Dung: "Hot" issue of illegal loggers elsewhere coming to deforest]. *Binh Thuan Online*, 30/10/2014. Retrieved January 20, 2015, from http://www.baobinhthuan.com.vn/phap-luat/phan-dung-nong-chuyen-lam-tac-noi-khac-den-pha-rungbr-70734.html

Nguyen, M. H., Vu, V. D., Nguyen, V. S., Hoang, V. T., Nguyen, H. D., Pham, N. T., et al. (2007). *Report on the review of Vietnam's wildlife trade policy*. Hanoi: CRES/FPD/UNEP/CITES/IUED.

Nguyen Nam. (2015). Khởi tố, bắt giam kiểm lâm viên huyện Hàm Tân [Charging and arresting a Kiem Lam staff in Ham Tan district]. *Tuoi Tre*, 28/3/2015. Retrieved March 29, 2015, from http://tuoitre.vn/tin/chinh-tri-xa-hoi/20150328/khoi-to-bat-giam-kiem-lam-vien-huyen-ham-tan/726599.html

Nguyen Thuong. (2011). Bao giờ kiểm lâm sống được bằng lương? [When can Kiem Lam live with their salary?]. *Con Nguoi & Thien Nhien*, 28/11/2011. Retrieved June 2, 2015, from http://www.thiennhien.net/2011/11/28/bao-gio-kiem-lam-song-duoc-bang-luong/

Nguyen, T. N. (2013a). Tham nhũng—nguyên nhân và biện pháp ngăn ngừa [Corruption—Causes and preventive measures]. *Communist Review—Tap chi Cong san*, 17/1/2013. Retrieved June 2, 2015, from http://www.tapchicongsan.org.vn/Home/PrintStory.aspx?distribution=19790&print=true

Nguyen, T. Q., & Tran, H. N. (2011). *How Viet Nam is prepared to meet legal requirements of timber export markets*. Hue City, Vietnam: Tropenbos International Viet Nam.

Nguyen, T. T. T., & Phan, M. T. (2014). *Wood processing industry—Activity code: No-5—Support business associations to prepare studies on development strategy for two selected industries*. Hanoi: EU-MUTRAP & Vietnam Chamber of Commerce and Industry. Retrieved from http://www.trungtamwto.vn/sites/default/files/publications/nghien_cuu_nganh_go.pdf

Nguyen, V. D. (2013b). *Muong livelihoods and the role of education in their development: A case study of a Muong community in Cam Thuy district, Thanh Hoa Province, Vietnam*. A thesis submitted to Victoria University of Wellington in fulfilment of the requirements for the degree of Doctor of Philosophy in Development Studies, Victoria University of Wellington.

Nguyen, V. S. (2003). *Vietnam's illegal trade in wildlife: Why it flourishes*. Hanoi, Vietnam: Economy and Environment Program for Southeast Asia.

Nguyen, V. S. (2008). Wildlife trading in Vietnam situation, causes, and solutions. *The Journal of Environment & Development, 17*(2), 145–165.

Nguyen Xuan. (2013). Hành trình sa ngã của một sĩ quan [The wrong journey of a officer]. *Khanh Hoa Online*, 23/11/2013. Retrieved June 2, 2015, from http://baokhanhhoa.com.vn/phap-luat/201311/hanh-trinh-sa-nga-cua-mot-si-quan-2277047/

Nhan Dan. (2004). Báo Nhân Dân: Người bạn tin cậy đồng hành cùng bạn đọc [Nhan Dan Newspaper: The trusted friend accompanying the readers]. *Nhan Dan*, 14/03/2004. Retrieved June 2, 2015, from http://www.nhandan.com.vn/bandoc/item/19795802-nguoi-ban-tin-cay-dong-hanh-cung-ban-doc.html

Nhan Dan. (2005). Vụ phá rừng lớn nhất ở Quảng Ngãi: Trùm Hai Chiên đã được dung túng? [The biggest case of illegal logging in Quang Ngai: Was Hai Chien magnate condoned?]. *Nhan Dan*, 25/10/2005. Retrieved June 2, 2015, from http://www.nhandan.com.vn/phapluat/thoi-su-phap-luat/item/4448702-.html

Nhien Di. (2013). Rừng Tây Nguyên teo tóp [The forests in Central Highland are atrophied]. 24/6/2013. Retrieved April 19, 2015, from http://nld.com.vn/thoi-su-trong-nuoc/rung-tay-nguyen-teo-top-20130624102358371.htm

Nong Nghiep Vietnam. (2011). Quảng Bình: Bảo vệ rừng liên tục bị hành hung [Quang Binh: Forest rangers are regularly assaulted]. *Bao Moi*, 25/2/2011. Retrieved June 2, 2015, from http://www.baomoi.com/Quang-Binh-Bao-ve-rung-lien-tuc-bi-hanh-hung/141/5763937.epi

Nong Nghiep Vietnam. (2012). Kiểm lâm—nghề đặc biệt [Kiem Lam—The special occupation]. *Nong nghiep Vietnam*, 22/05/2012. Retrieved June 2, 2015, from http://nongnghiep.vn/nongnghiepvn/72/1/15/94960/kiem-lamnghe-dac-biet.aspx

NORAD. (2011). *Joint evaluation of support to anti-corruption efforts—Viet Nam country report*. Oslo, Norway: Norwegian Agency for Development Cooperation.

Nsiah-Gyabaah, K. (2010). Human security as a prerequisite for development. In R. A. Matthew, J. Barnett, B. McDonald, & K. L. O'Brien (Eds.), *Global environmental change and human security* (pp. 237–259). Cambridge, MA: MIT Press.

Nurse, A. (2013a). *Animal harm: Perspectives on why people harm and kill animals*. Farnham, Surrey, UK; Burlington, VT: Ashgate.

Nurse, A. (2013b). Perspectives on criminality in wildlife. In R. Walters, D. Westerhuis, & T. Wyatt (Eds.), *Emerging issues in green criminology: Exploring power, justice and harm* (pp. 127–144). Houndmills, Basingstoke, UK: Palgrave Macmillan.

Obidzinski, K., Andrianto, A., & Wijaya, C. (2007). Cross-border timber trade in Indonesia: Critical or overstated problem? Forest governance lessons from Kalimantan. *International Forestry Review, 9*(1), 526–535.

OECD. (2012). *Illegal trade in environmentally sensitive goods*. Paris: Organisation for Economic Co-operation and Development.

Okubo, S. (2011). Glogalisation, human security and the right to live in peace. In S. Okubo & L. Shelley (Eds.), *Human security, transnational crime and human trafficking: Asian and Western perspectives* (pp. 9–35). Hoboken: Taylor and Francis.

Okubo, S., & Shelley, L. (2011). *Human security, transnational crime and human trafficking: Asian and Western perspectives*. Hoboken: Taylor and Francis.

Oldfield, S. (2003a). Chapter 12: regulation of the timber trade. In S. Oldfield (Ed.), *The trade in wildlife: Regulation for conservation*. London: Earthscan.

Oldfield, S. (2003b). *The trade in wildlife: Regulation for conservation*. London: Earthscan.

Olken, B. A., & Pande, R. (2012). Corruption in developing countries. *Annual Review of Economics, 4*, 479–509.

Ostrom, E. (1990). *Governing the commons: The evolution of institutions for collective action*. Cambridge: Cambridge University Press.

Ostrom, E. (1999). *Self-governance and forest resources*. Jakarta: Centre For International Forestry Research.

Ostrom, E. (2008). The challenge of common-pool resources. *Environment: Science and Policy for Sustainable Development, 50*(4), 8–21.

Paris, R. (2001). Human security: Paradigm shift or hot air? *International Security, 26*(2), 87–102.

Passas, N. (2005). Lawful but awful: 'Legal corporate crimes'. *The Journal of Socio-Economics, 34*(6), 771–786.

Pemberton, A. (2014). Environmental victims and criminal justice: Proceed with caution. In T. Spapens, R. White, & M. Kluin (Eds.), *Environmental crime and its victims: Perspectives within green criminology* (pp. 63–86). Surrey: Ashgate.

Pendleton, M. R. (2007). *The social basis of illegal logging and forestry law enforcement in North America*. In L. Tacconi (Ed.), *Illegal logging, law inforcement, livelihoods and the timber trade*. London: Earthcan.

Peters, J. L. (2000). Illegal trafficking of timber in Cambodia. *Colorado Journal of International Environmental Law and Policy, 11*, 102.

Pham, D. X. (2008). *Investigation delivered by the investigation police on crimes of economic management and official position over the violations against regulations on forest exploitation and protection*. The thesis of PhD in Law, The People's Police Academy (in Vietnamese).

Pham, Q. H. (2007). *The solutions for the police of investigation of economic crimes, Lang Son province public security, to enhance the effectiveness of the detection and investigations into cross-border smuggling in forest products*. The dissertation of Master in Law, The People's Police Academy (in Vietnamese).

Pham, Q. N. (2011a). Đấu tranh phòng, chống tội phạm và vi phạm pháp luật về bảo vệ môi trường [The work preventing and combatting environmental offences]. *Tap chi Cong san Communist Review*. Retrieved April 30, 2015, from http://www.tapchicongsan.org.vn/Home/Viet-nam-tren-duong-doi-moi/Phong-chong-tham-nhung/2011/11801/Dau-tranh-phong-chong-toi-pham-va-vi-pham-phap-luat-ve.aspx

Pham, S. D. (2011b). Điều gì khiến kỳ nam đắt khủng khiếp? [Why Ky Nam is extremely expensive?]. *Dan Tri*, 18/07/2011. Retrieved June 2, 2015, from http://dantri.com.vn/su-kien/dieu-gi-khien-ky-nam-dat-khung-khiep-499653.htm

Pham, T. L. (2006). *Characteristics of the criminal offense of violating the regulations on the exploitation and protection of forests in the province of Gia Lai*. Master dissertation, The People's Police Academy (in Vietnamese).

Pham, T. T., Moeliono, M., Nguyen, T. H., Nguyen, H. T., & Vu, T. H. (2012). *The context of REDD+ in Vietnam—Drivers, agents and institutions*. Occasional Paper 75, CIFOR, Bogor, Indonesia. Retrieved from http://www.cifor.org/publications/pdf_files/OccPapers/OP-75.pdf

Pham, T. T. T. (2010). Deforestation in Vietnam: Driving factors and some recommendations. *Economic Management, 35* (8+9/2010), 14–26 (in Vietnamese).

Phan, T. B. (2011). Lợi dụng chính sách xóa nhà tạm để tàn phá rừng đại ngàn [Taking advantage of the policy of eliminating temporary houses to ... destry massive forests]. *Cong An Nhan Dan*, 15/2/2011. Retrieved January 2, 2015, from http://cand.com.vn/Xa-hoi/Loi-dung-chinh-sach-xoa-nha-tam-de-tan-pha-rung-dai-ngan-173905/

Phan, X. S., & Pham, T. L. (2010). *Identifying corruption and solutions to prevent and combat corruption in Vietnam nowadays*. Hanoi: Chinh tri quoc gia (in Vietnamese).

Phap Luat Viet Nam. (2013). Nguyên Phó Văn phòng Cỏ quan CSĐT tù 8 năm vì nhận hối lộ [The former chief of investigative police office sentenced 8 years in prison for receiving bribe]. *Dantri*, 24/4/2013. Retrieved June 2, 2015, from http://dantri.com.vn/phap-luat/nguyen-pho-van-phong-co-quan-csdt-tu-8-nam-vi-nhan-hoi-lo-723047.htm

Phap Luat Viet Nam. (2014). Gia Lai: Bắt gỗ lậu, kiểm lâm bị côn đồ chém đứt gân tay [Gia Lai: Arresting illegal timber, Kiem Lam is slashed off their hand tendon by gangsters]. *Doi Song & Phap Luat*, 18/09/2014. Retrieved June 2, 2015, from http://www.doisongphapluat.com/phap-luat/an-ninh-hinh-su/gia-lai-bat-go-lau-kiem-lam-bi-con-do-chem-dut-gan-tay-a51269.html

Phung, V. M. (2011). Để việc thực hiện chủ trường, chính sách của Đảng và Nhà nước ở có sở không bị lợi dụng [To eliminate the grounds for abusing the guideline and policies of the Party and the State]. *Xay Dung Dang*,

25/10/2011. Retrieved January 2, 2015, from http://www.xaydungdang.org.vn/Home/dien-dan/2011/4332/De-viec-thuc-hien-chu-truong-chinh-sach-cua-Dang-va-Nha.aspx

Picarelli, J. T. (2008). Transnational organized crime. In P. Williams (Ed.), *Security studies: An introduction* (pp. 453–467). London: Routledge.

PMV. (2003). *Directive No.12/2003/CT-TTg dated 16/05/2003—Regarding "implementing urgent solutions in forest protection and development"*. Hanoi: The Prime Minister of Vietnam.

PMV. (2004). *Decision 134/2004/QĐ-TTg—A number of policies to support the households of poor ethnic minorities with production lands, residential lands, residential houses and clean water*. Hanoi: The Prime Minister of Vietnam.

Potter, G. (2010). What is green criminology? *Sociology Review, 20*, 8–12.

Potter, G. (2014). The criminogenic effects of environmental harm: Bringing a 'green' perspective in mainstream criminology. In T. Spapens, R. White, & M. Kluin (Eds.), *Environmental crime and its victims: Perspectives within green criminology* (pp. 7–21). Surrey: Ashgate.

Prieksat, J. (2009). *International trade and trafficking in wildlife: The value of biodiversity cites, and going further*. International Environmental Law.

Pulido, L. (2000). Rethinking environmental racism: White privilege and urban development in southern California. *Annals of the Association of American Geographers, 90*(1), 12–40.

Punch, K. (2005). *Introduction to social research: Quantitative and qualitative approaches*. Thousand Oaks, CA: Sage.

Pye-Smith, C. (2006). *Logging in the Wild East: China and the forest crisis in the Russian Far East*. Forest Trends & Rights and Resources.

Quang Binh Investigation Police Agency. (2012). *The investigative conclusion No.25/KLDT-PC46 dated 12/1/2012 on the case: Duong Van Tanh offence on violation of regulations on forest exploitation and protection*. Investigating Police Agency, Quang Binh Public Security Department, Quang Binh (in Vietnamese).

Quang Binh Public Security. (2013). *The investigative conclusion of the criminal case No.13/KLDT-PC46 dated 15/01/2013: Violation of he regulations on forest exploitation and protection taking place in the Phong Nha—Ke Bang National Park*. Investigating Police Agency, Quang Binh Public Security Department, Quang Binh (in Vietnamese).

Quang Long. (2014). "Báu vật" rừng già [The "treasure" of old forests]. *Tien Phong*, 4/5/2014. Retrieved January 21, 2015, from http://www.tienphong.vn/xa-hoi/bau-vat-rung-gia-700950.tpo

Quang Nhat. (2013). Bi kịch đời phu trầm [The life tragedy of agarwood hunters]. *Nguoi Lao Dong*, 30/03/2013. Retrieved June 2, 2015, from http://dantri.com.vn/xa-hoi/bi-kich-doi-phu-tram-713495.htm

Quang Thai. (2015). "Nóng" tình trạng khai thác lâm sản trái phép tại tỉnh Kon Tum ["Hot" situation of illegal harvest of forest products in Kon Tum]. 7/1/2015. Retrieved January 10, 2015, from http://www.vietnamplus.vn/nong-tinh-trang-khai-thac-lam-san-trai-phep-tai-tinh-kon-tum/300738.vnp

Quang Tri Environmental Police. (2012). Conference paper: The Quang Tri police of preventing and combatting environmental crimes and the work of preventing and combating crimes and violations of forest protection and natural reservation. In *Preventing and combatting violations against legislations on forest protection* (pp. 149–153). Hanoi: The Department of Preventing and Combating Environmental Crimes (in Vietnamese).

Quang Vu. (2013). Tình hình tội phạm đang diễn biến nghiêm trọng hơn [Criminal situation remains increasingly serious]. *Vietnam+*, 28/10/2013. Retrieved June 2, 2015, from http://www.vietnamplus.vn/tinh-hinh-toi-pham-dang-dien-bien-nghiem-trong-hon/227343.vnp

Quynh Anh. (2014). Kỳ II: Áp lực từ "cơn bão" di cư tự do [Volume II: Pressure from "storms" of free migration]. *Thanh Tra*, 25/11/2014. Retrieved January 20, 2015, from http://thanhtra.com.vn/xa-hoi/doi-song/ky-ii-ap-luc-tu-con-bao-di-cu-tu-do_t114c1159n82558

Ravenel, R. M., & Granoff, I. M. E. (2004). Illegal logging in the tropics: A synthesis of the issues. *Journal of Sustainable Forestry, 19*(1–3), 351–366.

Reboredo, F. (2013). Socio-economic, environmental, and governance impacts of illegal logging. *Environment Systems and Decisions, 33*(2), 295–304.

RECOFTC. (2014). *Vietnam's forestry reforms: A brief overview of forest reform in the country*. The Center for People and Forest. Retrieved December 1, 2014, from http://www.recoftc.org/country/vietnam/basic-page/vietnam%E2%80%99s-forestry-reforms

Reiman, J. (1995). *The rich get richer and the poor get prison: Ideology, crime, and criminal justice* (pp. 237). ISBN 0-02-439250-2.

Richards, M. (1997). Tragedy of the commons for community-based forest management in Latin America. *Natural Resource Perspectives, 22*, 1–9.

Richards, M., Wells, A., Gatto, F. D., Contreras-Hermosilla, A., & Pommier, D. (2003). Impacts of illegal ity and barriers to legality: A diagnostic analysis of illegal logging in Honduras and Nicaragua. *International Forestry Review, 5*(3), 282–292.

Roberton, S. (2013). Corruption along the Vietnam-China border permits massive smuggling of endangered animals. *Policy Innovation*, 7/8/2013. Retrieved June 2, 2015, from http://www.policyinnovations.org/ideas/briefings/data/000260

Rosander, M. N. (2008). *Illegal logging: Current issues and opportunities for Sida/SENSA engagement in Southeast Asia*. Bangkok, Thailand: RECOFTC & Sida.

Ruggiero, V., & South, N. (2013). Green criminology and crimes of the economy: Theory, research and praxis. *Critical Criminology, 21*(3), 359–373.

Ruiz Pérez, M., Nguinguiri, J.-C., Nzala, D., Toirambe, B., Yalibanda, Y., Ezzine De Blas, D., et al. (2005). Logging in the Congo Basin: A multi-country characterization of timber companies. *Forest Ecology and Management, 214*(1), 221–236.

Salo, R. S. (2003). When the logs roll over: The need for an international convention criminalizing involvement in the global illegal timber trade. *Georgetown International Environmental Law Review, 16*(1), 127.

Saunders, J. (2014). *Trade in illegal timber: The response in Vietnam—A Chatham House Assessment*. Retrieved from http://www.illegal-logging.info/sites/files/chlogging/CHHJ2362_Vietnam_Logging_Research_Paper_FINAL.pdf

Schloenhardt, A. (2008). *The illegal trade in timber and timber products in the Asia-Pacific region—Research and Public Policy Series No. 89*. Canberra: Australian Institute of Criminology.

Schlosberg, D. (2007). *Defining environmental justice: Theories, movements, and nature*. Oxford: Oxford University Press.

Schnaiberg, A. (1980). *The environment: From surplus to scarcity*. New York: Oxford University Press.

Schneider, J. L. (2008). Reducing the illicit trade in endangered wildlife. The market reduction approach. *Journal of Contemporary Criminal Justice, 24*(3), 274–295.

Sen, A. (2000). Conference paper: Why human security. In *International Symposium on Human Security*, Tokyo.

Seneca Creek Associates, LLC. (2004). *"Illegal" logging and global wood markets: The competitive impacts on the US wood products industry*. American Forest & Paper Association.

SGTT. (2011). Đại gia "chơi" gỗ quý tuyệt chủng tiền tỷ [The well-off enjoys specious timbers worth of about a billion VND]. *Báo Mới*, 26/02/2011. Retrieved June 2, 2015, from http://www.baomoi.com/Dai-gia-choi-go-quy-tuyet-chung-tien-ty/79/5772035.epi

Shearing, C. (2015). Criminology and the Anthropocene. *Criminology and Criminal Justice, 15*(3), 255–269.
Sikor, T., & To, P. X. (2011). Illegal logging in Vietnam: Lam Tac (forest hijackers) in practice and talk. *Society & Natural Resources, 24*(7), 688–701.
Silverman, D. (2011). *Interpreting qualitative data*. London: Sage.
Simon, D. R. (2000). Corporate environmental crimes and social inequality: New directions for environmental justice research. *American Behavioral Scientist, 43*(4), 633–645.
Simons, H. (2009). *Case study research in practice*. London: Sage.
Skinnider, E. (2011). *Victims of environmental crime: Mapping the issues*. International Centre for Criminal Law Reform and Criminal Justice Policy.
Smith, W. (2002). The global problem of illegal logging. *ITTO Tropical Forest Update*, 12/1.
Smith, W. (2004). Undercutting sustainability: The global problem of illegal logging and trade. *Journal of Sustainable Forestry, 19*(1–3), 7–30.
Smith, J., Obidzinski, K., Subarudi, S., & Suramenggala, I. (2003). Illegal logging, collusive corruption and fragmented governments in Kalimantan, Indonesia. *International Forestry Review, 5*(3), 293–302.
Sollund, R. (2013). The victimization of women, children and non-human species through traffcking and trade—Crime understood through an ecofeminist perspective. In N. South & A. Brisman (Eds.), *The Routledge international handbook of green criminology* (pp. 317–330). London: Routledge.
Solomon, S. (2007). *Climate Change 2007: The physical science basis: Contribution of working group I to the Fourth Assessment Report of the Intergovernmental Panel Climate Change*. Cambridge: Cambridge University Press.
South, N. (1998). A green field for criminology? A proposal for a perspective. *Theoretical Criminology, 2*(2), 211–233.
South, N. (2010). The ecocidal tendencies of the late modernity: Transnational crime, social exclusion, victims and rights. In R. White (Ed.), *Global environmental harm: Criminological perspective*. Cullompton, UK: Willan.
South, N. (2012a). Climate change, environmental (in)security, conflict and crime. In S. Farrall, T. Ahmed, & D. French (Eds.), *Criminological and legal consequences of climate change* (pp. 97–111). Oxford: Hart.
South, N. (2012b). Conference paper: Green criminology: Theories, concepts and connected meanderings. In *ESRC Green Criminology Seminar*, Northumbria University, Newcastle Upon Tyne.

South, N. (2015). Anticipating the Anthropocene and greening criminology. *Criminology and Criminal Justice, 15*(3), 270–276.

South, N., & Beirne, P. (2006). Introduction. In N. South & P. Beirne (Eds.), *Green criminology* (pp. xxvii–xxvii). Aldershot: Ashgate.

South, N., & Beirne, P. (2007). *Issues in green criminology: Confronting harms against environments, humanity and other animals*. Cullompton, UK; Portland, OR: Willan.

South, N., & Brisman, A. (2013). *The Routledge international handbook of green criminology*. London: Routledge.

South, N., Brisman, A., & Beirne, P. (2013). A guide to a green criminology. In N. South & A. Brisman (Eds.), *The Routledge international handbook of green criminology* (pp. 27–42). London: Routledge.

South, N., & Wyatt, T. (2011). Comparing illicit trades in wildlife and drugs: An exploratory study. *Deviant Behavior, 32*(6), 538–561.

Spapens, T. (2014). Invisible victimes: The problems of policing environmental crime. In T. Spapens, R. White, & M. Kluin (Eds.), *Environmental crime and its victims: Perspectives within green criminology* (pp. 221–236). Surrey: Ashgate.

Spapens, T., White, R., & Kluin, M. (2014). *Environmental crime and its victims: Perspectives within green criminology*. Ashgate Publishing.

Spencer, D. C., & Fitzgerald, A. (2013). Three ecologies, transversality and victimization: The case of the British Petroleum oil spill. *Crime, Law and Social Change, 59*(2), 209–223.

SRV. (2012). *Implementation of sustainable development in Vietnam—National report at the United Nations Conference on Sustainable Development (RIO+20)*. Socialist Republic of Vietnam, Hanoi.

Stahl, J. (2010). The rents of illegal logging: The mechanisms behind the rush on forest resources in Southeast Albania. *Conservation and Society, 8*(2), 140.

Stake, R. E. (2000). The case study method in social inquiry. In M. Hammersley, R. Gomm, & P. Foster (Eds.), *Case study method: Key issues, key texts* (pp. 19–26). London: Sage.

Stern, N. H. (2007). *The economics of climate change: The Stern review*. Cambridge: Cambridge University Press.

Steverson, B. K. (1994). Ecocentrism and ecological modeling. *Environmental Ethics, 16*(1), 71–88.

Stewart, D. (2014). Project leaf, and interpol's work on illegal logging and forest crime. In T. Spapens, R. White, & M. Kluin (Eds.), *Environmental crime and its victims: Perspectives within green criminology* (pp. 236–247). Surrey: Ashgate.

Stretesky, P., & Lynch, M. J. (1998). Corporate environmental violence and racism. *Crime, Law and Social Change, 30*(2), 163–184.

Stretesky, P. B., Long, M. A., & Lynch, M. J. (2014). *The treadmill of crime: Political economy and green criminology*. Oxon and New York: Taylor and Francis.

Sumrall, K. A. (2009). *Confronting illegal wildlife trade in Vietnam: The experience of education for nature—Vietnam*. Degree of Master of Science, University of Michigan.

Sunderlin, W. D., & Huynh, T. B. (2005). *Poverty alleviation and forests in Vietnam*. Subur Printing, Indonesia: Center for International Forestry Research—CIFOR.

Sygna, L., O'Brien, K., & Wolf, J. (2013). *A changing environment for human security: Transformative approaches to research, policy and action*. Routledge.

Tacconi, L. (2007a). The problem of illegal logging. In L. Tacconi (Ed.), *Illegal logging: Law enforcement, livelihoods and the timber trade* (pp. 1–16). London: Earthscan.

Tacconi, L. (2007b). Illegal logging and the future of the forest. In L. Tacconi (Ed.), *Illegal logging: Law enforcement, livelihoods and the timber trade* (pp. 275–289). London: Earthscan.

Tacconi, L., Boscolo, M., & Brack, D. (2003). *National and international policies to control illegal forest activities*. Report for the Ministry of Foreign Affairs, Government of Japan. Bogor, Indonesia: CIFOR.

Tadjbakhsh, S., & Chenoy, A. M. (2007). *Human security: Concepts and implications*. Oxon: Routledge.

Tadros, V. (2008). Crimes and security. *The Modern Law Review, 71*(6), 940–970.

Takemoto, H., Ito, M., Shiraki, T., Yagura, T., & Honda, G. (2008). Sedative effects of vapor inhalation of agarwood oil and spikenard extract and identification of their active components. *Journal of Natural Medicines, 62*(1), 41–46.

Tan Vu. (2014). Tan nát rừng đầu nguồn [Smashed watershed forests]. *Tuoi Tre*, 12/12/2014. Retrieved January 4, 2015, from http://tuoitre.vn/tin/ban-doc/20141212/tan-nat-rung-dau-nguon/683775.html

Telapak & EIA. (2007). *The thousand-headed snake: Forest crimes, corruption and injustice in Indonesia*. Bogor, Indonesia.

Thai, B. D., & Trung, T. (2012). Khởi tố vụ án phá rừng nghiêm trọng: Xử lý trách nhiệm nhiều cá nhân để xảy ra phá rừng [Prosecution of the serious case of illegal logging: Handling many individuals responsible for illegal logging]. *Tuoi Tre*, 27/08/2012. Retrieved June 2, 2015, from http://tuoitre.vn/Chinh-tri-xa-hoi/Phap-luat/509006/xu-ly-trach-nhiem-nhieu-ca-nhan-de-xay-ra-pha-rung.html

Thanh Chau. (2013). TAND tỉnh Hà Tĩnh xét xử vụ phá rừng đầu nguồn Sơn Hồng : Tuyên phạt 15 bị cáo 36 năm 6 tháng tù [Ha Tinh Province People's Court adjudicates the illegal logging case in the Son Hong watershed forest: 15 accused sentenced to 36 years and 6 months imprisonment. *Nhan Dan*, 31/01/2013. Retrieved June 2, 2015, from http://www.nhandan.com.vn/phapluat/thoi-su-phap-luat/item/225502-.html

Thanh Huyen. (2014). Người Việt chưa giàu nhưng "sang": Chỉ vì thích phô trương! [The Vietnamese are not yet rich but "posh": Because of exhibitionism! *Dan Tri*, 15/2/2014. Retrieved March 9, 2015, from http://dantri.com.vn/kinh-doanh/nguoi-viet-chua-giau-nhung-sang-chi-vi-thich-pho-truong-1008200.htm

Thanh Nguyen. (2014). Hung thần gỗ lậu [Deadly elfs of illegal timber]. *Thanh Nien*, 6/10/2014. Retrieved June 3, 2015, from http://www.thanhnien.com.vn/pages/20141005/hung-than-go-lau.aspx

Thanh Nien. (2013a). 3 arrested for brutally killing loggers in Vietnam. *Thanh Nien News*, 5/4/2013. Retrieved June 3, 2015, from http://www.thanhniennews.com/society/3-arrested-for-brutally-killing-loggers-in-vietnam-2982.html

Thanh Nien. (2013b). Khai thác trái phép gần 1.000 m3 gỗ dựng nhà tái định cư [Illegally harvesting almost 1,000 m^3 of timber for building resettlement houses]. *Con Nguoi & Thien Nhien—Thiennhien.Net*, 4/2/2013. Retrieved June 9, 2015, from http://www.thiennhien.net/2013/02/05/khai-thac-trai-phep-gan-1-000-m3-go-dung-nha-tai-dinh-cu/

Thanh Nien News. (2015). About us. *Thanh Nien News*. Retrieved June 3, 2012, from http://www.thanhniennews.com/about/about-us-27720.html

The Economist. (2010). Keeping it in the community—Well-organised locals often make the best forest managers, but they need help. *The Economist*, 23/9/2010. Retrieved February 4, 2015, from http://www.economist.com/node/17062703

The Lancet Commissions. (2009). *Managing the health effects of climate change.* Lancet and University College London Institute for Global Health Commission.

Thien Nhan. (2013). Hãi hùng xe máy chở gỗ lậu [Horrificly motorcycles carrying illegal timber]. *Thanh Nien*, 02/12/2013. Retrieved June 3, 2012, from http://www.thanhnien.com.vn/pages/20131201/hai-hung-xe-may-cho-go-lau.aspx

Thomas, C. (2000). *Global governance, development and human security: The challenge of poverty and inequality*. London: Pluto.

Thomas, C. (2001). Global governance, development and human security: Exploring the links. *Third World Quarterly, 22*(2), 159–175.

Thomas, G. (2011). *How to do your case study: A guide for students and researchers*. Los Angeles, CA; London: Sage.

Thuy Nguyen. (2014). Nghi báo kiểm lâm bắt gỗ, nhóm lâm tặc đốt nhà dân [Suspecting of informing Kiem Lam to arrest timber, Lam Tac burn the people house]. *Phap Luat & Doi Song*, 17/7/2014. Retrieved June 3, 2015, from http://www.doisongphapluat.com/dia-phuong/mien-nam/nghi-bao-kiem-lam-bat-go-nhom-lam-tac-dot-nha-dan-a41416.html

Tien Thanh. (2014). Bị phát hiện, lâm tặc đánh hạt phó kiểm lâm nhập viện [Being detected, Lam Tac beat a vice-captain of Kiem Lam to be hospitalised]. *Doi Song & Phap Luat*, 06/08/2014. Retrieved June 3, 2015, from http://www.doisongphapluat.com/dia-phuong/mien-trung/bi-phat-hien-lam-tac-danh-hat-pho-kiem-lam-nhap-vien-a44685.html

Tienphong. (2011). Bắt 15 toa tàu chở gỗ lậu quý hiếm [Arresting 15 train wagons illegally transporting rare and endanger timber]. *Tienphong*, 15/12/2011. Retrieved June 3, 2015, from http://www.tienphong.vn/Phap-Luat/561202/Bat-15-toa-tau-cho-go-lau-quy-hiem-tpp.html

To, P. X., Tran, L. H., & Nguyen, T. Q. (2014). *The overview of the Vietnamese timber sector: Current status and trends*. Forest Trends/VIFORES/FPA Binh Dinh.

To, X. P. (2012). Natural forests and its suffering from timber need [Rừng tự nhiên chịu sức ép từ nhu cầu sử dụng gỗ]. *Con Nguoi & Thien Nhien*, 6/11/2012. Retrieved June 3, 2015, from http://www.thiennhien.net/2012/11/06/rung-tu-nhien-chiu-suc-ep-tu-nhu-cau-su-dung-go/

To, X. P., & Canby, K. (2011). *Vietnam: Overview of forest governance and trade*. Kuala Lumpur: European Forest Institute—FLEGT Asia Regional Office.

To, X. P., & Tran, H. N. (2014a). *Forest land allocation in the context of forestry sector restructuring: Opportunities for forestry development and upland livelihood improvement*. Tropenbos International Viet Nam, Hue, Vietnam.

To, X. P., & Tran, H. N. (2014b). *Rubber expansion and forest protection in Vietnam*. Hue, Viet Nam. Retrieved from http://www.forest-trends.org/documents/files/doc_4671.pdf

Tombs, S., & Whyte, D. (2009). The state and corporate crime. In R. Coleman, J. Sim, S. Tombs, & D. Whyte (Eds.), *State, power, crime* (pp. 103–115). London: Sage.

Tombs, S., & Whyte, D. (2015). *The corporate criminal: Why corporations must be abolished.* Abingdon, Oxon; New York, NY: Routledge.

Ton, T. S. (2006). Kỳ nam và Trầm hương. *Tuoitre*, 04/09/2006. Retrieved June 6, 2015, from http://tuoitre.vn/Chinh-tri-xa-hoi/Moi-truong/159942/ky-nam-va-tram%c2%a0huong%c2%a0.html

TRAFFIC. (2008). *What's driving the wildlife trade? A review of expert opinion on economic and social drivers of the wildlife trade and trade control efforts in Cambodia, Indonesia, Lao PDR and Vietnam.* Washington, DC: TRAFFIC.

TRAFFIC. (2013). *Timber trade.* Retrieved March 6, 2015, from http://www.traffic.org/timber-trade/

Tran, C. D., Duong, T. N., Bui, V. Q., & Ly, V. T. (2010). Examining impacts of eco-social factors on forest resources of the Ba Be National Park—Ba Be distric, Bac Kan province. *The Ba Be National Park*, 199–210. Bac Kan: The Management Board of Ba Be National Park (in Vietnamese).

Tran, L. H. (2001). Overview of environmental offences and related issues. *Legal Review—Law University of Ho Chi Minh City*, 4/2001 (in Vietnamese).

Tran, N. Q. (2011). Tàn sát Vườn quốc gia Yók Đôn [Massacring the Yok Don National Park]. *Thanh Nien*, 18/04/2011. Retrieved June 3, 2015, from http://www.thanhnien.com.vn/pages/20110418/tan-sat-vuon-quoc-gia-yok-don.aspx

Tran Thu. (2014). Lợi dụng xóa nhà tạm để... phá rừng? [Taking advantage of the elimination of temporary houses to ... deforest]. *Dan Viet*, 7/3/2014. Retrieved January 2, 2015, from http://danviet.vn/dien-dan-ban-doc/loi-dung-xoa-nha-tam-de-pha-rung-196903.html

Tran, T. L. (2013a). *The operation of preventing the offence of illegal trade and transport in forest products delivered by the Investigation Police on the Offences of Position and Economic Management Order of the Quang Binh Province Department of Public Security.* Dissertation of Master in Law, People's Police Academy (in Vietnamese).

Tran, V. V. (2013b). "Rừng thiêng" của người Hà Nhì ["Holy forest" of Ha Nhi people]. *Nong Nghiep Viet Nam*, 25/9/2013. Retrieved January 21, 2015, from http://land.net.vn/info/Tin-tuc/Khong-de-Rung-Tay-Nguyen-tro-thanh-hoai-niem!-61-317.html

Transparency International. (2012). *Overview of corruption and anti-corruption in Vietnam No.315.* U4 & The Transparency International. Retrieved from

http://www.u4.no/publications/overview-of-corruption-and-anti-corruption-in-vietnam/
Trigger, D., & Mulcock, J. (2005). Forests as spiritually significant places: Nature, culture and belonging in Australia. *The Australian Journal of Anthropology, 16*(3), 306–320.
TTXVN. (2005). Băng nhóm XHĐ Hai Chi tham gia phá rừng nhú thế nào? [How does Hai Chi gang takes part in timber trafficking?]. *Viet Bao*, 10/7/2005. Retrieved June 3, 2015, from http://vietbao.vn/An-ninh-Phap-luat/Bang-nhom-XHD-Hai-Chi-tham-gia-pha-rung-nhu-the-nao/55073230/302/
Tuoi Tre. (2011). Kiểm lâm "làm luật"—Kỳ 2: Làm sai lệch hồ só [Kiem Lam "making laws"—Vol 2: Falsifying documentations]. *Tuoi Tre*, 4/10/2011. Retrieved June 3, 2015, from http://tuoitre.vn/Tuoi-tre-cuoi-tuan/Van-de-Su-kien/458808/Kiem-lam-%E2%80%9Clam-luat%E2%80%9D---Ky-2-Lam-sai-lech-ho-so.html
Tuoitre News. (2014). Infiltrating 'deforestation factory' in Vietnam jungle. *Tuoi Tre News*, 22/9/2014. Retrieved June 3, 2015, from http://tuoitrenews.vn/features/22636/infiltrating-a-deforestation-factory-in-a-vietnam-jungle
Tuong Vy. (2009). 9 "quan" hầu tòa vụ phá rừng Khe Diên [9 "leaders" brought into the court in an illegal deforestation case]. *Dan Tri*, 28/07/2009. Retrieved June 3, 2015, from http://dantri.com.vn/phap-luat/9-quan-hau-toa-vu-pha-rung-khe-dien-340122.htm
UK Customs. (2012). *The Forest Law Enforcement, Governance and Trade regulations No. 178/2012*. HM Revenue and Customs. Retrieved from http://www.legislation.gov.uk/uksi/2012/178/pdfs/uksi_20120178_en.pdf
Ullman, R. H. (1983). Redefining security. *International Security, 8*(1), 129–153.
UNDP. (1994). *Human Development Report 1994*. New York: United Nations Development Program.
UNDP. (2005a). *Human Development Report 2005; International cooperation at a crossroads: Aid, trade and security in an unequal world*. New York: United Nations Development Programme.
UNDP. (2005b). *Overcoming fear: Citizen (in)security and human development in Costa Rica (Summary)*. New York: United Nations Development Programme.
UNDP. (2013). *Justice index—Assessment of distributive justice and equality from a citizen-based survey in 2012*. Hanoi: United Nation Development Programme.
UNEP. (2007). *Global Environment Outllook, GEO 4—Environment for development*. Valletta, Malta: United Nations Environment Programme.

UNEP. (2012). Organized crime trade worth over US$30 billion responsible for up to 90 % of tropical deforestation. *UNEP Press Release*, 29/9/2012. Retrieved June 3, 2015, from http://www.unep.org/newscentre/default.aspx?DocumentID=2694&ArticleID=9286

United Nations. (2007). *Report on the drug and crime in Vietnam*. Hanoi, Vietnam: United Nations.

UNODC. (2004). *United Nations convention against transnational organized crime and the protocols thereto*. New York: United Nations.

UNODC. (2010). *The globalization of crime—A transnational organized crime threat assessment*. United Nations Office on Drugs and Crime.

UNODC. (2012a). *Vietnam country programme 2012–2017*. Vietnam: United Nations Office on Drugs and Crime. Retrieved from http://www.unodc.org/documents/southeastasiaandpacific//vietnam/publication/2014/CP_final_E.pdf

UNODC. (2012b). *Wildlife and forest crime analytic toolkit—Revised edition*. New York: United Nations Office for Drugs and Crime. Retrieved from http://www.unodc.org/documents/southeastasiaandpacific//Publications/wildlife/CJS_Response_-_VIETNAM_01_13_Dec_201.pdf

UNODC. (2013a). *Criminal justice responses to the illegal trade in timber in Vietnam*. Bangkok, Thailand: United Nations Office for Drugs and Crime, Regional Office for Southeast Asia and the Pacific. Retrieved from http://www.unodc.org/documents/southeastasiaandpacific//Publications/wildlife/CJS_Response_-_VIETNAM_01_13_Dec_201.pdf

UNODC. (2013b). *Transnational organized crime in East Asia and the Pacific*. Bangkok, Thailand: United Nations Office on Drugs and Crime—Regional Office for Southeast Asia and the Pacific.

UPM. (2013). *Confab provides inputs and collaboration to spur UPM's gaharu researchers*. Faculty of Forestry—Universiti Putra Malaysia. Retrieved March 4, 2014, from http://www.forr.upm.edu.my/ISSA_BI

USAID. (2013). *Vietnam tropical forest and biodiversity assessment*. Retrieved from http://www.usaid.gov/sites/default/files/documents/1861/Vietnam118_119FinalDraft%2015Oct2013%20(Public%20Version).pdf

Uyen Thu. (2013a). *Vì sao hơn 130,000 ha rừng Tây Nguyên bị bức tử: Kỳ 1 "Cơn lốc" cao su "tàn sát" rừng [Why more than 130,000 ha of forests in Central Highlands are forced to suicide: Vol 1: The rubber "cyclones" massacre the forests]*. Gia Lai, p. 3, released 16/9/2013: Gia Lai Newspaper.

Uyen Thu. (2013b). *Vì sao hơn 130,000 ha rừng Tây Nguyên bị bức tử: Kỳ 2: Nhiều bất cập trong quản lý rừng [Why more than 130,000 ha of forests in*

Central Highlands are forced to suicide: Vol 2: Many shortcomings in forest management]. Gia Lai. Gia Lai, p. 3, released 17/9/2013: Gia Lai Newspaper.

V.Tao. (2013). "Trảm" 2 sĩ quan huyện đội "ăn chặn kỳ nam" [Handling 2 officers of district army appropriating agarwood]. *Nguoi Lao Dong*, 19/06/2013. Retrieved June 3, 2015, from http://nld.com.vn/phap-luat/tram-2-si-quan-huyen-doi-an-chan-ky-nam-20130619080619l.htm

VAF. (2014). Workshop outcomes: "The shortcoming of the policy and practice in the work of forest and land allocation and post allocation"—Kết quả Tọa đàm: "Những bất cập, tồn tại về chính sách và thực tiễn trong công tác giao đất, giao rừng và sau giao đất, giao rừng", 8/5/2014. Hanoi: Vietnam Administration of Forestry.

Valverde, M. (2011). Questions of security: A framework for research. *Theoretical Criminology, 15*(1), 3–22.

Valverde, M. (2014). Studying the governance of crime and security: Space, time and jurisdiction. *Criminology and Criminal Justice, 14*(4), 379–391.

Van Thanh. (2014a). Nguy cơ tuyệt chủng cây thủy tùng trên đất Tây Nguyên [The risk of extinction of thuy tung trees in the Central Highlands]. *Cong An Nhan Dan Online*, 12/12/2014. Retrieved January 19, 2015, from http://cand.com.vn/Ban-doc-cand/Nguy-co-tuyet-chung-cay-thuy-tung-tren-dat-Tay-Nguyen-334264/

Van Thanh. (2014b). Rừng bảo tồn thiên nhiên Ea Sô bị xẻ thịt [The reservation forest Ea So is being slaughtered] *Cong An Nhan Dan*, 19/12/2014. Retrieved December 19, 2014, from http://cand.com.vn/Ban-doc-cand/Rung-bao-ton-thien-nhien-ea-So-bi-xe-thit-335128/

Vanrijckeghem, C., & Weder, B. (2001). Bureaucratic corruption and the rate of temptation: Do wages in the civil service affect corruption, and by how much? *Journal of development economics, 65*(2), 307–331.

Viet Ba. (2011). Thiếu tướng Nguyễn Xuân Lý: "Ý thức bảo vệ môi trường phải được nâng lên" [Major General Nguyen Xuan Ly: "The awareness of environmental protection have to be raised"]. *An Ninh The Gioi Online*, 25/08/2011. Retrieved June 3, 2015, from http://antg.cand.com.vn/Su-kien-Binh-luan-antg/Thieu-tuong-Nguyen-Xuan-Ly-Y-thuc-bao-ve-moi-truong-phai-duoc-nang-len-301312/

Viet Bao. (2005). Những kiểm lâm "ăn lương" của lâm tặc [Kiem Lam officers get salaries from Lam Tac]. *Viet Bao*, 10/11/2005. Retrieved June 3, 2015, from http://vietbao.vn/An-ninh-Phap-luat/Nhung-kiem-lam-an-luong-cua-lam-tac/55089059/302/

Viet Huong. (2014). Ngắm biệt thự bằng gỗ tiền tỷ của đại gia Việt [Seeing the billions-valued villas of the Vietnamese super rich]. *Phap Luat & Doi Song*, 8/9/2014. Retrieved December 8, 2014, from http://www.doisongphapluat.com/kinh-doanh/doanh-nhan/ngam-biet-thu-bang-go-tien-ty-cua-dai-gia-viet-a49822.html

Viet Quoc. (2015). Bình Thuận khó giữ được giáp ranh [It is hard for Binh Thuan to protect the adjacent forests]. *VOV*, 7/1/2015. Retrieved January 30, 2015, from http://vov.vn/phap-luat/binh-thuan-kho-giu-duoc-rung-giap-ranh-376264.vov

Vietnam Chamber of Commerce and Industry. (2011). *Extractive Industries Transparency Initiative (EITI)—Solution to sustainable development*. Retrieved April 04, 2013, from http://vccinews.com/news_detail.asp?news_id=23299

Vo, Q., & Vo, T. S. (2009). *Challenges on the environment and sustainable development in Vietnam*. Research Centre on Resources and Environment—Hanoi National University (in Vietnamese).

Voelkner, N. (2012). *Human security assemblages in global politics: The materiality and instability of biopolitical governmentality in Thailand and Vietnam*. Submitted for the degree of DPhil, Department of International Relations, University of Sussex.

VOV. (2014). Cảnh báo tình trạng phá đá "săn" Hoàng Đàn [Warning the situations of quarrying to "hunt" Hoang Dan timber]. *Lang Son Television and Radio*, 31/10/2014. Retrieved January 19, 2015, from http://langsontv.vn/node/46079

VTC News. (2012). Đi tìm lời giải về "thần dược"nấm lim xanh chữa ung thủ [Looking for answer on the "magic drug" green Lim mushroom]. *VTC*, 20/4/2012. Retrieved December 8, 2014, from http://vtc.vn/di-tim-loi-giai-ve-than-duocnam-lim-xanh-chua-ung-thu.321.330157.htm

VTV2. (2013). *Timber villages in Vietnam in the context of international integration. TV Programme of Green Vietnam*. Education Science Board, Vietnam Television Broadcast.

VTV. (2014). Xưởng mộc—Nơi ẩn náu và tiếp tay cho lâm tặc [The carpentry workshops that shelter and abailt illegal loggers]. *Dai Truyen Hinh Viet Nam*, 27/10/2014. Retrieved December 22, 2014, from http://vtv.vn/trong-nuoc/xuong-moc-noi-an-nau-va-tiep-tay-cho-lam-tac-20141027161722054.htm

Vu, D. T. (2014). Khai thác rừng trồng phá luôn rừng phòng hộ [Harvesting production forests, the destroy protection forests]. *Nong Nghiep Viet Nam*, 6/11/2014. Retrieved January 4, 2015, from http://nongnghiep.vn/khai-thac-rung-trong-pha-luon-rung-phong-ho-post134185.html

Vu, T. H. (2010). *The offence of violating regulations on forest exploitation and protection in the Vietnamese criminal law*. Master dissertation, School of Law, Hanoi National University (in Vietnamese).

Walker, G. P. (2012). *Environmental justice: Concepts, evidence and politics*. London: Routledge.

Walters, R. (2010a). *Crime, political economy and genetically modified food* (1st ed.). Hoboken: Taylor and Francis.

Walters, R. (2010b). Toxic atmospheres air pollution, trade and the politics of regulation. *Critical Criminology, 18*(4), 307–323.

Walters, R., Wyatt, T., & Westerhuis, D. (2013). *Emerging issues in green criminology: Exploring power, justice and harm*. Houndmills, Basingstoke, UK: Palgrave Macmillan.

Wells, A., Gatto, F. D., Richards, M., Pommier, D., & Contreras-Hermosilla, A. (2007). Rural livelihoods, forest law and the illegal timber trade in Honduras and Nicaragua. In L. Tacconi (Ed.), *Illegal logging: Law enforcement, livelihoods and the timber trade* (pp. 139–167). London: Earthscan.

Westerhuis, D. S. (2013). A Harm analysis of environmental crime. In R. Walters, D. Westerhuis, & T. Wyatt (Eds.), *Emerging issues in green criminology: Exploring power, justice and harm* (pp. 197–217). Houndmills, Basingstoke, UK: Palgrave Macmillan.

Weston, B. H. & Bollier, D. 2009. *Green governance: Ecological survival, human rights, and the law of the commons*.

Wheatley, M. (1997). Social and cultural impacts of mercury pollution on aboriginal peoples in Canada. *Water, Air, and Soil Pollution, 97*(1–2), 85–90.

Wheeler, T., & von Braun, J. (2013). Climate change impacts on global food security. *Science, 341*(6145), 508–513.

White, R. (2005). Environmental crime in global context: Exploring the theoretical and empirical complexities. *Current Issues in Criminal Justice, 16*(3), 1.

White, R. (2007). Green criminology and the pursuit of social and ecological justice. In P. Beirne & N. South (Eds.), *Issues in green criminology: Confronting harms against environments, humanity and other animals*. Cullompton, UK: Willan.

White, R. (2008a). *Crimes against nature: Environmental criminolgy and ecological justice*. Cullompton, UK: Willan.

White, R. (2008b). The criminalisation of environmental harm. *Criminal Justice Matters, 74*(1), 35–37.

White, R. (2009a). Introduction: Environmental crime and eco-global criminology. In R. White (Ed.), *Environmental crime: A reader* (pp. 1–8). Cullompton, UK: Willan.

White, R. D. (2009b). *Environmental crime: A reader*. Cullompton, UK: Willan.
White, R. (Ed.). (2010a). *Global environmental harm—Criminological perspectives*. Cullompton, UK: Willan.
White, R. (2010b). *Global environmental harm: Criminological perspectives*. Cullompton, UK: Willan.
White, R. (2010c). Globalisation and environmental harm. In R. White (Ed.), *Global environmental harm: Criminological perspective*. Cullompton, UK: Willan.
White, R. (2011). *Transnational environmental crime: Toward an eco-global criminology*. London: Routledge.
White, R. (2013a). The conceptual contours of green criminology. In R. Walters, D. Westerhuis, & T. Wyatt (Eds.), *Emerging issues in green criminology: Exploring power, justice and harm* (pp. 17–33). Houndmills, Basingstoke, UK: Palgrave Macmillan.
White, R. (2013b). Eco-global criminology and the political economy of environmental harm. In N. South & A. Brisman (Eds.), *The Routledge international handbook of green criminology*. London: Routledge.
White, R. (2013c). Guest editor's introduction. *Journal of Social Criminology*, Green Criminology Issue.
White, R. (2014). Environmental insecurity and fortress mentality. *International Affairs, 90*(4), 835–851.
White, R., & Heckenberg, D. (2014). *Green criminology: An introduction to the study of environmental harm*. Oxon, UK; New York: Routledge.
Williams, C. (1996). An environmental victimology. *Social Justice, 23*(4), 16–40.
Williams, P. (2008). *Security studies: An introduction*. In P. Williams (Ed.), *Security studies: An introduction*. London: Routledge.
Wolfers, A. (1952). 'National security' as an ambiguous symbol. *Political Science Quarterly, 67*(4), 481–502.
World Bank. (2002). *A revised forest strategy for the World Bank Group*. World Bank Group.
World Bank. (2005). *Vietnam environment monitor 2005—Biodiversity*. Hanoi: The Vietnamese Ministry of Environment and Natural Resources, the World Bank, and the Swedish International Development Cooperation Agency. Retrieved from http://siteresources.worldbank.org/INTEAPREGTOPENVIRONMENT/Resources/VN_Env_Monitor_05.pdf
World Bank. (2007). Weak forest governance costs US$15 billion a year. *News Release No. 2007/86/SDN*. Retrieved June 3, 2015, from http://web.world-

bank.org/WBSITE/EXTERNAL/COUNTRIES/AFRICAEXT/EXTAFRSUMESSD/EXTFORINAFR/0,,contentMDK:21055716~menuPK:2493506~pagePK:64020865~piPK:149114~theSitePK:2493451,00.html

World Bank. (2009). *Country social analysis: Ethnicity and development in Vietnam—summary report*. Washington, DC: World Bank. Retrieved from http://siteresources.worldbank.org/INTEAPREGTOPSOCDEV/Resources/499760ESW0Whit1C10VietnamSummary1LR.pdf

World Bank. (2012). *Well begun, not yet done: Vietnam's remarkable progress on poverty reduction and the emerging challenges*. Hanoi: World Bank. Retrieved from http://www.worldbank.org/en/news/feature/2013/01/24/poverty-reduction-in-vietnam-remarkable-progress-emerging-challenges

Wright, G. (2011). Conceptualising and combating transnational environmental crime. *Trends in Organized Crime, 14*(4), 332–346.

WWF. (2008). Destruction of Sumatra forests driving global climate change and species extinction. *WWF*. Retrieved June 3, 2015, from http://worldwildlife.org/press_releases/destruction-of-sumatra-forests-driving-global-climate-change-and-species-extinction-wwf

Wyatt, T. (2012). *Green criminology & wildlife trafficking: The illegal fur and falcon trades in Russia Far East*. Saarbrücken Lap Lambert Academic Publishing.

Wyatt, T. (2013a). The Russian Far East's illegal timber trade: An organized crime? *Crime Law and Social Change, 1*(61), 15–35.

Wyatt, T. (2013b). The security implications of the illegal wildlife trade. *The Journal of Social Criminology*, August, 130–158.

Wyatt, T. (2013c). Uncovering the significance of and motivation for wildlife trafficking. In N. South & A. Brisman (Eds.), *The Routledge international handbook of green criminology*. London: Routledge.

Wyatt, T. (2013d). *Wildlife trafficking—A deconstruction of the crime, the victims and the offenders*. Basingstoke, UK: Palgrave Macmillan.

Wyatt, T., & Cao, N. A. (2015). *Corruption and wildlife trafficking* (pp. 1–54). Bergen, Norway: U4 Anti-Corruption Resource Centre.

Wyatt, T., Westerhuis, D., & Walters, R. (2013). Introduction. In R. Walters, D. Westerhuis, & T. Wyatt (Eds.), *Emerging issues in green criminology: Exploring power, justice and harm* (pp. 1–13). Houndmills, Basingstoke, UK: Palgrave Macmillan.

Wyler, L. S., & Sheikh, P. A. (2008). *International illegal trade in wildlife: Threats and US policy*. DTIC Document.

Xuan Dung—Trong Loi. (2014). Giữ bình yên cho những cánh rừng giáp ranh [Keep the peacefulness for the adjacent forests]. *Binh Dinh Online*, 26/10/2014. Retrieved January 21, 2015, from http://baobinhdinh.com.vn/viewer.aspx?macm=5&macmp=5&mabb=30307

Xuan Hoa. (2011). Vụ lật xe: Hé lộ bí ẩn kinh hoàng trên cung đường gỗ lậu [The case of flipped truck: Revealing the horrifying secrets on the curve of illegal timber road]. *Giao duc Viet Nam*, 14/12/2011. Retrieved June 3, 2015, from http://giaoduc.net.vn/Xa-hoi/Vu-lat-xe-He-lo-bi-an-kinh-hoang-tren-cung-duong-go-lau-post26592.gd

Xuan Hoai. (2014). Về vụ khai thác lâm sản trái phép tại Quảng Nam: Khởi tố bị can trong tháng này [The case of illegal logging in Quang Nam—Initiating the accused with criminal procedures in this month]. *Cong an THANH PHO HO CHI MINH*, 6/10/2014. Retrieved June 3, 2015, from http://trangtin.congan.com.vn/?mod=detnews&catid=1101&p=&id=526393

Xuan Hong. (2012). Khi chủ rừng và kiểm lâm cùng nhau... phá rừng [When forest owner and Kiem Lam together... deforest]. *Nguoi Dua Tin*, 27/12/2012. Retrieved May 29, 2015, from http://www.nguoiduatin.vn/khi-chu-rung-va-kiem-lam-cung-nhau-pha-rung-a42177.html

Xuan Huy & An Bang. (2011). Nỗi lo "hung thần" chở gỗ lậu [The concern of deadly elfs transporting illegal timber]. *Phu Yen Online*, 18/8/2011. Retrieved January 11, 2015, from http://www.baophuyen.com.vn/89/68101/noi-lo--hung-than--cho-go-lau.html

Xuan Minh. (2014). Khai thác cây lâm sản ngoài gỗ: Gắn phát triển với bảo tồn [Harvesting non-timber forest products: Linking development with conservation]. *Vietnam+*, 22/11/2014. Retrieved June 3, 2015, from http://www.vietnamplus.vn/khai-thac-cay-lam-san-ngoai-go-gan-phat-trien-voi-bao-ton/292572.vnp

Xuan Tho. (2014). Rừng bảo tồn Mường La đang bị "rút ruột" [Muong La Reservation forest is "being gutted"]. *VOV—Dai Tieng Noi Viet Nam*, 24/11/2014. Retrieved December 23, 2014, from http://vov.vn/xa-hoi/rung-bao-ton-muong-la-dang-bi-rut-ruot-366763.vov

Xuan Trach Comnune. (2013). *Report on the poverty status in the 2005–2012 period and a number of directions in the new context*. Xuan Trach Commune People's Committee (in Vietnamese).

Yin, R. K. (2009). *Case study research: Design and methods*. Thousand Oaks, CA: Sage.

Yokoyama, M. (2007). Environmental pollution by corporations in Japan. In H. N. Pontell & G. Geis (Eds.), *International handbook of white-collar and corporate crime* (pp. 327–346). New York: Springer.

Zedner, L. (2007). Pre-crime and post-criminology? *Theoretical Criminology, 11*(2), 261–281.

Zedner, L. (2009). *Security*. London: Routledge-Cavendish.

Zilney, L. A., McGurrin, D., & Zahran, S. (2006). Environmental justice and the role of criminology an analytical review of 33 years of environmental justice research. *Criminal Justice Review, 31*(1), 47–62.

The manufacturer's authorised representative in the EU is Springer Nature Customer Service Centre GmbH, Europaplatz 3, 69115 Heidelberg, Germany. If you have any concerns regarding our products, please contact ProductSafety@springernature.com

Printed and bound by CPI Group (UK) Ltd, Croydon, CR0 4YY

06/10/2026

02228468-0007